Thin Film Structures in Energy Applications

Suresh Babu Krishna Moorthy
Editor

Thin Film Structures in Energy Applications

Springer

Editor
Suresh Babu Krishna Moorthy
Centre for Nanoscience and Technology
Pondicherry University
Kalapet, Pondicherry, India

ISBN 978-3-319-14773-4 ISBN 978-3-319-14774-1 (eBook)
DOI 10.1007/978-3-319-14774-1

Library of Congress Control Number: 2015933997

Springer Cham Heidelberg New York Dordrecht London

Printed on acid-free paper

Springer International Publishing AG Switzerland is part of Springer Science+Business Media (www.springer.com)

Preface

In the past decades, advancement in thin film technology has led to the miniaturization of device dimensions which resulted in applications in various fields. On the other end, energy demands are growing fast in proportion to human population and industrialization. Integration of thin films to improve the energy efficiency of the existing tools as well as in energy harvesting and storage devices of the future is necessary to meet the challenges in energy sector. Most of the available books deal with either the fundamental aspects of deposition or one particular area of energy such as photovoltaic energy. A book that gives an overview on utilization of thin films across broad spectrum of energy is useful for a new researcher as well as scientist working in a specialized field. In order to facilitate the readers, each chapter has a brief introduction required for understanding the chapter content.

Chapter 1 provides an overview of the widely used thin film fabrication and characterization techniques. Chapter 2 focuses on coatings used in various applications ranging from solar cells, fuel cells, batteries to hydro, wind, and geothermal energy technologies. Ternary and quaternary inorganic semiconductor compounds used in thin film solar cells are discussed in Chap. 3. Semiconductors based on organic compounds that are becoming popular are discussed with respect to the fundamentals as well as application in the areas of solar cells and light emitting diodes (LEDs) in Chap. 4. The potential of one-dimensional nanostructures, in particular titanium oxide for photocatalytic and dye-sensitized solar cells (DSSCs), is discussed in Chap. 5. An overview of solid oxide fuel cell (SOFC) operation and ionic conductivity single and multilayer electrolyte thin film is presented in Chap. 6. Though there are numerous techniques to generate energy, still efficient conversion of one form of energy to the other remains a challenge. In Chap. 7, the fundamentals of thermoelectric materials and its application as sensor are discussed. Thin film phosphors in electroluminescent devices are detailed in

Chap. 8. Cutting tools are used everyday for shaping objects, and improving the durability of the tool life saves energy which is given in Chap. 9.

The objective of the book is to provide an overview of the current status of energy harvesting in the form of thin films for both beginners and advanced level researchers.

Kalapet, Pondicherry, India
December 2014

Suresh Babu Krishna Moorthy

Acknowledgements

First I am thankful to Springer International Publishing AG, Switzerland, for providing this opportunity and scope to write the book on a challenging topic. I am especially grateful to Ms. Ania Levinson and the editorial and production staff for being supportive from the start to completion of the work and for providing a pleasant experience to work.

The task of preparing the book with wide spectrum of focus areas cannot be feasible with single person's effort but only with the contribution of authors having specific domain of knowledge. I thank all the authors for their technical contribution as well as their timely responses. I wish to thank the following authors and their team (in the order of appearance in the book): Anup Kumar Keshri (IIT, Patna), Malar (SRM University, Chennai), Parameswar Krishnan Iyer (IIT, Guwahati), Tom Mathews (IGCAR, Kalpakkam), Manjula I. Nandasiri and Suntharampillai Thevuthasan (PNNL, USA), Ramesh Chandra Mallik (IISc, Bangalore), Samuel Paul David and Romain Gaume (UCF, USA), and Hisayuki Suematsu and Tsuneo Suzuki (Nagaoka University of Technology, Japan). I am also thankful to the publishers who have provided copyright permission for the figures used in various chapters of the book.

Also I am indebted to my research students Preethi, Arunkumar, and Sushil Kumar Kuanr for updating the chapter contents, formatting, checking the copyright permissions, preparing indices, etc. in addition to their regular task as researchers. I wish to acknowledge the financial assistance provided by SERB (SR/FTP/ETA-0107/2011) fast-track scheme for young scientist, India, to carry out the research on thin films.

I am thankful to all the family members, friends, and colleagues for extending their support. Last, but not the least, I am grateful to my son Lingesh Pranav and wife Anitha for sparing their family time in order to permit me to concentrate on the work. Without their constant encouragement and support, completion of this book would have been a distant dream.

Kalapet, Pondicherry, India Suresh Babu Krishna Moorthy

Contents

Contributors

Rajini P. Antony School of Materials Science and Engineering, Nanyang Technological University, Singapore, Singapore

P. Arunkumar Centre for Nanoscience and Technology, Madanjeet School of Green Energy Technologies, Pondicherry University (A Central University), Puducherry, India

K. Suresh Babu Centre for Nanoscience and Technology, Madanjeet School of Green Energy Technologies, Pondicherry University (A Central University), Puducherry, India

Ashoka Bali Thermoelectric Materials and Devices Laboratory, Department of Physics, Indian Institute of Science, Bangalore, Karnataka, India

Raju Chetty Thermoelectric Materials and Devices Laboratory, Department of Physics, Indian Institute of Science, Bangalore, Karnataka, India

Dipjyoti Das Centre for Nanotechnology, Indian Institute of Technology Guwahati, Guwahati, Assam, India

Samuel Paul David College of Optics and Photonics (CREOL), University of Central Florida, Orlando, FL, USA

Anamika Dey Centre for Nanotechnology, Indian Institute of Technology Guwahati, Guwahati, Assam, India

Romain Gaume College of Optics and Photonics (CREOL), University of Central Florida, Orlando, FL, USA

Parameswar Krishnan Iyer Department of Chemistry and Centre for Nanotechnology, Indian Institute of Technology Guwahati, Guwahati, Assam, India

Anup Kumar Keshri Department of Materials Science and Engineering, Indian Institute of Technology Patna, Navin Government Polytechnic Campus, Patna, Bihar, India

Sushil Kumar Kuanr Centre for Nanoscience and Technology, Madanjeet School of Green Energy Technologies, Pondicherry University (A Central University), Puducherry, India

P. Malar Research Institute, SRM University, Chennai, Tamil Nadu, India

Ramesh Chandra Mallik Thermoelectric Materials and Devices Laboratory, Department of Physics, Indian Institute of Science, Bangalore, Karnataka, India

Tom Mathews Thin Films and Coatings Section, Surface and Nano-Science Division, Indira Gandhi Centre for Atomic Research, Kalpakkam, Tamil Nadu, India

Manjula I. Nandasiri Environmental Molecular Sciences Laboratory (EMSL), Pacific Northwest National Laboratory (PNNL), Richland, WA, USA

Ashish Singh Centre for Nanotechnology, Indian Institute of Technology Guwahati, Guwahati, Assam, India

M. Sribalaji Department of Materials Science and Engineering, Indian Institute of Technology Patna, Navin Government Polytechnic Campus, Patna, Bihar, India

Hisayuki Suematsu Extreme Energy-Density Research Institute, Nagaoka University of Technology, Nagaoka, Japan

Tsuneo Suzuki Extreme Energy-Density Research Institute, Nagaoka University of Technology, Nagaoka, Japan

Suntharampillai Thevuthasan Environmental Molecular Sciences Laboratory (EMSL), Pacific Northwest National Laboratory (PNNL), Richland, WA, USA

Chapter 1
Thin Film: Deposition, Growth Aspects, and Characterization

P. Arunkumar, Sushil Kumar Kuanr, and K. Suresh Babu

Abstract Thin film science and technology plays an important role in the development of devices in the future ranging from energy-efficient display devices to energy-harvesting and storage devices such as solar cell, fuel cell, batteries, super capacitor, etc. Thin films have properties that can be different from that of their corresponding bulk structures. A film is considered as thin, as long as its surface properties are different from its bulk behaviour. Thin films have larger surface to volume ratio, hence the surface and near surface characteristics decide the properties of the thin film. As a result thin film properties generally depend on the thickness of the film which extends from few micrometre to nanometre, substrate nature on which the films are grown and deposition methodology/conditions used in the fabrication of thin films. Thin film fabrications are generally carried out by depositing the required material in the atomistic deposition (atom by atom) over the required substrate, which may result in either single crystalline, polycrystalline, or amorphous structure depending on the deposition conditions. Thin film technology has the potential to engineer the various properties such as porosity, surface morphology, surface roughness, and crystallite size. These advantages in thin film assist in the development of new products and minimize the waste as in the conventional manufacturing techniques. This chapter provides an overview of various thin film processing methods, mechanism behind the growth and important tools used for the characterization of thin films.

1.1 Introduction

Thin film science and technology plays an important role in the development of devices in the future ranging from energy-efficient display devices to energy-harvesting and storage devices such as solar cell, fuel cell, batteries, super capacitor, etc. Thin films have properties that can be different from that of their

---

P. Arunkumar • S.K. Kuanr • K.S. Babu (✉)
Centre for Nanoscience and Technology, Madanjeet School of Green Energy Technologies, Pondicherry University (A Central University), Puducherry 605 014, India
e-mail: sureshbabu.nst@pondiuni.edu.in

S. Babu Krishna Moorthy (ed.), *Thin Film Structures in Energy Applications*, DOI 10.1007/978-3-319-14774-1_1

corresponding bulk structures. A film is considered as thin, as long as its surface properties are different from its bulk behaviour. Thin films have larger surface to volume ratio, hence the surface and near surface characteristics decide the properties of the thin film. As a result thin film properties generally depend on the thickness of the film which extends from few micrometre to nanometre, substrate nature on which the films are grown and deposition methodology/conditions used in the fabrication of thin films. Thin film fabrications are generally carried out by depositing the required material in the atomistic deposition (atom by atom) over the required substrate, which may result in either single crystalline, polycrystalline, or amorphous structure depending on the deposition conditions. Thin film technology has the potential to engineer the various properties such as porosity, surface morphology, surface roughness, and crystallite size. These advantages in thin film assist in the development of new products and minimize the waste as in the conventional manufacturing techniques.

Development of thin film-based solid state devices is an economically mean to enhance the commercialization of various renewable energy sources. Thin films technology finds large scale applications in power generation in photovoltaic power system and solid oxide fuel cells [1]. For example, the 'First Solar' company Japan to install the 100 MW power system through thin film based cadmium telluride photovoltaic module, which has an increased efficiency of about 8 % compared to conventional silicon modules [2]. On the other hand, using thin film technology in solid oxide fuel cell (SOFC) current researches have been focused on reducing the high operating temperature (~1,000 °C) of SOFC to intermediate range (500–700 °C) [3]. Thus, in the energy field thin film technology has vast opportunity to make enormous difference in the development of various products in a cost-effective way. Understating the fundamentals of the thin film deposition process is essential to tailor new thin film-based devices with enhanced properties and higher efficiency.

1.2 Classification of Deposition Technique

Depending on the deposition nature, the thin film techniques can be classified into (1) physical vapour deposition (PVD) and (2) chemical vapour deposition (CVD) techniques. Generally in PVD technique, the deposition is carried out through the vaporization of target material by either thermal or athermal process. In thermal process, target material undergoes sublimation or evaporation through the supply of thermal energy, while in athermal process the vaporization takes places physically by striking the ionized gas molecules over the target material. In CVD technique, the gas/liquid phase of target material undergoes chemical reaction during deposition process. The classification of thin film technique based on the nature of the deposition process is shown in Fig. 1.1 and discussed below.

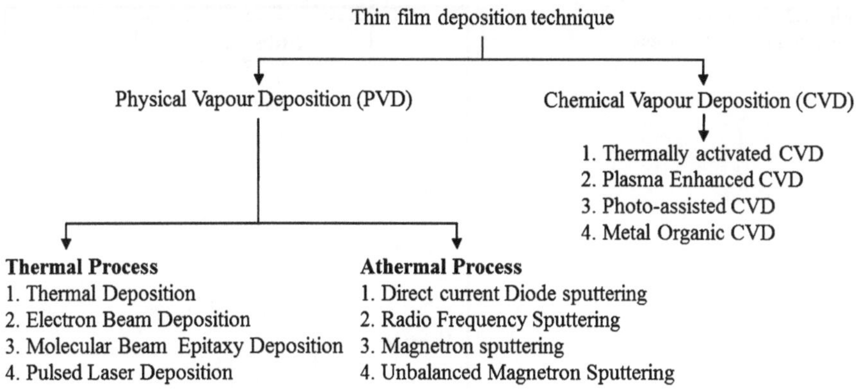

Fig. 1.1 Classification of thin film deposition technique

1.3 Physical Vapour Deposition Technique

PVD technique is used to deposit the material from target over a substrate in the range of few angstrom to several micrometres. The deposition process in PVD involves various stages, in which the first stage is the transfer of high energy to the target material by means of filament heating or bombardment of electron/ion over the target material. In the second stage, the target material vaporizes due to the transfer of energy through evaporation or sublimation or physical sputtering to the target atom under high vacuum or low pressure conditions. At the third stage the transportation of vaporized target material travels towards substrate surface. The fourth stage involves the reaction between the vaporized target material and gas present in the chamber during the transportation. This fourth stage is applicable only for the specialized coatings such as nitride, oxide, carbide, etc., in which the final coating differs from the chemistry of the target material. In the fifth stage, the vaporized material transported to the substrate surface undergoes condensation process and induces the growth of the thin film.

Hence, the overall PVD process involves the transportation of material in atomistic level to the substrate surface leading to the formation of thin film. Figure 1.2 explains all the five stages of PVD process during the deposition.

Following are the associated advantages and disadvantages of the PVD technique.

Advantages

- The composition of the deposited thin film can be controlled by the selection of the target material.
- Formation of different compounds or alloys is possible in PVD technique by reactive or co-deposition process.

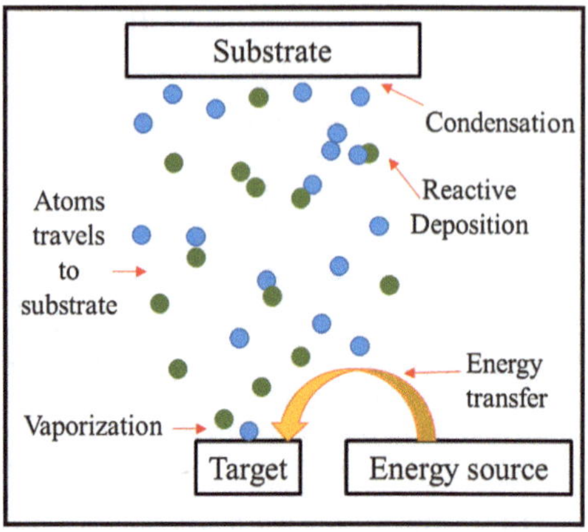

Fig. 1.2 Basic steps involved in PVD process

- Almost all the inorganic material and some organic material can be deposited as thin film in PVD.
- As most of the deposition in PVD technique carried out under high vacuum, contamination is low.
- By controlling the rate of deposition, the required morphology, porosity, texture, etc. can be tailored.

Disadvantages

- High Sophisticated instruments such as filaments, vacuum pumps, and vacuum chamber are basically needed for PVD technique which increases the overall initial cost.
- Line of sight process which results in thickness variation on moving from the centre to the edges.
- Suffers from poor step coverage and shadowing.

1.3.1 Vaporization Process

PVD involves vaporization of the target material and subsequent condensation of vapour over the substrate which results in the formation of thin film. Usually the vaporization of the target material has been carried out under high vacuum in order to avoid the unwanted collision with the foreign gaseous matters to maintain the purity in the thin film. High vacuum increases the mean free path of the vaporized atom which enables the movement in a straight line without the gas-to-gas collision towards the substrate. The evaporation or sublimation depends on the vapour pressure of the material which is to be coated. The thermodynamic equilibrium between the gaseous and condensed solid or liquid phases denotes the vapour pressure of the material being evaporated. Knudsen effusion method is used to

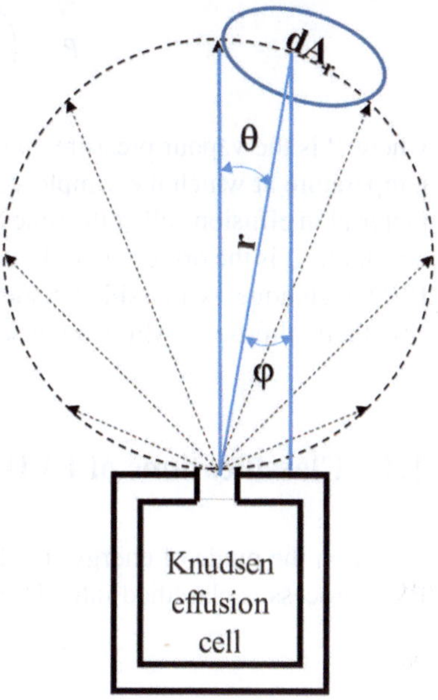

Fig. 1.3 Knudsen effusion cell with cosine law of distribution of deposition geometry

determine the vapour pressure of the target material [4–6]. This method consists of an isothermal Knudsen effusion cell with an orifice (Fig. 1.3), which is smaller than that of target surface. Through this orifice opening, the effusion of the evaporated material spreads in an isotropic distribution. This isotropic distribution of evaporated material obeys the cosine law of emission. Cosine law provides the relation between the evaporated target material with the angular distribution during deposition. The cosine law for the vaporization of the target material can be represented by the following equation.

$$J = \frac{J_o \cos\theta \cos\varphi}{r^2} dA_r, \tag{1.1}$$

where J is the receiving flux over the area dA_r away from the orifice at the distance r, θ is the polar angle between the normal to the orifice and the direction of the receiving flux, φ is the azimuthal angle between the normal to the dA_r and the direction of the receiving flux as shown in Fig. 1.3. $J_o = KQ/r^2$, where Q is the evaporation rate and K is the geometrical factor which depends on the surface of the evaporation source. For point source the $K = 1/4\pi$ and for source which is evaporated from crucible or boat the $K = 1/\pi$.

The evaporated material which follows the cosine law undergoes isotropic distribution of sphere. The vapour pressure of the evaporated material which is in thermodynamic equilibrium with its solid or liquid phase is measured by Knudsen effusion method (Eq. 1.2) [7]

$$P = \left(\frac{m}{t\ W_o A_o}\right)\sqrt{\frac{2\pi RT}{M}}, \tag{1.2}$$

where P is the vapour pressure, m mass rate of the sample at given time t, T is the temperature at which the sample starts evaporating or sublimating from the source material in effusion cell at the time t, M is the molar mass of the source, R is the gas constant, A_o is the orifice area, W_o is the transmission probability of the vapour. The PVD technique is classified under two broad categories: (1) thermal process, (2) athermal process which are discussed in Sect. 1.4.

1.4 Classifications of PVD

Based on the mode of energy supplied for the vaporization of the source material, PVD process is classified into (1) thermal and (2) athermal process.

1.4.1 Thermal Process

In thermal process, the thermal energy is involved in the vaporization of the target material. Generally large volume of the target material is required for the vaporization and the temperature raised should be high above the equilibrium point of the vapour pressure in order to initiate the vaporization process of the target material. The thermal energy is supplied to the target material by means of resistive heating or high energy electron beam or photons for the evaporation/sublimation process. Depending on the thermal source used for the vaporization of the target material, the thermal process can be classified into

1. Thermal deposition technique
2. Electron beam deposition technique
3. Molecular beam epitaxy
4. Pulsed laser deposition technique

1.4.1.1 Thermal Deposition Technique

Thermal Deposition technique uses the resistive heating to evaporate the target material. The target material is loaded in a suitable holder and heated by passing the high current through it, which transfers the thermal energy to the target material resulting in the evaporation or sublimation. Figure 1.4 represents the working principle of thermal deposition technique. Generally high melting point materials such as tungsten, molybdenum, and graphite are used as holder material. In order to evaporate the target material, holder is heated resistively using the joule heating principle.

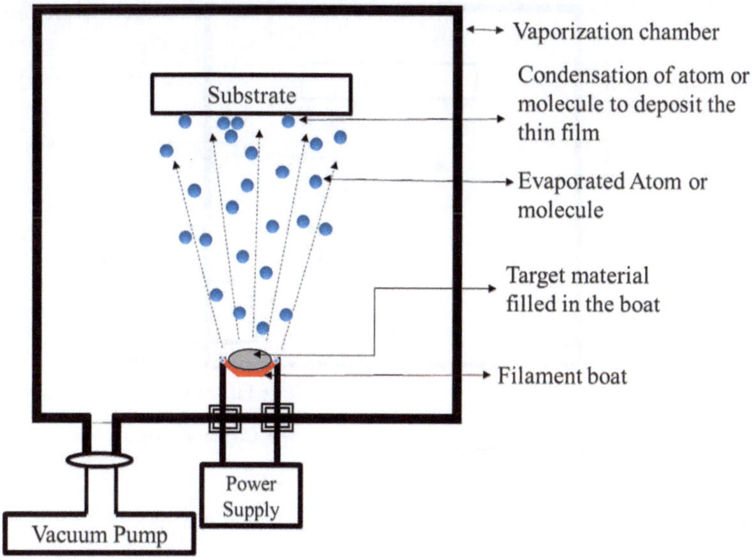

Fig. 1.4 Schematic representation of thermal evaporation system

The vaporization of the target is achieved by supplying the adequate power to the holder to achieve sufficient thermal energy *t*. The power required to achieve the sufficient temperature in the holder by joule heating is given by Eq. (1.3)

$$P = I^2R, \tag{1.3}$$

where I is the current, R is the resistance of the holder. The resistance R can be calculated from Eq. (1.4)

$$R = \rho(L/A), \tag{1.4}$$

where L is the length, A is the cross-sectional area of the material, and ρ is the electrical resistivity of the holder material.

Depending on the nature of the target material and vacuum pressure inside the chamber, the target material either undergoes evaporation or sublimation at higher temperature. Following conditions need to be analysed before carrying out thermal evaporation. (1) The chamber pressure should be better than 10^{-3} m bar to prepare high quality thin film and ensure that the vaporized atom travel from the target to the substrate in a trajectory path without collision. (2) The holder should not react with the target material at higher temperature. (3) The holder should be resistively heated up to transfer the thermal energy to the target material which should be high enough to vaporize the material. (4) The holder material should possess high melting and low vapour pressure in order to evaporate the wide range of the target materials by resistive heating.

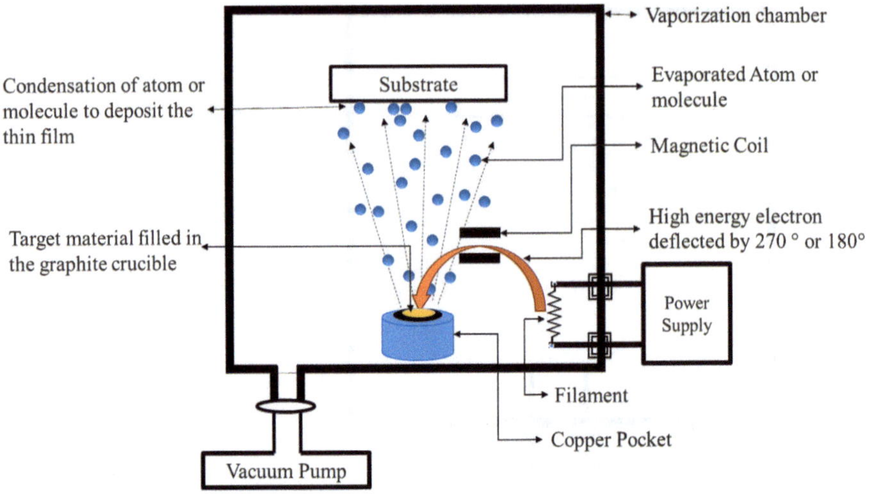

Fig. 1.5 Schematic representation of electron beam deposition technique

1.4.1.2 Electron Beam Deposition Technique

In electron beam deposition technique, an intense high energy electron beam is used to hit the target material locally to induce the vaporization and subsequent formation of thin films. Here the source producing the electron and target acts as the cathode and anode, respectively. When a high voltage of 6–40 KV is applied to the filament, high density of electrons are ejected to evaporate the wide range of target materials differing in melting point. The high energy electron deflected by 270° or 180° are precisely focussed towards the target material by the applied electric and magnetic field as shown in Fig. 1.5. The electron beam can also be scanned in x and y direction at high speed over a specified area on the target by coupling the frequency generator with the magnetic field. This sweeping control provides an additional advantage of increasing the electron beam width over wide area on the target material during vaporization. In electron beam the target material used for vaporization is kept in the water-cooled crucible. The high energy electron beam travels at high speed and strikes the surface of the target. The high kinetic energy of the traveling electron gets converted to thermal energy over the target and results in melting or sublimation and produces sufficient vapour flux for the deposition of the thin film. The electron beam deposition technique helps to vaporize the target materials with high melting point around 2,800 °C. In the electron beam system the electron-producing filament is kept away from the target material to maintain the purity of the coating during deposition process.

The vaporization of the target material depends on the velocity of the striking electron on the surface of the target. Increase in velocity of the electron in turn increases the heat energy generated over the target (anode). The velocity of the electron depends on the applied potential to the electron-producing filament (cathode).

Usually the cathode and anode is kept at high negative and positive potentials, respectively; hence, the accelerated electron produced from the cathode directly strikes over the positive anode.

By applying the potential V, the energy of the accelerated electron is given by, $V = eV$, where e is the charge of the accelerated electron. Hence, these energy in the accelerated electron get converted into kinetic energy as given in Eq. (1.5)

$$Ve = \frac{1}{2}\left(mv^2\right), \tag{1.5}$$

where v is the velocity of the accelerated electron, m is the mass of the electron. By rearranging, Eq. (1.6), the velocity of the accelerated electron can be calculated for applied potential V.

$$v = \sqrt{\frac{2eV}{m}}. \tag{1.6}$$

In the electron beam deposition system the accelerated electron is deflected by the applied magnetic field, B. The electron deflection from the cathode to the anode forms a circular path by 180° or 270°. This circular path of the accelerated electron induced by the Lorentz force undergoes a centripetal force which gives the radius of the orbit travelled by the electron.

$$eVB = \frac{mv^2}{r}, \tag{1.7a}$$

where r is the radius of the orbit travelled by the accelerated electron under applied magnetic field. The electron beam deposition chamber should be under high vacuum (>10^{-2} m bar); hence the accelerated electron can avoid the dissipation of its kinetic energy by inelastic collision with other gas molecules. In electron beam deposition system the acceleration voltage, applied beam current, focusing of the electron, vacuum pressure, and electron beam diameter incident on the target surface influence the deposition process.

1.4.1.3 Molecular Beam Epitaxy

The word 'epitaxial' means the growth of monocrystalline structure and 'molecular beam' denotes the movement of molecule or atom to form the epitaxial thin film. In molecular beam epitaxy (MBE), the atoms or molecules produced by the heating of the target material travel in an ultra-high vacuum chamber without collision between each other or with other gas molecules. These atoms or molecules impinge on the substrate surface and undergo diffusion to form thin film. MBE technique is as schematically represented in Fig. 1.6. The pressure maintained inside the MBE

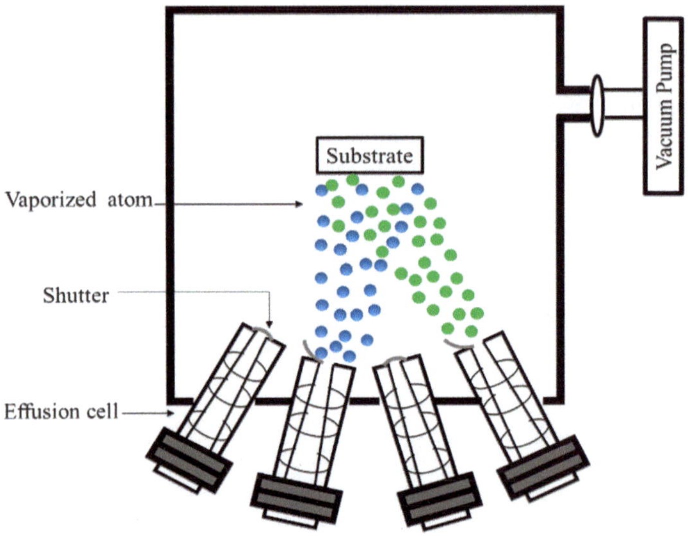

Fig. 1.6 Schematic representation of molecular beam epitaxy

chamber is typically in the range of ~10^{-9} m bar. Hence, the mean free path between the vaporized atom or molecule increases, which in turn avoids the collision among them. Generally the lattice parameter variation between the substrate and target has to be kept below 10 % in order to obtain the monocrystalline thin film. The target material vaporized from the effusion cell is placed inside the MBE chamber. Effusion cell is also called the Knudsen cell in which the target material undergoes thermal vaporization under controlled vapour pressure. The vaporized vapour flux escapes through the orifice opening and provides the flux stability to provide uniformity in the deposited epitaxial layer. The effusion cell is focussed towards the heated substrate and the temperature in the effusion cell should be precisely maintained within the associated error of 1 % for the uniform distribution of the vaporized flux.

The schematic representation of the effusion cell is as shown in Fig. 1.7, which consists of (1) crucible in cylindrical or conical shape, (2) heating filament, (3) heat shielding, (4) thermocouple. Hence, the effusion cell parameters such as crucible geometry, positioning angle towards substrate, and heating filament decide the final quality of the target vapour flux, which in turn influences the crystalline nature of the thin film.

1.4.1.4 Pulsed Laser Deposition Technique

Pulsed laser deposition technique (PLD) established by Smith and Turner in 1965 uses the laser beam to vaporize the target material for the deposition of thin film [8]. A pulsed laser beam is focused on the surface of the target material and generates

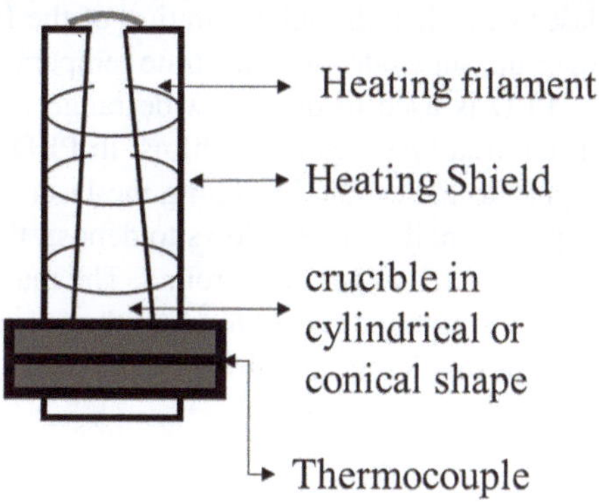

Fig. 1.7 Schematic representation of the effusion cell

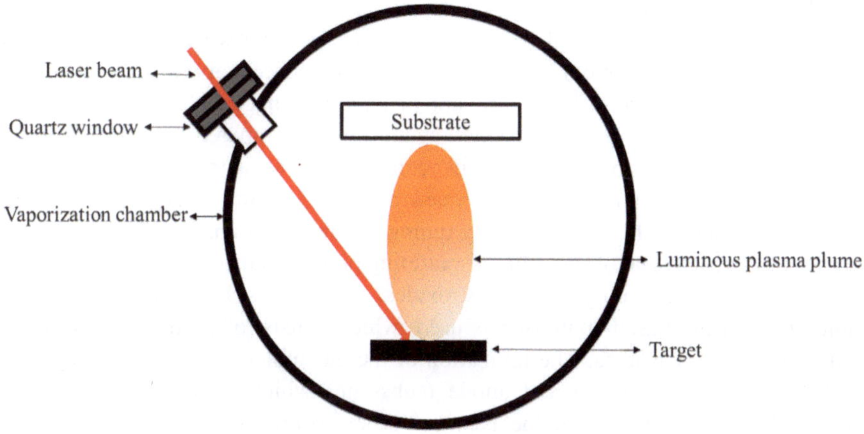

Fig. 1.8 Schematic representation of the pulsed laser deposition technique

the required thermal energy by the strong absorption of the electromagnetic wave inducing the vaporization of the target material. Excimer laser (YAG or ARF) is used in the PLD due to their high energy and efficiency. The schematic representation of the basic set up of PLD is shown in Fig. 1.8. In PLD high energy laser is focussed on the target at the short pulse duration and transfers the heat energy to vaporize the target material to form the thin film. Due to the laser ablation, the vaporized target material generates luminous plasma plume near to the surface of the target. The vaporized plume consists of various ionized and excited species that affect the transfer rate of the evaporated/sublimated target material. The ablated material has the same stoichiometry as that of the target material. By principle the PLD is the simplest technique; however, in practical phenomenon the situation is so complex. The uniformity of the thin film in PLD depends on the spot size of the laser, plume temperature, and target to the substrate distance. The growth of the thin film in PLD is influenced by several parameters such as energy density of the laser,

laser wave length, pulse duration of the laser, condensation of vaporized material over the substrate, and substrate temperature.

PLD is used to deposit wide range of oxides, nitrides, carbides, metallic, and different alloys-based thin films. In PLD the deposition rate can be changed from higher to lower rate by tuning the laser power and pulse rate. The laser source is kept externally which allows to deposit the thin film in a clean atmosphere both in inert or reactive gas background. The main challenge of the PLD is that the plasma plume is highly directed towards the forward direction, which results in uniformity only in smaller area limiting the large scale production.

1.4.2 Athermal Process

In athermal process physical energy is (non-thermal) involved in the vaporization process of the target material by ejecting the surface atoms (target) through the bombardment of the accelerated gaseous ions. This physical vaporization process in PVD is called sputtering. Large amount of energy is needed for the ejection of atoms from the target material through sputtering as compared to that of thermal process. This ejection of target atom depends on several factors such as chemical bonding within the target material, mass and acceleration energy of incident-charged ion, and applied potential. By tuning the above parameters the stoichiometry of the target to that of thin film in sputtering process can be obtained for various compound materials, alloys, mixtures, oxides, and nitrides. In the sputtering technique the plasma-based sputtering is used, which consist of partially ionized gas with all charged particles. The ionization of the gas molecule occurs between the cathode (sputtering target) and anode (substrate) which is maintained at high potential under vacuum. On the basis of power source used to ionize the gas molecule for the generation of the plasma, sputtering techniques are classified into

1. Direct current diode (DC) sputtering
2. Radio frequency (RF) sputtering
3. Magnetron sputtering
4. Unbalanced magnetron sputtering

1.4.2.1 Direct Current Diode (DC) Sputtering

DC sputtering is the simplest system, in which the plasma is created by applying the high DC potential between the anode and cathode in order to ionize the sputtering gas. Argon gas is preferred as sputtering gas due their inert nature, high mass, and low cost. The pressure in the sputtering chamber should be better than 10^{-7} m bar to minimize the impurity level in the thin film. The sputtering gas should be allowed inside the chamber during the deposition process. The partial pressure of sputtering gas should be maintained between ~10^{-3} and 10^{-2} m bar. After achieving the

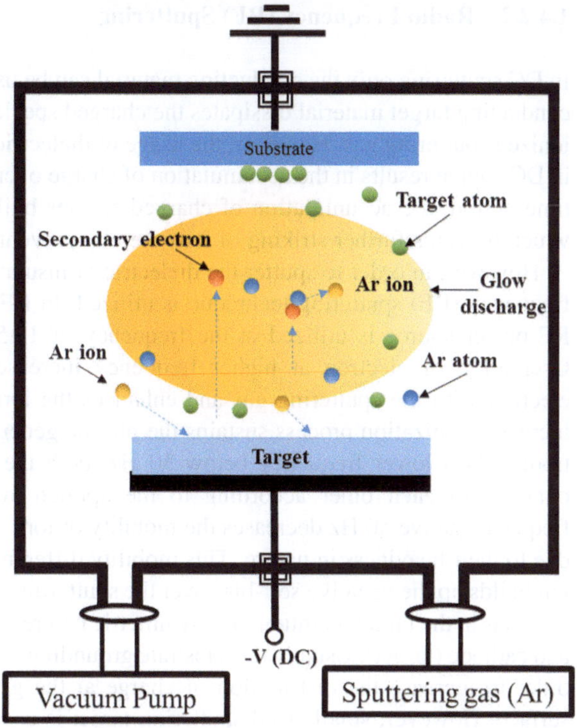

Fig. 1.9 Schematic representation of the direct current diode (DC) sputtering

desired partial pressure of the sputtering gas, negative DC bias with high voltage is applied to the cathode (sputtering Target). However, the anode (substrate) is usually grounded with respect to the cathode. The high potential in the cathode generates the plasma near to the surface of the cathode. The high voltage near the cathode accelerates the electrons towards the anode collide with the outer shell of the sputtering gas atom results in ionization process of the gas molecule. These positively ionized cations (sputtering gas atom) accelerated at high energy towards the negative cathode (target) lead to the collision of charged species with the sputtering target. This collision of charged species ejects the secondary electron from the sputtering target, which accelerates back to the plasma and increases the collision process with the sputtering gas to form more ionized species. These secondary electrons play a major role in generation of ions which keeps the plasma in a steady state and leads to the sustained plasma. This sustained plasma increases the collision process of ionized gas and in turn ejects the target atom towards the substrate. The number of target atom ejected from the surface by the collision of ionized gas decides the sputtering yield (deposition rate). The sputtering yield depends on the energy of accelerated ions, incident angle of ion bombardment over the target material, and binding energy within the target atom (Fig. 1.9). Hence, the required quality of the thin film can be obtained by tuning the variables such as sputtering gas nature, working pressure, base pressure, applied voltage to the target, substrate temperature, and deposition rate.

1.4.2.2 Radio Frequency (RF) Sputtering

In DC sputtering only the conducting material can be used as sputtering target. This conducting target material dissipates the charged species induced by the striking of ionized sputtering gas. However, the usage of dielectric or insulting target material in DC source results in the accumulation of charge over the target. Over a period of time period, the accumulation of charged species builds up on the target surface which prevents further striking of ions thereby preventing further sputtering.

However, in order to sputter the dielectric or insulated target material, the radio frequency (RF) sputtering technique is utilized. In RF sputtering, the commercial RF power source is utilized at the frequency of 13.56 MHz for the deposition. Oscillation of electron at higher frequency increases the collision process of electron with the sputtering gas and enhances the ionization process. Hence, the increased ionization process sustains the plasma generation between the two electrodes. At a lower frequency below 50 Hz, both the electron and ions oscillate opposite to each other according to the applied voltage. However, increased frequency above 50 Hz decreases the mobility of ions compared to that of electron due to their heaviness in nature. This mobility difference between the electron and ion builds up the negative self-bias over the sputtering target. This induced negative self-bias in the target maintains an asymmetric nature between the anode (substrate) and cathode (target). Usually the substrate grounding is connected to the chamber in order to increase the distribution of charge at the given area over the substrate surface. However, small-sized sputtering target is used compared to that of large substrate. Hence, the charge density at the given area is higher in the sputtering target compared to that of substrate surface. Accordingly most of the ions travel towards the negatively self-biased sputtering target where the concentration of electron is high (Fig. 1.10). Thus, the target undergoes collision with the ionized gas deposits over the substrate and results in the formation of thin film.

1.4.2.3 Magnetron Sputtering

In DC sputtering the formation of secondary electron from the target acts as the main source for ionizing the sputtering gas. This secondary electron plays a major role in order to sustain the plasma during sputtering process. The secondary electron ejected from the target is not efficiently utilized for ionizing the sputtering gas which leads to degradation of the plasma. Hence, in the magnetron sputtering appropriate usage of magnetic arrangement parallel to the sputtering target deflects the ejected secondary electron within the limited boundary near to the target surface. If the applied electric field, E, is perpendicular to the applied magnetic field, B, then a perpendicular drift developed in the traveling electron. Thus, the applied $E \times B$ leads the electron to travel in a circular path near to the cathode surface and increases the ionization process of the sputtering gas. These increased ionization process in turn increases the plasma flux density near to the surface of the

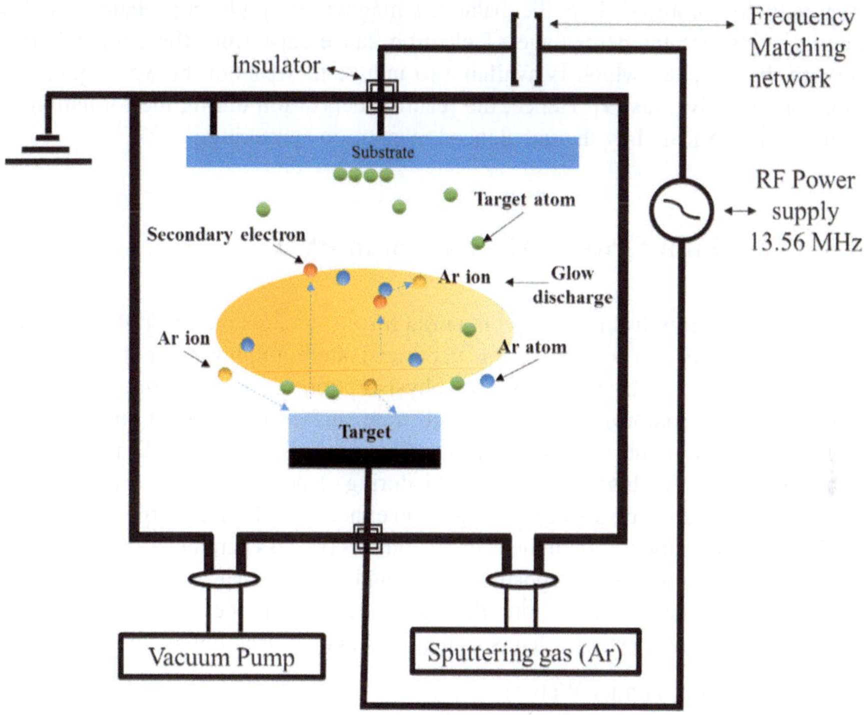

Fig. 1.10 Schematic representation of the radio frequency (RF) sputtering

target. The increased plasma density effectively sputters the target material. This circular motion of the electron significantly increases the ionization rate in the magnetron sputtering which effectively deposits the thin film at lower pressure as compared to DC sputtering. The major disadvantage of the conventional magnetron sputtering is the non-uniform sputtering nature over the target surface due to the localization of the electrons. These localized electrons preferentially sputter the target material leading to the non-uniform utilization of target material. In order to improve the target utilization, the magnets or target should be physically moved in certain sequence to change the sputtering location which leads to increased utilization of the target material. Thus, in magnetron sputtering the secondary electrons are effectively utilized to ionize the sputtering gas which in turn increases the sputtering yield during the formation of the thin film.

1.4.2.4 Unbalanced Magnetron Sputtering

In magnetron sputtering the plasma confined near to the sputtering target makes it unavailable for the reactive gas for the reactive sputtering deposition. In unbalanced magnetron sputtering the applied magnetic field near the sputtering target is relatively less compared to the balanced magnetron sputtering. Hence, in this configuration greater percentage of electron can escape from the confined field near to the cathode, which is available to induce the reaction between sputtered atom and reactive gas [9]. Hence, the reactive deposition during the formation of thin film is fabricated by the unbalanced magnetron sputtering.

1.5 Thin Film Growth Mechanism in PVD

The vaporized atom from the target material travels towards the substrate inside the vacuum chamber. Accordingly, the vaporized atom impinges on the substrate surface either by chemisorption or physisorption. At the initial stage, the adsorbed-atom (adatom) reaching the substrate surface has its own energy transferred to the atom during vaporization. Also in most of the PVD process the substrate is kept at higher temperatures during deposition. Hence, the adatom arriving on the substrate undergoes migration either by its initial energy or substrate energy or combination of both the energy. Various process parameters of PVD such as substrate temperature, deposition rate, and residual gas pressure influence the migration energy of adatom during the growth process of the thin film [10, 11]. Different growth mechanisms involved in the formation of the thin film are

1. Island growth (Volmer Weber)
2. Layer-by-layer growth (Frank–van der Merwe)
3. Stranski–Krastanov (mixed growth)

1.5.1 Island Growth (Volmer Weber)

Island growth of the thin film occurs when the vaporized atoms (source) have strong bonding with each other compared to the substrate surface. This strong bonding within the atoms leads to the formation of localized clusters (Island growth) on the substrate surface. This localized cluster continues to nucleate with the deposition until the exposure of the substrate surface exists. After completely occupying the entire substrate surface, these nucleated clusters merge with each other leading to the coalescence by the minimization of the surface energy to form continuous thin film as shown in Fig. 1.11.

1.5.2 Layer-by-Layer Growth (Frank–van der Merwe)

Layer-by-layer growth of the thin film occurs when the vaporized atom has strong bonding with the substrate surface compared to the bonding level between each other. This growth mode undergoes a sequential formation of layer driven by the

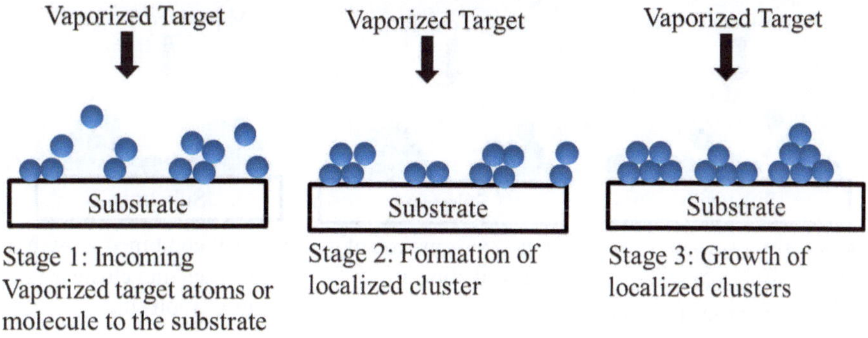

Fig. 1.11 Growth mechanism of thin film deposition by Island growth (Volmer Weber) mode

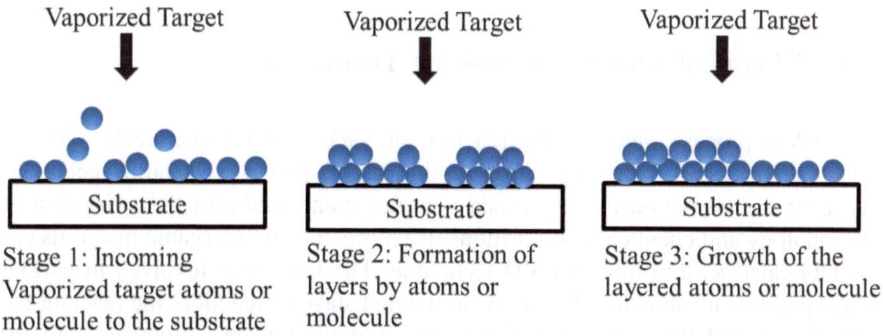

Fig. 1.12 Growth mechanism of thin film deposition by layer-by-layer growth (Frank–van der Merwe) mode

lower surface energy of substrate with the adatom. Hence, the formation of new layer begins after the completion of first one (Fig. 1.12).

1.5.3 Mixed Growth (Stranski–Krastanov)

The combination of both layer-by-layer and island growth leads to Stranski–Krastanov growth mode in the formation of the thin film. At the initial stage, the adatom nucleates to form the thin film by means of two-dimensional layer-by-layer growth mode and then the thin film transforms into three-dimensional island growth mode. The mechanism of Stranski–Krastanov growth mode undergoes three stages [12]. At the first stage, the adatom reaching the substrate undergoes layer-by-layer growth mode and leads to the formation of metastable island over the substrate. In the second stage these metastable islands undergo further growth and stabilize after exceeding the critical thickness. At the third stage further arrival of adatom leads to the formation of island over the stable island and continues this process to form the thin film as represented in Fig. 1.13.

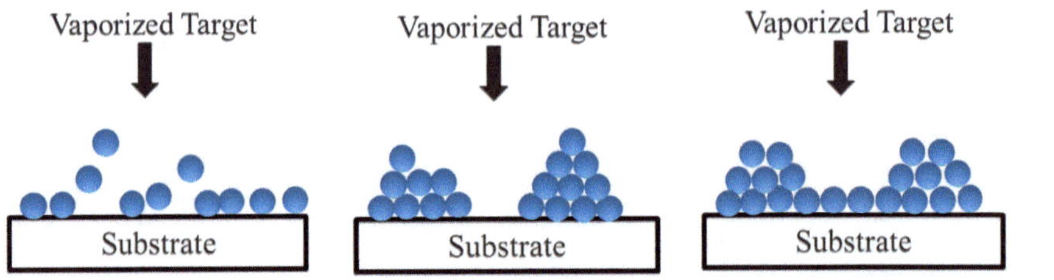

Fig. 1.13 Growth mechanism of thin film deposition by Stranski–Krastanov (mixed growth) mode

1.6 Chemical Vapour Deposition Technique

The CVD process involves the reaction of vapour phase chemicals over the substrate to result in the deposition of thin film. CVD is a versatile technique, used for the formation of simple and complex compound thin film with desired morphology and chemical compositions. Both organic and inorganic materials can be fabricated as thin film in CVD technique. The four steps involved in various CVD process are shown in Fig. 1.14. In the first stage of reaction, the reactant gas (precursor) and the carrier gas (inert gas, if required) are introduced into the reaction chamber in a required flow rate. In the second stage, the adsorption of reactant gas takes places over the substrate over which the reactant gases undergo a chemical reaction activated by the thermal energy or ionized plasma. In the third stage these gases undergo surface diffusion and decomposition on the substrate leads to the formation of nucleation sites. With time the nucleation sites grow to form the required thin film on the substrate. In the fourth stage, the by-product formed on the substrate undergoes desorption and carried out through the outlet of the reaction chamber. The quality of the thin film formation is determined by the thermodynamics and kinetics of the chemical reaction, flow rate of the gaseous reactant, substrate temperature, and chemical–physical properties of the substrate.

CVD technique possesses following advantages and disadvantages.

Advantage

- Both organic and inorganic material can be fabricated as thin film in CVD technique.
- High pure and dense coatings can be obtained from the CVD technique.
- Low operating temperature.
- Initial cost is lower compared to PVD.
- Generally large number of substrates can be coated at the same time with unique property.

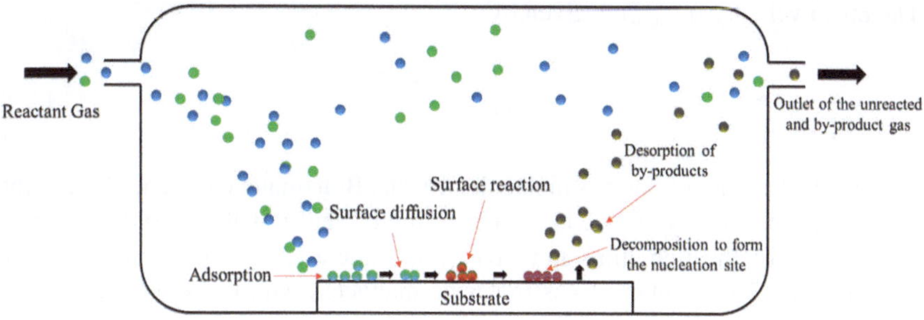

Fig. 1.14 Schematic representation of the basic steps in CVD process involved in the formation of thin film

Disadvantage

- Many of the by-products released for the reactant chamber are toxic and corrosive in nature and lead to the environmental pollution.
- Often the reaction is incomplete, which results in the wastage of precursor material.
- The reaction should be heterogeneous between the gas phase and substrate. However, the undesirable homogenous reaction within the gas phase results in the contamination of the deposited film.
- During the CVD process, the chamber wall is also prone to the deposition; hence periodic cleaning of reactant chamber is required.

1.6.1 Gas Transportation Process

The thin film fabrication through chemical vapour deposition depends on the transportation of gas into the reaction chamber. The transportation of gas depends on the various parameters such as kinetic theory of the gases and the vapour pressure inside the chamber. The gas momentum transportation can be represented through ideal behaviour of gas. The ideal behaviour of gas for all the temperature obeys the Boyle's law.

$$PV = nRT, \tag{1.7b}$$

where P is the pressure, V is the volume, n is the number of gas molecule in terms of moles, R is the universal gas constant (8.31 J K^{-1} mol^{-1}), and T is the absolute temperature. The velocity of an ideal gas obeys the max-well distribution theory.

$$\frac{\delta N}{\delta \nu} = N4\pi\nu^2 e^{\left[\frac{-mV^2}{2KT}\right]}\sqrt{\left[\frac{m}{2\pi KT}\right]}. \tag{1.8}$$

The mean velocity (C_{mean}) is given by

$$C_{mean} = \sqrt{\frac{8RT}{m}}, \tag{1.9}$$

where m is the mass of one gas molecule, K is the Boltzmann constant, $\delta N/\delta \nu$ is the probability of velocity, N is the number of moles, and ν is the absolute velocity. The growth of the thin films initiated by the interaction of gas molecule with the substrate. The number of gas molecule interacted with the substrate is given by the cosine law (J).

$$J = \frac{nC_{mean}}{4}, \tag{1.10}$$

where n is the number of gas molecule per unit volume. The precursor gas molecule transported to the substrate follows the above theory. The transportation of the precursor gas molecule is known as mass transport. The deposition rate is either controlled by the mass transport or the reaction rate of the gas molecule. The deposition phenomenon also depends on the temperature in the chamber or substrate, because it affects the rate of gas phase chemical reaction during thin film formation. The CVD process is classified on the basis of operating pressure, heating method to induce the gas phase reaction, and the nature of precursor as given below

1. Thermally activated chemical vapour deposition (TACVD)
2. Plasma-enhanced chemical vapour deposition (pECVD)
3. Photo-assisted chemical vapour deposition (PCVD)
4. Metal organic chemical vapour deposition (MOCVD)

1.6.2 Thermally Activated Chemical Vapour Deposition

In thermal CVD process heat energy is supplied for the precursor gas molecule to induce the reaction between the precursor gas and the substrate. Different source of heat energy (resistive, radiation, or inductive) is used in the thermally activated chemical vapour deposition. Based on the deposition pressure, the TACVD is classified into low pressure chemical vapour deposition technique (LPCVD) and atmospheric pressure chemical vapour deposition technique (APCVD). The schematic representation of TACVD is shown in Fig. 1.15. In the thermally activated CVD process, the precursor gas passes through the chamber and reacts with the substrate assisted by the thermal energy resulting in the formation of thin film (Fig. 1.15). The above basic principle of both LPCVD and APCVD was same; however, they differ in the mass transfer rate of the precursor gas and the substrate reaction. The mass transfer rate of the precursor gas and substrate reaction influence the quality of the thin film deposited over the substrate. In LPCVD the mass transfer

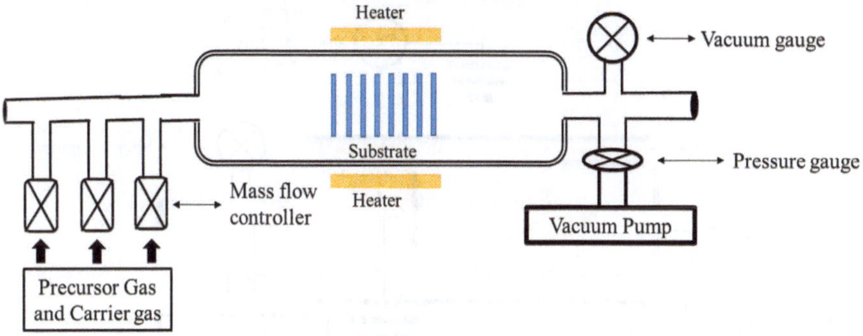

Fig. 1.15 Schematic representation of thermally activated chemical vapour deposition

rate of precursor gas is low as compared to that of surface reaction rate, while in APCVD both are in the same magnitude. In LPCVD the pressure maintained inside the deposition chamber is below ~10^{-1} m bar; hence the diffusion process of transported precursor gas reduces as compared to that of APCVD. The reduction in the diffusion process reduces the mass transport velocity of the precursor gas; hence most of the gas molecule interacts with the substrate which in turn increases the surface reaction rate. Hence, in LPCVD the deposition rate is controlled by the substrate reaction rate and leads to the uniformity, good step coverage, and purity of the thin film as compared to that of APCVD [13].

1.6.3 Plasma-Enhanced Chemical Vapour Deposition

In conventional CVD process high temperature is required to obtain the acceptable deposition rate during the growth process of the thin film. However, this high temperature during the deposition process affects the thermally sensitive substrate leading to the formation of defects in the deposited thin film. Hence, in PECVD the plasma energy is used to activate the reaction of the precursor gas molecule during the thin film deposition. The plasma energy enhances the inelastic collision process of the charged species (electron) with the precursor gas molecule to activate the chemical reaction during the thin film deposition and it also enhances the surface energy of the substrate to induce the nucleation process of the deposited atoms. Usually high voltage radio frequency (RF) power at reduced pressure is used to generate the charged species to ionize the precursor gas molecule by inelastic collision at reduced temperature. Hence, the ionized gas molecule is more reactive at reduced activation energy and is utilized for the growth of the thin film. The PECVD is used in two modes: (1) direct and (2) remote PECVD (RPECVD).

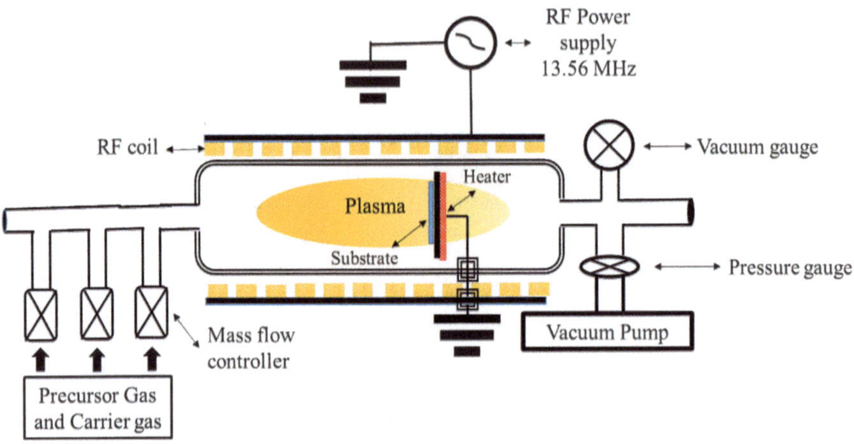

Fig. 1.16 Schematic representation of direct PECVD

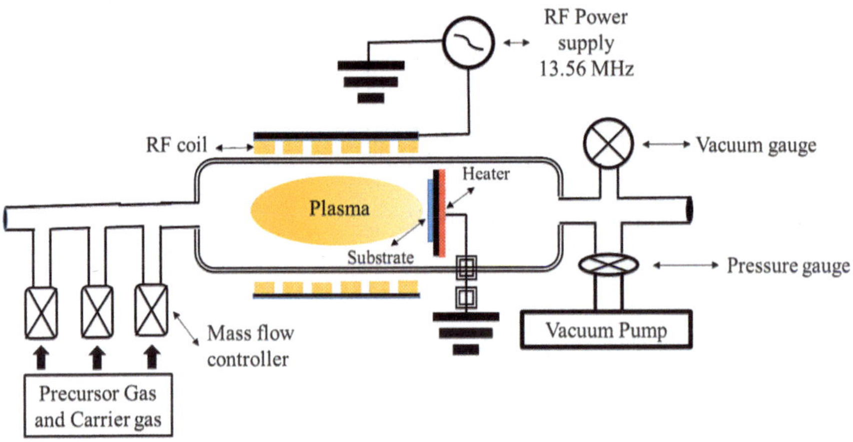

Fig. 1.17 Schematic representation of remote PECVD

In the direct mode, generally the RF power is used to generate the glow discharge of the precursor gas molecules and the substrate is placed between the RF electrodes as shown in Fig. 1.16, where the plasma generation is concentrated. The presence of substrate in the plasma region during the deposition process undergoes the bombardment of the charged species over the deposited thin film. This bombardment of energetic particle leads to the formation of pinholes, electrical defects and increases the complexity in the chemical reaction. However, in RPECVD the substrate is spatially separated from the plasma region and only a small quantity of reactant gas undergo the reaction and lead to the controlled deposition of the thin film (Fig. 1.17). This controlled deposition minimizes the formation of pin holes and defects over the deposited thin film.

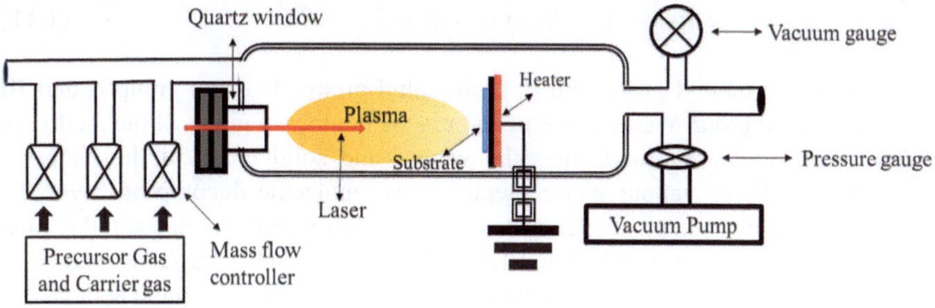

Fig. 1.18 Schematic representation of PCVD

1.6.4 Photo-Assisted Chemical Vapour Deposition

In order to obtain the acceptable deposition rate, the PCVD uses the low substrate temperature as compared to the conventional thermal CVD. Basic principle involved in PCVD process is to stimulate the deposition process over the substrate through photothermal energy. The photo-assisted CVD process uses the pulsed or scanning laser as photo energy source. The photo energy source focused locally over the substrate to raise the temperature, which in turn induces the chemical reaction of precursor gas molecule over the substrate and leads to the growth of the thin film. The PCVD set up consists of focused laser with the optical access provision in the deposition chamber. Hence, the laser beam is scanned across the *x*–*y* plane in the required pattern over the substrate. Thus, in PCVD the selected area deposition can be achieved through the *x*–*y* sweeping control of focused laser. Unlike PECVD, the PCVD process has the flexibility to perform the deposition process either under atmospheric or lower pressure. In PCVD, controlling the laser power source, pulse rate, and deposition atmosphere decides the final quality of the deposited thin film (Fig. 1.18).

1.6.5 Metal Organic Chemical Vapour Deposition

MOCVD is used to grow the epitaxial layer of compound semiconductors, oxides, and super conductors for the advanced high performance electronics and optoelectronic applications. The epitaxial deposition process of MOCVD undergoes thermal decomposition of metal alkyls, nitrides, and hydrides. The precursor gas is introduced inside the deposition chamber. Usually the deposition chamber is kept at room temperature during the deposition process. However, the substrate to be coated is kept at elevated temperature to form the desired chemical reaction for the thin film growth. The chemical decomposition occurs during the growth of the epitaxial layer and is given by Eq. (1.11).

$$R_XM(v) + NR_n^{'}(v) \rightarrow MN(s) + XRR^{'}(v), \qquad (1.11)$$

where R is the methyl group and R' is the ethyl group, M is the group II and III metals, N is the group V and VI elements, x is the number of moles of the methyl or ethyl group, and v and s denotes the vapour and solid phase of the material, respectively. These vapour phase precursor gas molecule decomposes over the substrate, which is kept at elevated substrate temperature to deposit the final product. The basic set up of MOCVD is similar to that of TACVD as shown in Fig. 1.15. The precursor gas is transported to the reaction chamber held at the room temperature along with the carrier gas (H_2). MOCVD deposition can be performed at atmospheric or low pressure. The low pressure deposition of thin film helps to maintain the purity and uniformity over the large area of the substrate. The Lewis acid and base reaction is one of the major problems in MOCVD, which occurs between the II/III group and V/VI group material. This problem is minimized by physically separating the flows of II/III and V/VI group precursor gas molecule until they reach the area where the thin film growth starts. Thus, in MOCVD the quality of the epitaxial layer depends on the factors such as substrate temperature, flow rate of the gas, design of the deposition chamber, and angle of the mounted substrate.

1.7 Thin Film Characterization

1.7.1 Importance of Thin Film Characterization

Optical, electrical, mechanical, and chemical composition of the thin film are highly dependent on their physical phenomenon such as thickness, microstructures, density, and surface roughness, which in turn depends on the deposition conditions. Various characterization tools are employed to study the properties in order to optimize the thin film properties for specific applications. Though numerous techniques are used for investigating the thin film properties, widely used structural, mechanical, and electrical characterizations are discussed below.

1.8 Structural Characterization

The structural properties such as crystal structure, grain orientation, surface morphology, surface roughness, thickness, defects, etc. have major effect in determining the performance of the thin film towards applications. Various techniques are involved to analyse the structural characterization of thin film and some of the important techniques are discussed below.

1.8.1 X-Ray Diffraction

X-ray diffraction technique is used for the analysis of the crystalline nature of materials. X-rays are part of the electromagnetic spectrum in the wavelength range of 0.1–100 Å. When X-rays fall on a material, they may undergo either elastic (Thomson) or inelastic (Compton) scattering primarily through the electron present in the atoms of the material. A periodically arranged crystalline material elastically scatters in the form of spherical waves. The generated waves cancel one another through destructive interference, while they may constructively interfere in few directions. The intensity ratio between the incident and scattered light can be derived as follows: Consider the case of a single free electron of mass 'm', charge 'e', and at a position 'R_0' interacts with the incoming monochromatic X-ray beam with wave vector K_0 and intensity I_0. The interaction of X-ray induces oscillation of the free electron which results in scattering by creating an electromagnetic field at a distance R away from the electron. The scattering of the electromagnetic wave from the free electron changes the intensity to $I(R)$ as given by Eq. (1.12)

$$I(R) = I_0 r_e^2 \frac{K_1 + K_2 \cos^2(2\theta)}{R^2}, \tag{1.12}$$

where, $r_e = \frac{e^2}{4\pi\varepsilon_0 mc^2}$ is the classical radius of the electron whose value is 2.82×10^{-15} m and 2θ is the scattering angle. K_1 and K_2 are fractions of the incident field amplitude perpendicular and parallel, respectively, to the scattering plane spanned by the incident and scattered beams. In the case of scattering from an atom, all the electrons are considered and r_e may be replaced by Zr_e in Eq. (1.12) to give a quantitative narration regarding X-ray diffraction by an atom.

However, the X-ray diffraction by crystal system is little complicated since they consist of periodic group of unit cells and atomic planes. For discussion, let us consider the case of simple cubic crystal structure which has eight atoms at each corner and is irradiated by a monochromatic X-rays of wavelength λ and intensity I_0 as shown in Fig. 1.19. Let K_0 and K be the wave vector of the primary incident beam and outgoing scattered beam, respectively, and Q be the wave vector difference, i.e., $K - K_0$. The necessary condition to obtain the maximum intensity at a position R can be set by Laue's equations for integer values of a reciprocal lattice indices (h, k, l) are as follows

$$\text{For } I(R) \to \max, \quad \begin{aligned} aQc_1 &= 2\pi h, \\ aQc_2 &= 2\pi k, \\ aQc_3 &= 2\pi l, \end{aligned} \tag{1.13}$$

where c_1, c_2, and c_3 are three unit vectors of three orthogonal directions in space and the values of integers h, k, and l may lie between $-\infty$ and $+\infty$.

A peak appears in XRD only when scattered waves interfere constructively. Bragg derived a relation between incident wavelength and the appearance of Bragg

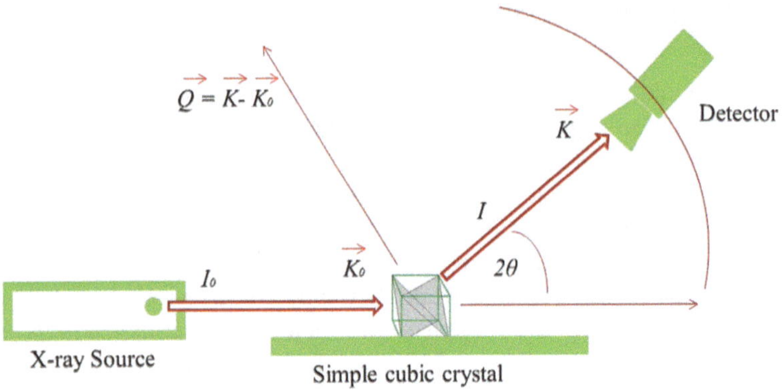

Fig. 1.19 Scattering of X-rays by a simple cubic crystal

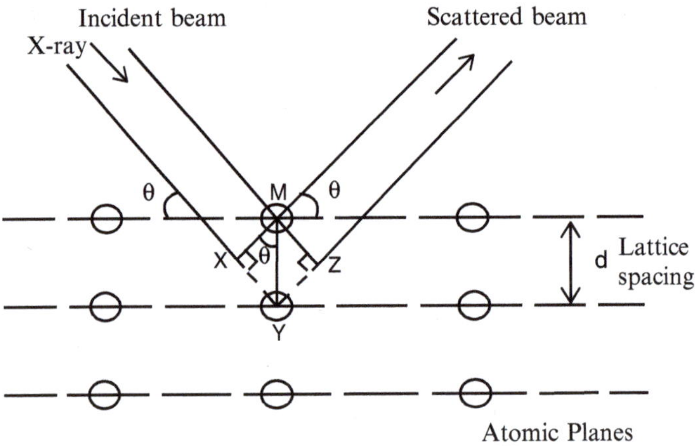

Fig. 1.20 Bragg's model of diffraction of X-rays

peaks through simple geometric calculations. Consider a monochromatic X-ray beam with wavelength λ is incident at an angle θ on a set of crystallographic lattice planes with spacing d_{hkl}. The atomic planes of a crystal cause an incident beam of X-rays to interfere with one another as they leave the crystal as given in Fig. 1.20.

Constructive interference occurs only when,

$$\begin{aligned} n\lambda &= XY + YZ \\ \text{Since}, XY &= YZ = MY \sin\theta \ \text{ and } \ MY = d_{hkl} \\ \Rightarrow n\lambda &= 2d_{hkl} \ \sin\theta. \end{aligned} \tag{1.14}$$

Equation (1.14) represents the most popular Bragg's equation, where n refers to reflection order.

X-ray diffractometer is an instrument used for measuring the Bragg reflection and consists of a radiation source, a monochromator beam adjusting slits, a detector, and sample holder. Generally goniometer is used to adjust the positions of the sample and the detector. An impenetrable beam stopper is usually used to block the X-rays which are not diffracted by the sample to save the detector from being damaged. A computer system is attached with the diffractometer to get the digital diffraction data in the form of intensity vs 2θ. The diffraction peaks vary with the size and shape of the crystallites present in the sample. The intensities of the diffracted waves depend on the orientation of the crystalline planes in the materials. However, most of the materials are composed of different oriented planes called as polycrystalline materials. These polycrystalline planes in X-ray diffractometer show multiple diffracted peaks.

The crystalline phase of the deposited material is identified from X-ray diffraction pattern. The analytical analysis of chemical phase from the X-ray diffraction pattern is identified by comparing with the standard diffraction database. The standard diffraction database consists of thousands of diffraction patterns for organic and inorganic solids, which were experimentally recorded and calculated using known material. International Centre for Diffraction Data (ICDD) maintains the standard diffraction database. The position of 2θ, the width, and the intensity of the diffracted peak is used to estimate the qualitative information such as mean crystallite size, lattice parameter, lattice stain, and crystalline nature.

In 1918, Scherrer related the width of a diffracted peak to the crystallite size of the samples and derived a popular equation known as Scherrer's equation.

$$D = \frac{K\lambda}{\beta \cos \theta}, \tag{1.15}$$

where D is the mean crystallite size, β is the full width at half maximum (FWHM), θ is the Bragg angle of the peak centroid, and K is the dimensionless shape factor known as Scherrer's constant, whose typical value is close to the unity ($1.0 > K > 0.89$).

1.8.2 Grazing Incidence X-Ray Diffraction

As the thin film thickness is (1–1,000 nm) extremely small, the primary beam of X-ray in the diffraction technique penetrates the film as well as the substrate under conventional 2θ scanning method. Accordingly the conventional X-ray scanning method of thin film gives additional intense signal from the substrate compared to that of thin film material. In order to obtain the strong X-ray diffraction signal from the thin film, the X-ray source is fixed at certain glazing angle to perform the 2θ scanning so that the X-ray penetrates through larger distance in the thin film instead of penetrating the substrate as in the case of conventional method. Thus, the Primary beam is allowed to enter into the thin film at a very small angle and increases the travel path inside the thin film. This increased travel path in turn increases the intensity of the signal from the thin film material compared to the substrate. This technique is known as Glazing Incidence X-ray diffraction technique (GIXRD). The glazing angle is fixed slightly above the critical angle based on the material nature of thin film to obtain the complete reflection.

Let the incidence angle taken here be denoted by α. The maximum path travelled (l) by the X-rays within the thin film layer can be estimated from the relation

$$l = \mathrm{t}/\sin\alpha, \tag{1.16}$$

where t is the thickness of the thin film. The reliable value of α can be chosen by comparing the average path length with the inverse attenuation coefficient where t is the thickness of the thin film. The reliable value of α can be chosen by comparing the average path layer with the inverse attenuation coefficient, i.e., $1 \approx 1/\mu$.

To derive the new absorption factor, a configuration factor k is introduced, i.e.,

$$k_\alpha = \frac{1}{\sin\alpha} + \frac{1}{\sin(2\theta - \alpha)}, \tag{1.17}$$

where α is used as subscript of k in order to characterize it as the appropriate GI-XRD factor. The GI-XRD absorption factor A_α can accordingly be written

$$A_\alpha = 1 - \exp(-\mu t k_\alpha). \tag{1.18}$$

The value of μt product may be determined from the measurement of a substrate reflection before and after the deposition and hence the intensity correction can be performed.

For example, the diffraction pattern of a nickel oxide (NiO) thin film of thickness ~800 nm at a glazing incidence mode is shown in Fig. 1.21. The NiO crystal structure can be identified as cubic type (ICDD reference code 01-089-7130) for which the reflection planes are indexed as shown in Fig. 1.21.

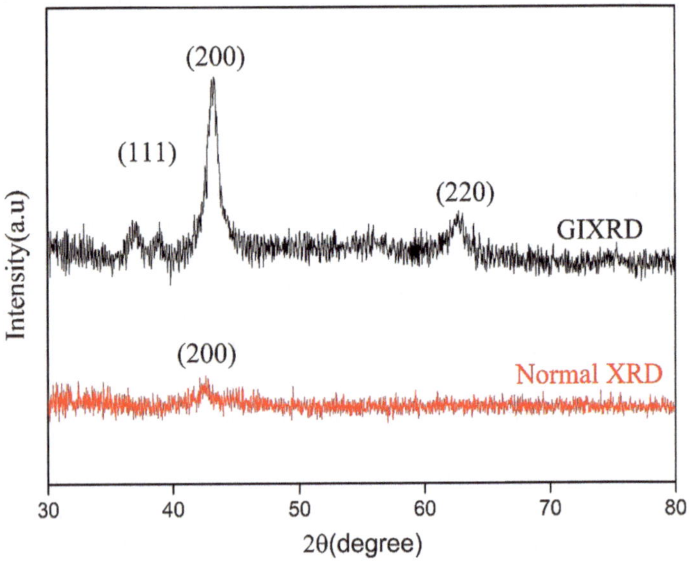

Fig. 1.21 GI-XRD pattern of NiO thin film with $\alpha = 1.5°$

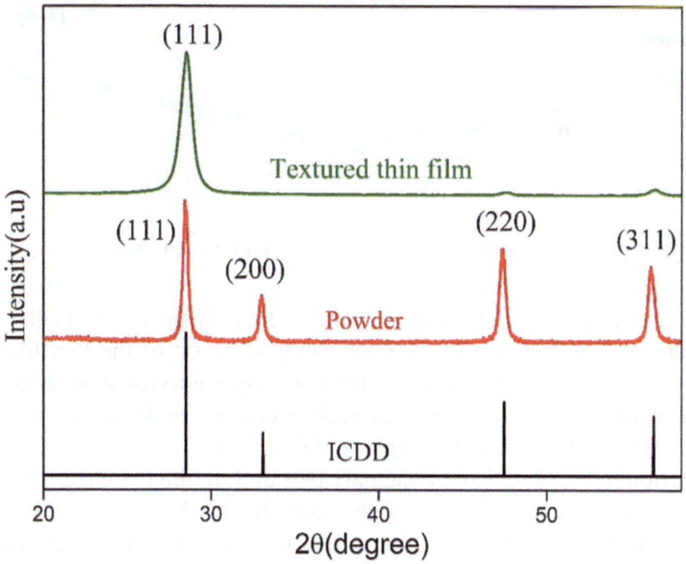

Fig. 1.22 XRD pattern of polycrystalline powder and textured thin film of ceria

1.8.3 Textured Film Analysis

In polycrystalline thin films, the crystalline grains are randomly oriented in all directions depending on the crystal structure of the material. In polycrystalline films, the grains are oriented towards out of plane and in-plane directions which are perpendicular and parallel to the substrate, respectively. However, the thin film may consist of preferred in-plane with different out-plane or vice versa, which denotes the preferential orientation of thin film. Symmetric $\theta/2\theta$ diffraction technique is used to analyse the preferential orientation of the thin films. In the $\theta/2\theta$ diffraction pattern, the preferentially oriented thin films exhibit an enhanced intensity of a particular Bragg peak in comparison to others.

For example, Fig. 1.22 demonstrates the effect of texture on a $\theta/2\theta$ diffraction pattern for a ceria thin film deposited by electron beam deposition. The powder pattern of X-ray diffraction follows the intensity ratio of the reflection peaks as per ICDD data which denotes the random orientation of grain. However, in the thin film, (111) plane is highly intense compared to other reflection plane due to the preferential orientation in thin films.

1.8.4 X-Ray Reflectivity

X-ray reflectivity (XRR) is a surface-sensitive analytical technique used to measure the thickness, density, and roughness of the thin film. XRR monitors the intensity of the reflected X-rays from the sample surface at the grazing angles. In the grazing

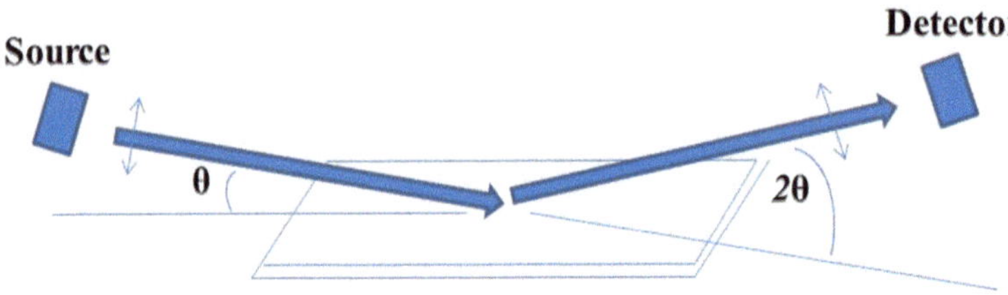

Fig. 1.23 Schematic of X-ray reflectivity where $\theta < \theta_c$ for total reflection

incident X-rays process, the incident angle (θ) is maintained smaller than the critical angle of reflection (θ_c) to avoid the penetration of the thin film surface. The intensity of the reflected beam is monitored by a detector at an angle 2θ and is used for analysis (Fig. 1.23). It uses $\theta/2\theta$ operation mode and ensures that the incident angle is always half the angle of diffraction.

Refraction index of real and imaginary part of thin film obtained from the XRR is used to calculate the density of the film. In the X-ray region the complex refractive index (n) of the thin film is less than 1, which is represented by the variables δ (real component) and β (imaginary component) as shown in Eq. (1.19).

$$n = 1 - \delta - i\beta. \tag{1.19}$$

The scattering cross-section correction (f') which depends on the material chemistry of the thin film is an important parameter to give the precise value of δ. Without appropriate correction factors for a given energy, the calculated density can be off by more than 10 %. Hence, the value of δ is determined by the following equation (1.20) for a given film.

$$\delta = \frac{r_0 \lambda^2}{2\pi} \cdot \left(Z + f'\right) \cdot n_a, \tag{1.20}$$

where λ is the X-ray wavelength, r_0 is the atomic Bohr radius, Z is the atomic number, and n_a is the atomic concentration, given by

$$n_a = \frac{N_A}{A} \cdot \rho, \tag{1.21}$$

where, N_A is the Avogadro's number, A is the atomic weight and ρ is the thin film density. From Eqs. (1.20) and (1.21) we can reduce as,

$$\delta \propto \rho \lambda^2. \tag{1.22}$$

Using Fresnel equations, it can be expressed that the real component of the refractive index is related to the critical angle as,

$$1 - \delta = \cos \theta_c. \tag{1.23}$$

Equations (1.22) and (1.23) are used to calculate the density of the material for several energies.

In XRR technique the thin film thickness is measured by maintaining the incident X-ray at higher angle compared to the critical angle of reflection ($\theta > \theta_c$). Thus, the X-ray beam penetrates inside the film by refraction and reaches the substrate. Therefore, the reflection of beam takes places from both the thin film and substrate results in an angle-dependent interference fringes. According to the modified Bragg equation, the values of θ from the maxima and minima of the interference fringes are related to the thickness of the film (d) as given in Eq. (1.24) [14],

$$d = \frac{\lambda}{2} \frac{n_i + \Delta n}{\sin^2\theta_i - \theta_c^2}, \tag{1.24}$$

where θ_i is the ith maximum or minimum position of the interference fringe, n_i is an integer, and $\Delta n = 1/2$ and 0 for the maximum and minimum, respectively. Using Eq. (1.24) the thickness of the thin film can be calculated with precision better than 1 Å.

1.8.5 Scanning Electron Microscopy

Scanning electron microscopy (SEM) uses electron as the probe to image the surfaces at high resolution. The magnification of SEM is greater than 100,000× in comparison to optical microscopy. SEM is used for analysing the microscopic characteristics of the thin film such as surface morphology, grain boundaries, and porosity. These microscopic characteristics have strong influence on the basic property of the thin film such as mechanical, electrical, and optical property.

A schematic representation of a typical SEM is shown in Fig. 1.24. In SEM electron beams under high vacuum are emitted from an electron gun by thermionic heating of tungsten filament or by field emission cathode. The electron produced from the electron gun accelerated towards the anode and is subsequently focused by the condenser and objective lenses before falling on the sample surface. The energy of the accelerated electrons can be varied from 100 eV to 30 keV. The incident electron travelling down the chamber is deflected in the raster scanning pattern for imaging the sample surface using the scanning coils and the typical spot size of an incident electron is around 10 Å.

The incident electron travelling at high energy towards the sample surface produces both elastic and inelastic scattering of the electrons. When the incident electron interacts with the nucleus of the sample, it results in elastically scattered electron as shown in Fig. 1.25. If the energy of the elastically scattered electron is equivalent to that of incident electron, it is referred to as backscattered electron. The probability of generating backscattered electrons depends on the atomic number and density of the sample. Thus, the difference in the generation of the

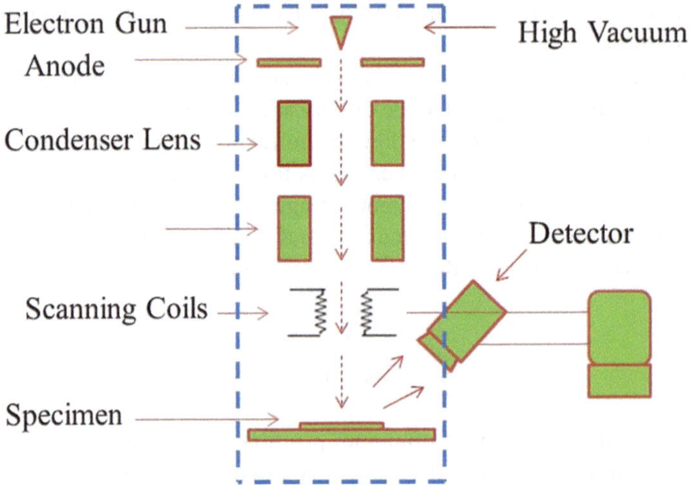

Fig. 1.24 Schematic diagram of typical SEM

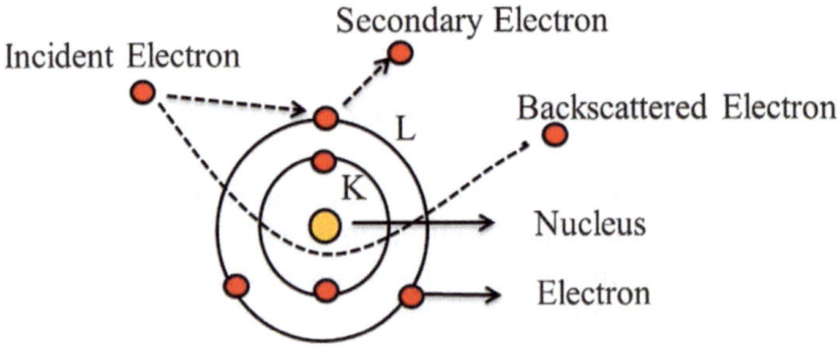

Fig. 1.25 Emission of secondary and backscattered electrons

backscattered electron depending on atomic number creates a contrast known as atomic number contrast. As a result, higher atomic number elements appear brighter than the lower atomic number elements.

The scattered electrons generated due to the inelastic collision of incident primary electron with the sample surface are known as secondary electrons. Secondary electron emitted from the near surface of the sample possesses a low energy of less than 50 eV. Production of secondary electrons depends on the topography of the sample surface which generates the topographic contrast. In order to obtain the image from SEM, the incident electrons from the cathode source are scanned across the sample in raster scanning pattern. During the scanning process, the electron scattered from the sample source are detected by the cathode ray tube (electron detector). The electron detected by the cathode ray tube appears as a bright or dark spot depending on intensity of electron beam ejected from the particular position on the sample surface. In general, electrons emitted from the specimen surface facing the detector are collected preferentially and appear bright in the image.

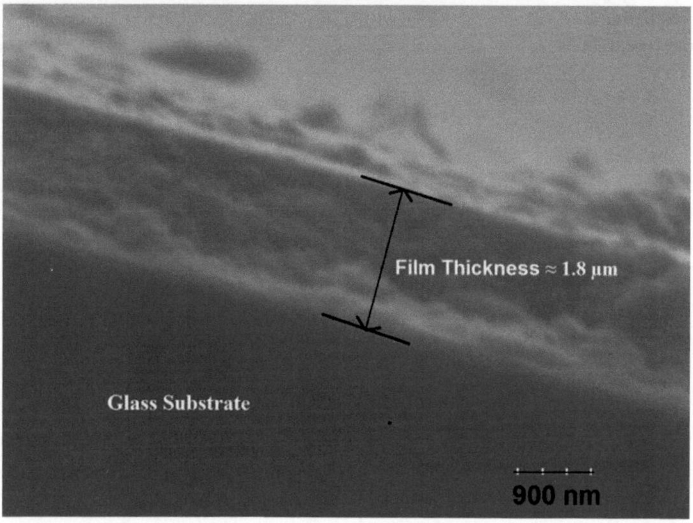

Fig. 1.26 SEM cross-sectional image of a samarium-doped ceria (SDC) thin film

If the sample surface is not facing the detector, the ejected electrons have difficulty in reaching the detector and the corresponding spot appears as dark. Figure 1.26 shows the cross-sectional SEM image recorded by secondary electron mode for nano-structured samarium-doped cerium oxide over glass substrate deposited by electron beam deposition method. The edges of the thin film and its thickness (approximately 1.8 μm) can be measured from the SEM image.

Currently the modified SEM like environmental scanning electron microscope (ESEM), field emission scanning electron microscope (FESEM), etc. are being used as a powerful analytical tool for surface analysis. The ESEM allows even creating specific gaseous atmosphere to collect electron micrographs of various samples and FESEM is used especially for high resolution, i.e., from sub-micron to <10 nm lateral spatial resolution for the analysis of nano-sized samples.

1.8.6 Transmission Electron Microscopy

Transmission electron microscopy (TEM) is a powerful technique used for investigating the morphology, crystallographic information, interface effect between individual layers, grain size distribution, and chemistry of the materials. In TEM, the electron gun accelerates a stream of monochromatic electron beam which reaches the condenser lens. Condenser lens focuses the electrons as thin coherent beam to a small spot size as shown in Fig. 1.27. The coherent electron beam strikes the sample surface and partly transmitted through sample. The electrons incident through the sample undergo both elastic and inelastic scattering process. If the energy of the incident electron beam remains the same after interacting with the specimen, the scattering process is referred as elastic scattering. These elastically scattered electrons give rise

Fig. 1.27 Schematic diagram of typical TEM

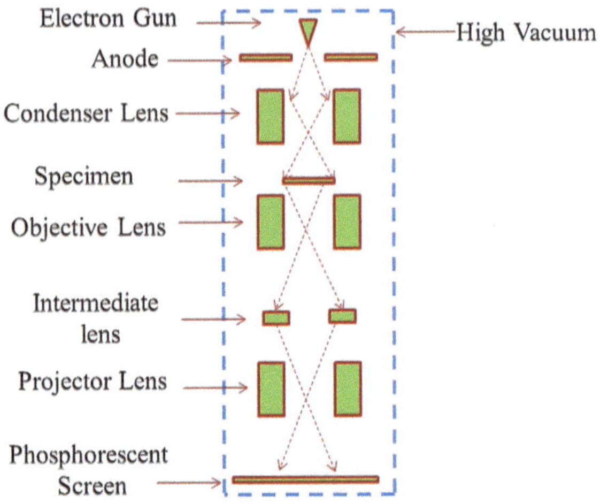

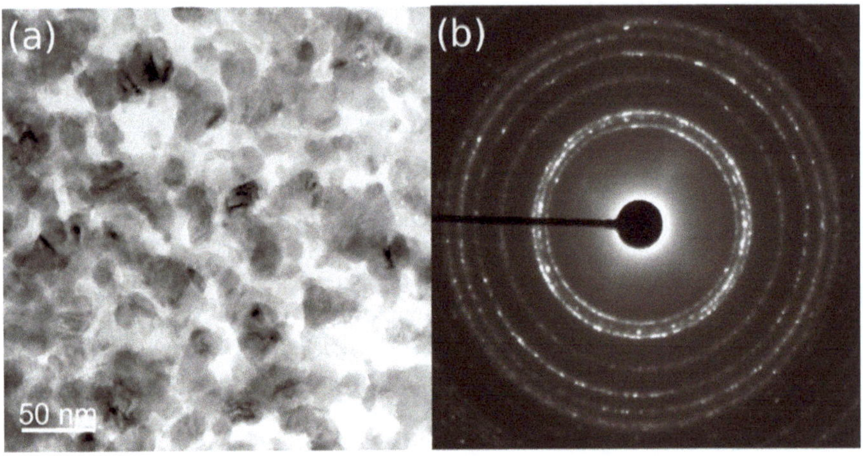

Fig. 1.28 (**a**) Bright-field TEM image of a ZnO thin film and its associated. (**b**) Electron diffraction pattern showing te rings from the ZnO wurtzite structure. Reprinted with permission from [15], B.S. Straumal et al., Beilstein. Journal of Nanotechnology, 4 (2013). © 2013, Beilstein

to various diffraction pattern of the sample similar to X-ray diffraction following Bragg's law. The representative electron diffraction pattern of zinc oxide thin film with TEM image is shown in Fig. 1.28. The incident electrons which are transmitted through the thin specimen without any interaction are known as unscattered electrons. The transmission of unscattered electrons is inversely proportional to the sample thickness. Thicker areas block the transmission of electrons and appear darker while thinner areas appear brighter due to more transmission. These electrons can be used for the analysis of the grain boundaries, dislocation, secondary phase formation, defect chemistry, and density variations. Primary and diffracted electron beams that emerge from the sample are made to pass through a series of objective and projector lenses.

The objective lens produces the first image of the sample. Finally, the projector lenses further magnify the image and projects it on a phosphorescent screen.

Unscattered electrons are selected using the objective aperture to form the image in bright field microscopy, while dark field uses the elastically scattered electrons. The resolution of the microscope denotes the ability to distinguish between two adjacent points located closely.

High resolution transmission electron microscopic (HRTEM) is used to obtain the structural information at the lattice level. Here both the direct and the diffracted beam are allowed through objective aperture to capture the image. The spatial resolution of HRTEM should be high enough (~0.2 nm) to obtain the lattice level imaging of the crystalline material. The HRTEM image is used to analyse the interfacial defects, lattice parameter, grain boundaries, and nanocrystalline features.

1.8.7 Elemental Identification and Quantification

Apart from structural analysis electron microscopy can also be useful for the identification and quantification of elements. Electron energy loss spectroscopic (EELS) technique in TEM is used to analyse the structural and chemical information of the sample. In EELS technique, the electrons with known kinetic energy are allowed to strike the sample; after hitting the electrons undergo inelastic collision by losing their kinetic energy with the electron cloud of the sample. The inelastic scattered electron were monitored though electron spectrometer and used for further analytical analyses of the sample. Generally the EELS spectrum along with electron diffraction pattern and imaging is used to analyse the atomic composition, oxidation states of the elements, electronic property, and local thickness of the sample during TEM analyses.

Energy dispersive X-ray spectroscopy (EDX) is a micro-analytical X-ray technique used to identify the elemental composition of the sample. It can be coupled with the instruments like SEM and TEM. The X-ray emitted from the sample during the interaction of electrons beam focused either from SEM or TEM over the sample surface is used in EDX for semi-quantitative analysis of chemical composition. The energy dispersive detector is used to detect the X-ray energy characteristics and its intensity produced from the sample. The X-ray energy characteristics are used to identify the chemistry of the element and the intensity count is used to quantify the concentration of the element.

Cross-sectional transmission electron microscopy (XTEM) technique is used for the analysis of the multi-layered structures of thin film. [16] The sample preparation for the cross-section view in XTEM undergoes difficult process. The processes involved in the sample preparation are (1) transverse cleaving of sample, (2) glue the transversely cleaved sample, face to face together at the area of interest, and (3) thinning the sample by grinding and polishing or dimpling.

1.8.8 Atomic Force Microscopy

A branch of microscopy that forms the image by scanning over the sample surface using a probe is known as scanning probe microscopy. One among the many popular scanning probe techniques is known as atomic force microscopy (AFM). AFM uses a small probe to scan across the sample surface to provide the surface topographical images as well quantifies the local atomistic properties such as friction, magnetic domains, and chemical nature. The probe consists of cantilever beam with sharp tip in the range of 15–40 nm. This probe is scanned across the sample in raster scanning mode across the selected area to measure in *x*–*y* direction as shown in Fig. 1.29. The interatomic force between the probe tip and the sample surface deflects the cantilever beam vertically and laterally with respect to the change in surface topography of the sample. These changes in the cantilever beam are monitored by the laser beam. The change in the surface topography of the sample during scanning process via probe is reflected in the laser beam. The change in the laser beam is detected by photo detector which is recorded through the computer feedback and generates the surface topography of the sample. The scanning process of AFM can be carried out in contact, non-contact, or tapping modes.

In contact mode the AFM probe is kept at constant force with the sample surface. This force between the probe and surface of the sample is maintained throughout the scanning process of AFM. When the probe encounters a topographical change over the surface of the sample, it comes back to the original position. These changes in the cantilever beam are recorded using computer feedback to obtain the topographical image with the lateral and vertical resolution of <1 nm and <1 Å, respectively. In the tapping mode the AFM probe oscillates at the resonance frequency and is used to scan the sample surface and the amplitude of oscillation is higher than 10 nm. During scanning process, the oscillating probe is maintained

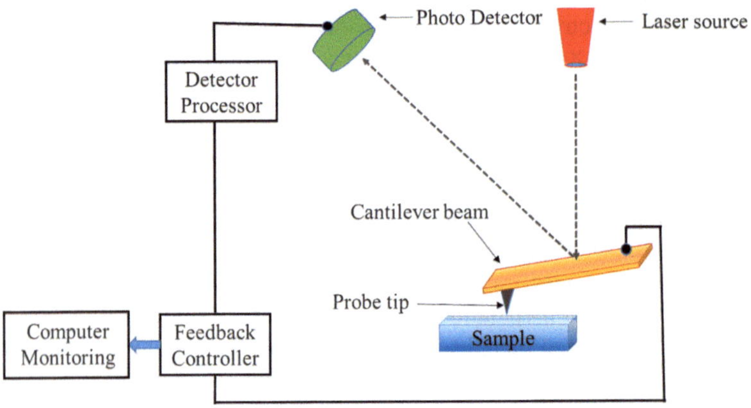

Fig. 1.29 Schematic diagram of AFM instruments

at certain distance, which barely taps or closely moves over the sample surface. The position of the cantilever beam and the oscillation frequency is continuously monitored by the laser to obtain the topographical image of the sample. The lateral and vertical resolution of image obtained through tapping mode is around 50 and 1 Å, respectively. In non-contact mode, cantilever is oscillated at its resonant frequency or just above and the amplitude of oscillation is less than 10 nm. Since there is no contact between AFM tip and the sample surface, non-contact does not suffer from tip or sample degradation.

AFM provides three-dimensional surface profile for both conducting as well as nonconducting samples with high resolution.

1.8.9 Profilometer

Profilometer is also called surface profiling system and is used to quantify the roughness and thickness of the thin film. The measurement in the profilometer is either done in contact or non-contact mode (Fig. 1.30).

In case of contact mode, a diamond stylus is used to move vertically and laterally in contact across the thin film surface with a specific force for a given distance. The vertical displacement of the stylus as a function of position with respect to the surface of the thin film is recorded. The displacement of the stylus from its normal position with respect to the thin film surface is used to calculate the thickness and roughness of the thin film.

The non-contact mode Profilometer is also called an optical Profilometer, in which a beam of light is focused on the sample surface and the variation in reflection time is monitored. Depending on the surface of the thin film, the reflection process of the light induces the constructive and destructive interference. This interference pattern is used for the calculation of thickness and roughness of the sample.

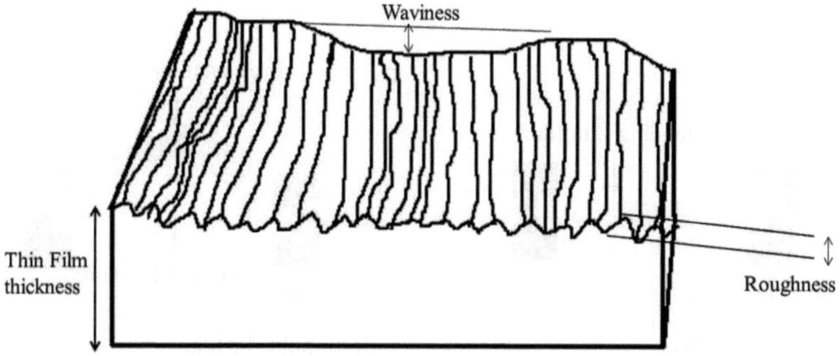

Fig. 1.30 Schematic of surface analysis by profilometer

1.9 Mechanical Characterization

The mechanical properties of the thin films differ from its bulk counterparts. Various factors can be attributed to the changes in mechanical properties of thin films.

1. The surface area to volume of thin film increases in proportion to the dimension. As a result size effect influences the mechanical properties than the macroscopic properties.
2. The processing methodology differs from the bulk structure formulation which modifies the phase, composition, and microstructure of thin films. The thickness of the film extends from few micrometre to nanometre, the conventional mechanical testing systems are not amenable for testing the thin film.

However, mechanical testing is needed to ensure the performance of the thin film. Various techniques are available for the evaluation of mechanical properties of thin films and the most important techniques are discussed below.

1.9.1 Microtensile Testing

Microtensile testing is used to evaluate the information of elasticity and plasticity of the material. The experimental challenges in microtensile testing of thin films are (1) extreme delicacy to detach the film from the substrate, (2) mounting it in a proper alignment, (3) careful loading and unloading process to measure the mechanical response (Fig. 1.31a). Electromagnetic transducer is used for loading and unloading process and the associated stress–strain curve for the thin film is measured. The following equations are used to estimate the stress, strain, and Young's modulus for thin film through microtensile testing.

$$\sigma = \frac{P}{A}, \tag{1.25}$$

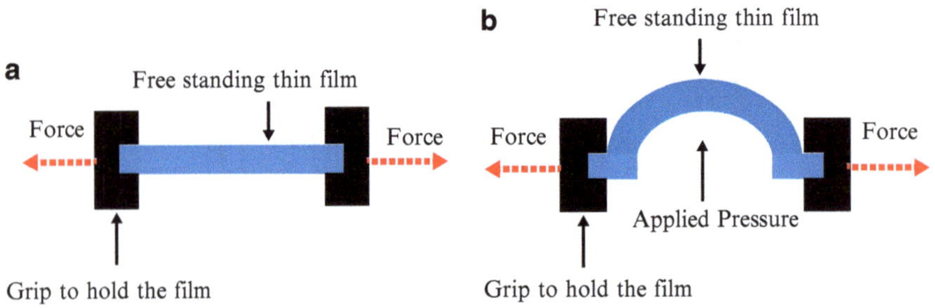

Fig. 1.31 Methods for mechanical testing of thin films: (**a**) tensile testing, (**b**) bulge testing

$$\varepsilon = \frac{L_o - L}{L}, \tag{1.26}$$

$$E = \frac{\sigma}{\varepsilon}, \tag{1.27}$$

where σ is the stress, ε is the strain, A is the original cross-section area of the thin film, L is the original length of the thin film, L change in length of the thin film, E is the calculated Young's modulus for the thin film.

1.9.2 Bulge Testing

Bulge testing in the thin film is used to characterize the elastic modulus, fracture toughness, residual stress, and yield strength. The experimental steps followed in the bulge testing are (1) detaching the thin film from the substrate, (2) attaching the free standing film at the end of the hollow tube, (3) gas pressure is applied from the other end of the hollow tube. Increase in gas pressure results in hemispherical bulge in the free standing film. This deflection is measured with respect to the applied gas pressure using optical interferometric technique in order to evaluate the associated strain as shown in Fig. 1.31b. The relation between the applied gas pressure (P) and bulge height (h) is given by Eq. (1.28) and is used to evaluate the Young's modulus (Y) of the thin film [17]

$$P = \frac{4dh}{r^2}\left\{\sigma_i + \frac{2Yh^2}{3(1-v)r^2}\right\}, \tag{1.28}$$

where d and r be the film thickness and radius, respectively, h is the bulge height from normal level, σ_i is the residual stress in the film under a zero applied pressure, v is the Poisson's ratio, and Y is the Young's Modulus.

1.9.3 Nanoindentation

Nanoindentation is used to analysis the mechanical properties of the thin film when the thickness is typically less than 1 μm. In nanoindentation testing, a hard tip of a mechanically known material is used to apply the load over the specimen to determine the mechanical property. The applied load on the thin film is typically lower than milliNewton and varied with respect to the material nature and thickness of the film. During loading and unloading process over the thin film, simultaneous measurement of intender displacement is recorded. The contact area of the indentation is measured using the penetration depth and indenter displacement for the evaluation of the hardness of the thin film using Eq. (1.29)

$$H = \frac{P_{\max}}{A}, \tag{1.29}$$

where H is the hardness, $P_{\max}$ is the maximum applied load, and A is the intender contact area. During the measurement of hardness in thin film, the influence of the substrate can't be ignored. However, the effect of substrate can be ignored when the penetration depth of the indentation doesn't exceed above 10–20 % of film thickness.

1.10 Optical Characterization

Understanding the thin film optical property is important in order to select materials for various energy applications such as in solar cells, photocatalysis, energy efficient display devices, etc.

1.10.1 Ultraviolet–Visible (UV–Vis) Spectroscopy

UV–Visible spectroscopic technique is used for studying the optical properties of materials including thin films. Reflectance, transmittance, or absorbance can be followed as a function of wavelength regime of 200–800 nm as shown in Fig. 1.32.

Upon passing UV–Visible waves, the atoms or molecules undergo electronic transitions by absorbing certain energy from an incident light and excites the electrons from lower energy state to higher energy state. When a beam of light is incident on a thin film surface, the beam may be reflected, transmitted, diffused, absorbed, or refracted. The possibilities of these phenomena depend on the angle of incidence of the incident beam with the thin film surface. Hence, any variations such as thickness, porosity, and inhomogeneities significantly affect the UV–visible optical properties of the thin films [18].

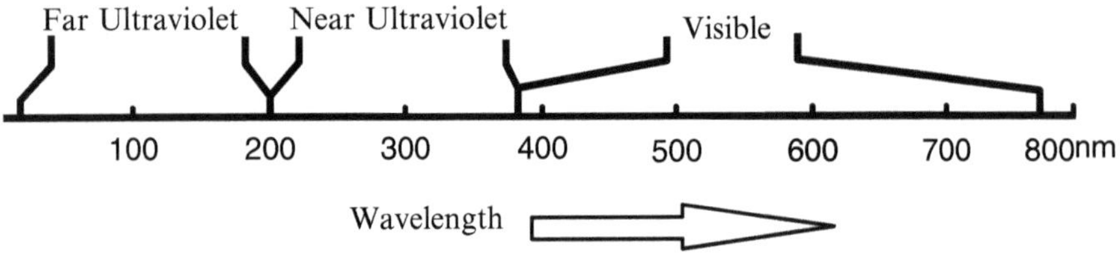

Fig. 1.32 UV–Vis electromagnetic spectrum

In a typical UV–Visible spectrophotometer, the initial intensity (I_0) and the intensity after passing through the thin film (I) are compared to get the transmittance (I/I_0) and expressed as percentage ($\%T$) of the incident beam of light transmitted by the sample. In turn the absorbance (A) is related to transmittance by the following equation (1.30).

$$A = \log(1/T). \tag{1.30}$$

For probing the band structure of semiconductor thin films material absorption study is primarily recorded. Generally, the optical band gap (E_g) and absorption coefficient α could be evaluated from transmittance or absorbance spectra. The absorption coefficient can be determined from the equation below as Tauc equation (1.31).

$$(\alpha h\upsilon)^{\frac{1}{n}} = A\left(h\upsilon - E_g\right) \tag{1.31}$$

where A is a constant and E_g is the band gap energy, n value depends on the nature of semiconductor material (direct or indirect) and has values of 1/2, 3/2, 2, and 3. From the graph of $(\alpha h\upsilon)^{1/n}$ versus $h\upsilon$, E_g and A parameter can be obtained.

Swanepoel method is employed for the determination of refractive index, absorption coefficient, band gap energy, and thickness of the thin films using transmittance data of the UV–visible Spectroscopy.

The refractive index of the thin film can be expressed by the equation given below as

$$n = \sqrt{N + \sqrt{(N^2 - s^2)}}, \tag{1.32}$$

where s is the refractive index of the substrate and

$$N = 2s\frac{T_{\mathrm{M}} - T_{\mathrm{m}}}{T_{\mathrm{M}}T_{\mathrm{m}}} + \frac{s^2 + 1}{2}, \tag{1.33}$$

where T_{M} and T_{m} are corresponding maxima and minima positions in the fitted transmission curve as shown by the example in Fig. 1.33.

Thickness of the films can be calculated by Swanepoel method by using the formula,

$$d = \frac{\lambda_1\lambda_2}{2(\lambda_1 n_2 - n_1\lambda_2)}, \tag{1.34}$$

where n_1 and n_2 are the refractive indices calculated from two consecutive maxima or minima corresponding to two wavelengths of λ_1 and λ_2.

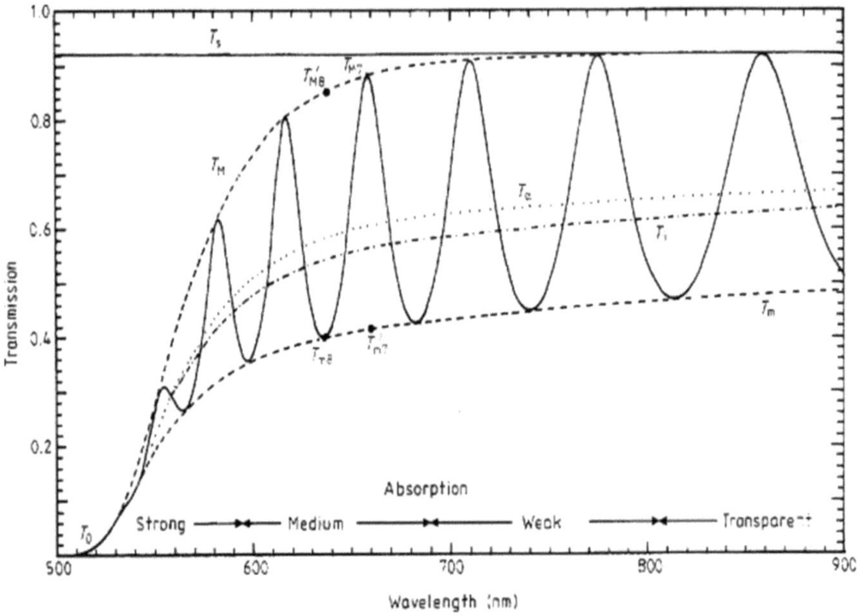

Fig. 1.33 Simulated transmission spectrum (full curve) for a 1 μm film of α-Si: H on a finite glass substrate with transmission T_s. Reprinted with permission from [19], Swanepol R, J. Phys. E: Sci. Instrum., 16 (1983). © 1983, IOP Publishing

1.10.2 Photoluminescence Spectroscopy

Photoluminescence spectroscopy is a complementary technique of UV–Vis spectroscopy for probing the optical emission properties of a material. Analysis of emitted light from the excited state to the ground state upon light-induced excitation is investigated. The emitted light is collected and analysed by a spectroscope with regard to space and time.

The photoluminescence excitation spectroscopy (PLE) is more suitable for thin films despite of its weak absorption spectrum. It can also measure the absorption coefficient of thin film on an opaque substrate. A tunable laser light dispersed from an incandescent lamp by an excitation monochromator is used to vary the exciting photon energy E_{ex} and setting the emission monochromator at a constant emission energy E_m to obtain the total or spectrally integrated photoluminescence intensity. Thus, PLE spectroscopy gives photoluminescence excitation energy spectra and analyses absorption and emission mechanism of emitting excitons.

1.11 Electrical Characterization

1.11.1 Electrical Conductivity

Electrical properties of the thin films deviate from their corresponding bulk forms. Consider a piece of material as shown in Fig. 1.34, of which electrical resistivity (ρ) is to be determined as given by Eq. (1.35)

$$\rho = R\frac{A}{l}, \tag{1.35}$$

where, R is the electrical resistance in ohm (Ω); l is the length of the specimen, A is the cross-sectional surface area of the material, and ρ is the electrical resistivity in ohm-m (Ω-m). Electrical conductivity σ is defined as the inverse of resistivity, ($1/\rho$) with the unit of Siemens per meter (S/m).

Electrical resistivity and conductivity of the thin films can be measured by various techniques. One of the techniques, namely four-point probe method, consists of four probes which are placed collaterally with equal spacing 's' between them and the tips of the probes touch the surface of the film. The current is passed through the two outer probes and the potential is measured between the two inner probes as shown in Fig. 1.35.

In case of thin film samples, where the distance between probes (s) is larger than its thickness 'd', the electrical resistivity (ρ) can be obtained by formula as expressed below. [20]

$$\rho = \frac{\pi}{\ln 2}\frac{V}{I}d, \tag{1.36}$$

$$\rho \approx 4.532\frac{V}{I}d, \tag{1.37}$$

where V is the potential difference between inner probes and I is the current through the outer pair of probes. The errors due to electrical contacts can be minimized

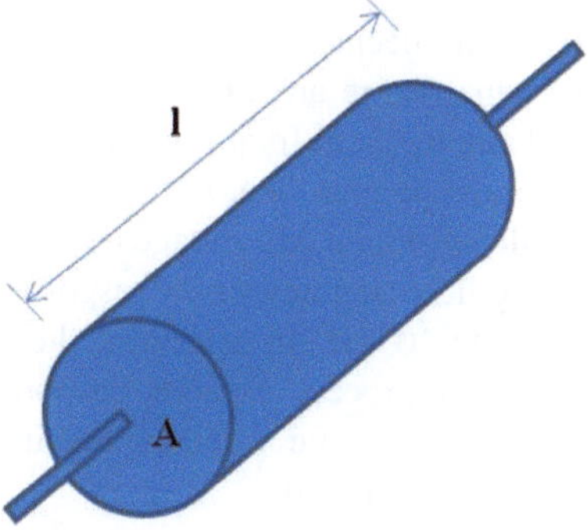

Fig. 1.34 A piece of resistive material specimen

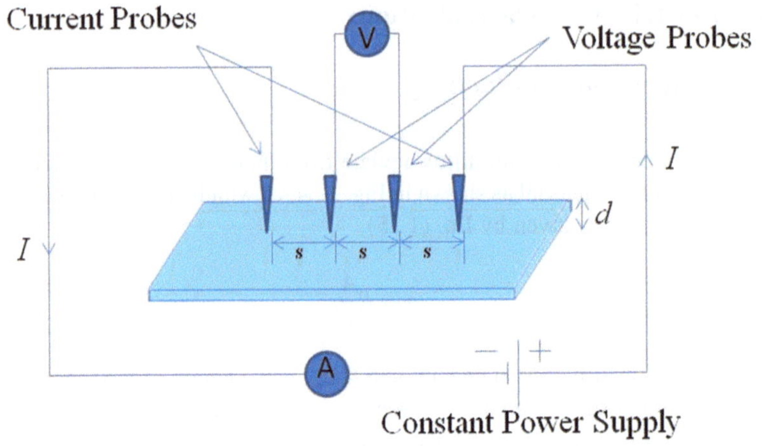

Fig. 1.35 Schematic of four-probe method

because of the separate current and voltage leads. This is a widely used technique to measure the electrical resistivity by superficial contact. This technique is mainly used in the semiconductor industry, research, and manufacturing field. The four-probe method is preferred over two-probe method because of negligible interface impedance.

1.11.2 Electrochemical Impedance Spectroscopy

Electrochemical impedance spectroscopy (EIS) is an important technique to study the charge transfer parameters of electrodes and interfaces by measuring the impedance $Z(\omega)$ (ω-angular frequency) over a range of frequency. Various properties such as frequency response of the system with respect to energy storage and dissipation properties can be investigated. The ionic and electronic transport properties in the thin film can be analysed. As a non-destructive technique this technique assumes lot of importance, particularly in energy applications in deducing the reaction mechanism of an electrochemical process in a fuel cell and battery systems. There are a number of instrument configurations available for EIS study and among them frequency response analyser (FRA) connected to a potentiostat is the most popular one. A FRA is used while measuring the impedance to impose a small amplitude AC voltage (typically 5–50 mV) to a fuel cell via the load as shown in Fig. 1.36. It analyses the response of voltage and current in a range of frequencies (10^{-4} to 10^{9} Hz) to determine the impedance of the fuel cell. Since each physico-chemical process exhibits response at different but characteristic frequencies, the exact specific value of the impedance associated with individual processes can be determined.

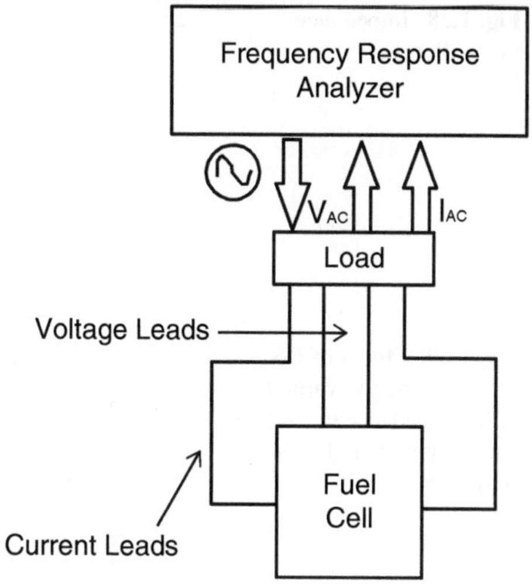

Fig. 1.36 Schematic for EIS of fuel cells

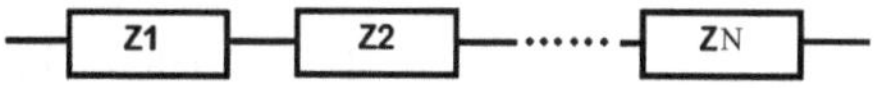

Fig. 1.37 Impedances in series

EIS measurements can also be carried out under real fuel cell operating conditions e.g. open circuit voltage (without load) or under load (DC voltage or current) and parameters such as permittivity, conductivity, and permeability can be determined. Bulk as well as interfacial properties of films and electrocatalysts can also be studied.

Equivalent circuit modelling is used to analyse the EIS data in terms of electrical circuits, which is the combination of ideal elements like capacitors, resistors, and inductors. These elements can be arranged in a series or parallel combination based on the application.

The impedance of elements in series is additive (Fig. 1.37),

$$Z_{\text{Total}} = Z_1 + Z_2 + \cdots + Z_N. \tag{1.38}$$

The impedance of elements in parallel is the inverse of the sum of the inverse of impedances (Fig. 1.38),

$$\frac{1}{Z_{\text{Total}}} = \frac{1}{Z_1} + \frac{1}{Z_2} + \cdots + \frac{1}{Z_N}. \tag{1.39}$$

The data obtained by EIS is represented graphically by a Bode plot or Nyquist plot as shown in Fig. 1.39. Bode plot is used to represent the impedance magnitude,

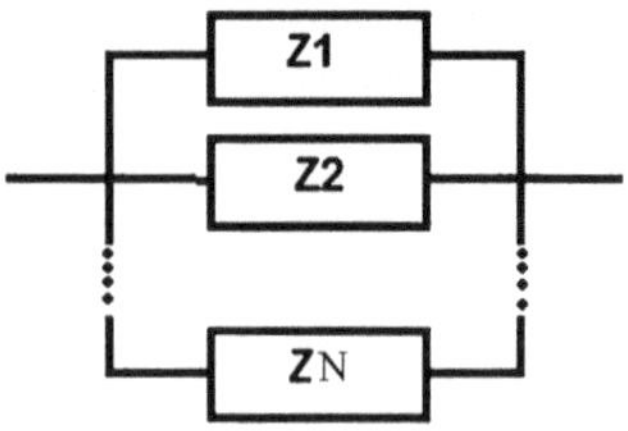

Fig. 1.38 Impedances in parallel

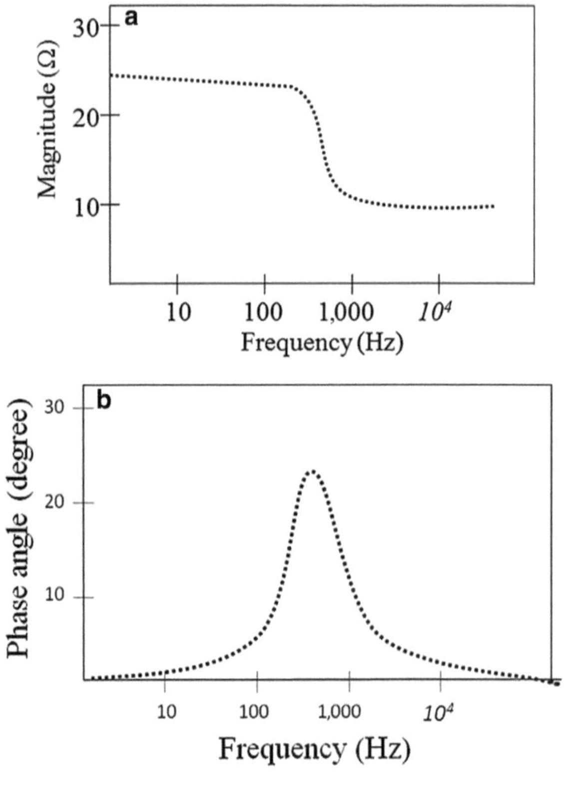

Fig. 1.39 Model of Bode Plot showing the variation of (**a**) impedance magnitude and (**b**) phase angle with frequency

i.e., the real or imaginary components of the impedance and phase angle as a function of frequency to show the system's frequency response.

A Nyquist plot is obtained by plotting imaginary impedance against the real impedance of the cell. Figure 1.40 shows a representative Nyquist plot in which the real and imaginary parts are plotted in x and y-axis of the graph, respectively. The advantage of Nyquist plot is that the activation controlled processes with distinct time constants appear as unique impedance arcs and the shape of the curve provides information about possible mechanism.

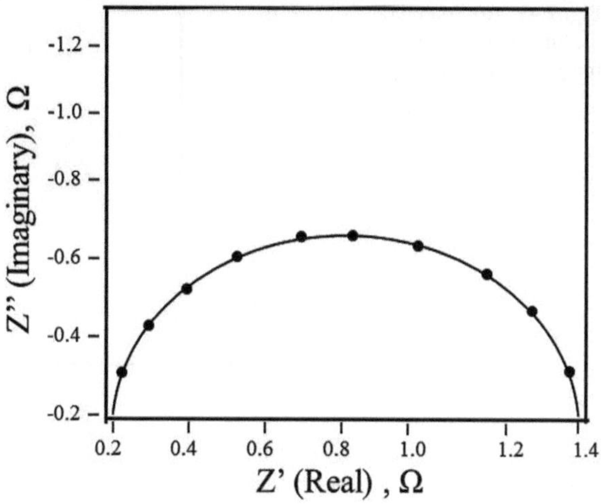

Fig. 1.40 Model of Nyquist Plot showing real and imaginary impedance, for example three frequencies (10^1, 10^2 and 10^3 Hz) are taken in the Nyquist plot

1.11.3 Cyclic Voltammetry

Cyclic Voltammetry (CV) is a potentiodynamic technique used to evaluate the electrochemical performance of the thin films. In this process the electric potential between two limit points is scanned in a cyclic manner using a cyclic voltammogram as shown in Fig. 1.41. Three electrode set up consisting of working electrode (WE), reference electrode (RE), and counter electrode (CE) is required for CV testing of thin films. The potential is measured between the WE and RE and current flow is measured between WE and CE to obtain the current-voltage polarization curve. The principle of this cyclic voltammetric curve is shown in Fig. 1.41. A CV scan starts from an initial potential (V_{Ini}), proceeds to a final potential (V_{Fin}) which is called forward scan, and then returns back to the initial potential known as reverse scan and accordingly this cycle is repeated. The flow of current is also changing with time and hence is measured and it is plotted with applied electrode voltage as shown in Fig. 1.41b

Consider the following redox reaction

$$M^{+} + e^{-} \rightleftharpoons M.$$

In Fig. 1.41b the reduction process occurs from initial potential (a) to switching potential (b) by scanning the potential negatively. After that the potential starts scanning positively from (b) to (c) and oxidation occurs.

The current–voltage curve obtained from this cyclic Voltammetry test helps to determine the redox potential. This also helps to get the information regarding the kinetics of electron reactions and also finds out the presence of intermediate oxidation–reduction reactions in many energy systems.

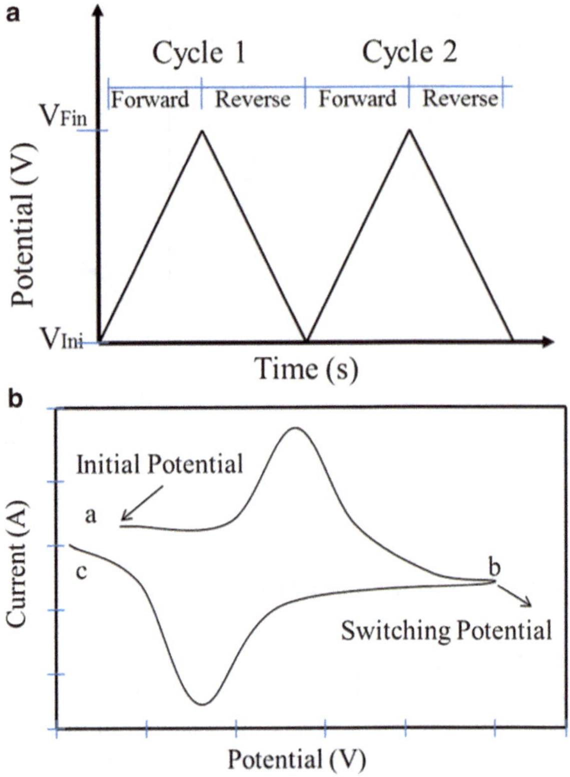

Fig. 1.41 (**a**) Potential vs. time and (**b**) current for cyclic voltammetry

References

1. V. Fthenakis, Renew. Sustain. Energy Rev. **13**, 2746–2750 (2009)
2. First Solar, First Solar, XSOL Establish Supply Agreement in Japan, Will enable broader application of safe, clean and renewable energy (2014), http://investor.firstsolar.com/releasedetail.cfm?ReleaseID=862193. Accessed 28 July 2014
3. J. Santiso, M. Burriel, J. Solid State Electrochem. **15**, 985–1006 (2011)
4. M. Knudsen, Ann. Phys. **28**, 75–130 (1909)
5. M. Knudsen, Ann. Phys. **28**, 999–1016 (1909)
6. M. Knudsen, Ann. Phys. **29**, 179–193 (1909)
7. M.A.V. Ribeiro da Silva, M.J.S. Monte, L.M.N.B.F. Santos, J. Chem. Thermodyn. **38**, 778–787 (2006)
8. P.K. Dutta, H. Wilman, J. Phys. D: Appl. Phys. **3**, 839–853 (1970)
9. P. Arunkumar, R. Ramaseshan, S. Dash, J. Basu, T.R. Ravindran, S. Balakumarand, K. Suresh Babu, RSC Adv. **4**, 33338–33346 (2014)
10. A. Baskaran, P. Smereka, J. Appl. Phys. **111**, 044321 (2012)
11. H.M. Smith, A.F. Turner, Appl. Opt. **4**, 147 (1965)
12. B. Windows, N. Savvides, J. Vac. Sci. Technol. A **4**, 196 (1986)
13. A. Stoffel, A. Kovács, W. Kronast, B. Müller, J. Micromech. Microeng. **6**, 20–33 (1996)
14. P. Scherrer, Nachr. Ges. Wiss. Göttingen **26**, 98–100 (1918)
15. B.S. Straumal, S.V. Protasova, A.A. Mazikin, T. Tietze, E. Goering, G. Sehutz, P.B. Straumanand, B. Baretzky, Beilstein J. Nanotechnol. **4**, 361–369 (2013)

16. L. Rivaud, J. Electron Microsc. Tech. **2**, 577–580 (1985)
17. R.W. Hoffman, in *Physics of Thin Films*, ed. by G. Hass, R.E. Thun, vol. 3 (Academic, New York, 1966)
18. P. Rouard, A. Meessen, Prog. Opt. **15**, 7 (1977)
19. R. Swanepol, J. Phys. E. Sci. Instrum. **16**, 1216–1222 (1983)
20. K.L. Chopra, *Thin Film Phenomena* (McGraw Hill, New York, 1969)

Chapter 2
Coatings for Energy Applications

Anup Kumar Keshri and M. Sribalaji

Abstract This chapter aims at providing an understanding about the potential applications of various types of coatings in energy sector. As the energy demands are growing day by day, there is need of enhancing the efficiency of energy systems, which can be enhanced using the advanced coatings. This chapter summarizes about the application of thin films and thick coatings of conventional/nanomaterials in both renewable and non-renewable energy sectors. A comparison between the efficiencies of systems with and without coatings has also been addressed. The importance and challenges associated with adding nanomaterials like carbon nanotubes (CNT), graphene, and various nanostructures with conventional coating material have also been discussed. This chapter can lead to better fundamental understanding about the coatings, which ensures new designs, high efficiency, and large application of coatings in energy sector.

2.1 Introduction

Energy has been the part of our life, especially electricity plays a vital role in our modern society. With an ever-increasing population, energy demand also growing faster for the past few decades. To meet the increased demand, every country has started to explore different energy systems to produce and store electricity in an efficient and most importantly in the eco-friendly way [1]. Hence, by increasing energy production, emission of greenhouse gases like carbon dioxide (CO_2), methane (CH_4), and nitrous oxide (N_2O) also rises to its peak. In order to decrease the emission of greenhouse gases, a worldwide initiation has also been in progress. Energy conservation steps have been introduced, either by reducing CO_2 emission from non-renewable sources like coal, natural gas, petroleum, and nuclear sources; or by avoiding such emissions using renewable energy sources as wind, water, sunlight, or biomass.

A.K. Keshri (✉) • M. Sribalaji
Department of Materials Science and Engineering, Indian Institute of Technology Patna, Navin Government Polytechnic Campus, Patliputra Colony, Patna 800 013, Bihar, India
e-mail: anup@iitp.ac.in

S. Babu Krishna Moorthy (ed.), *Thin Film Structures in Energy Applications*,
DOI 10.1007/978-3-319-14774-1_2

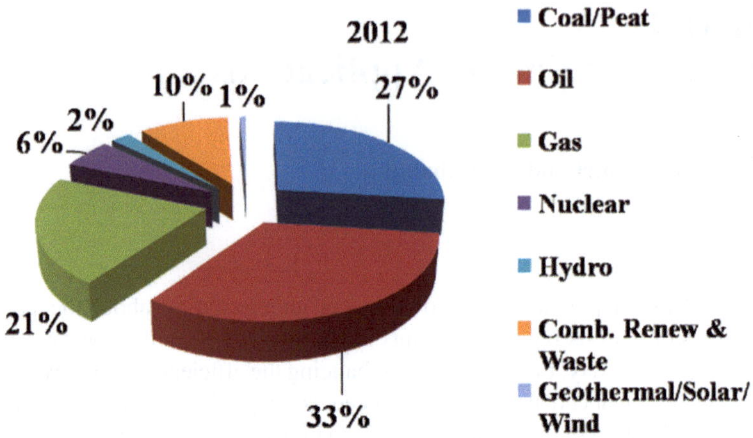

Fig. 2.1 Global energy generation by various sources during the year 2012 [2]

Looking at the present global energy intake in Fig. 2.1 [2], the conventional energy sources like coal, oil, and natural gas are more dominant with more than ~80 % of the world consumption. While the renewable energy sources like geothermal, solar, wind, hydro, and biomass contribute to less than ~20 % of the energy usage [2]. This current scenario for conventional energy sources would still increase for the next two decades; however, the fraction of renewable energy sources will also increase considerably. To fulfil this increased energy demand, it is essential to improve the efficiencies of all energy generation processes. This includes the development of advanced materials, its operation at higher temperatures, increased electricity storing capacity, and electrochemical performance. These specific material properties can be achieved by means of the high-performance multi-scale advanced coatings on its surface.

Such coatings with its unique properties can be commercially exploited for several energy systems in the form of thin films and thick coatings. Thin films are essentially important in solar energy system, fuel cells, batteries, and super capacitors, as light trappers, anti-reflectors, and electrochemical layers [3–5]. While thick coatings are useful for thermal and nuclear energy sources, power plant equipments like turbines, boilers, heat exchangers, etc., which gets frequently exposed to harsh environment such as high thermal exposure, wear, erosion, and corrosion. In order to protect them, various thick coatings like thermal barrier, corrosion-resistant, wear-resistant, abradable, high emissivity coatings were being developed on these structures [6].

Throughout this chapter, advanced coating materials, its fabrication, significance and potential applications will be discussed for various forms of energy generation and storage as shown in Fig. 2.2. This includes the coatings for solar energy, fuel cells, batteries, super capacitors, wind energy, tidal energy, conventional fossil fuel-based energy, bioenergy, nuclear energy, geothermal energy, etc. The forthcoming section discusses about the advanced coating materials for above-mentioned energy applications.

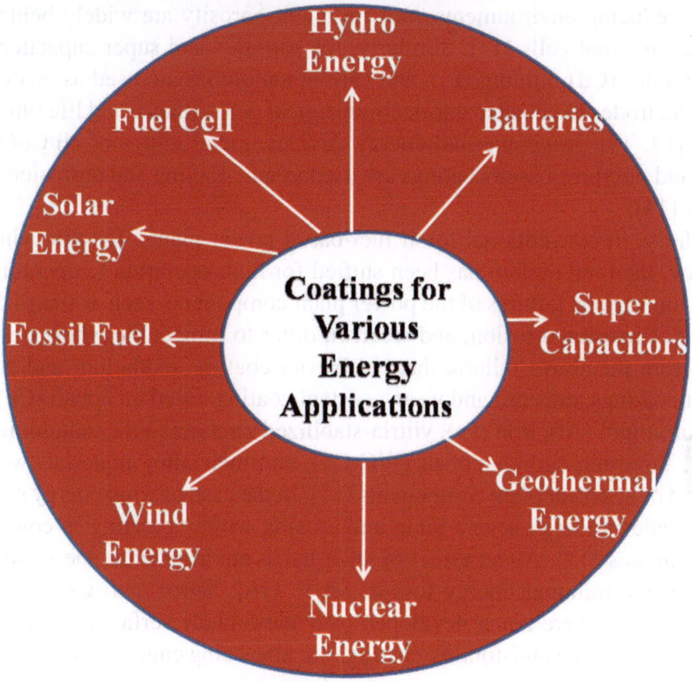

Fig. 2.2 Illustration of the potential applications of coatings for various energy applications

2.2 Advanced Coating Materials for Energy Applications

Coatings are generally formed over a bulk substrate in order to achieve the properties that are not easily attainable or unattainable with the substrate alone. Therefore, the range and requirement of coatings are very broad for various energy systems. Coatings in the form of thin films are prominently used for solar-based energy systems. This has large benefit over the first-generation solar energy system which is based on wafer-based cells and have several constrains in it, such as, it requires high purity silicon crystals, high fabrication temperature, and also needs large amount of material [7]. These limitations had led to the second generation of thin film-based solar cells, where varieties of thin film were established for different purposes. Several researchers have reported the thin film solar cell of silicon (Si) type [8], cadmium telluride (CdTe), copper indium selenide (CIS), copper zinc tin sulphide (CZTS), and copper indium gallium selenide (CIGS) [9]. Lee et al. had coated a layer of ZnO as anti-reflecting coatings, which enhances the efficiency of solar cells [10].

For fuel cells, fluorite-structured oxide materials such as yttria-stabilized zirconia, rare-earth-doped ceria and rare-earth-doped bismuth oxide have been widely investigated as electrolytes because these materials enhance the ionic conductivity of fuel cells [11]. Metals like nickel (Ni), cobalt (Co), and lanthanum (La) which are stable at reducing environments with sufficient porosity are widely being used as electrodes for fuel cells [11]. Similarly, for batteries and super capacitors, nickel (Ni)/cadmium (Cd), lithium (Li), and carbon nanotubes are used as coating materials in electrodes to enhance the electrochemical performance and life time of these devices [12, 13]. In geothermal energy devices, metal alloys of Ni/Cu/Al, metal oxides, and polymer-based coatings are used as anti-fouling and corrosion-resistant coatings [14].

Similarly, in conventional fossil fuel-based power plants, for achieving higher efficiency, the total system has been shifted for high operating temperature which leads to the various failures of the power plant components such as thermal fatigue, oxidation, corrosion, erosion, and wear. In order to protect the whole power plant system from the above failure, thermal barrier coating, oxidation- and corrosion-resistant coatings, erosion- and wear-resistant coating based on cobalt (Co), cerium (Ce), aluminium (Al), iron (Fe), yittria-stabilized zirconia (YSZ), aluminium oxide (Al_2O_3), Tungsten carbide-cobalt (WC-Co), and their alloy materials were being employed over power plant components [15]. In the case of wind energy system, the turbine blades undergo severe icing and erosion while operating in cold [16] and humid climates [17]. As an effect of such harsh environments, the wind turbines were facing a minimal energy loss of ~5 % [16]; hence, anti-wetting and anti-fouling coatings were being developed over these blade surfaces to enhance their efficiency [18]. Such anti-fouling coatings are also being engaged on tidal- or wave-based energy systems. Mostly, polymer (silicone)-based materials were utilized for such systems, because they are super hydrophobic in nature.

The performance of these energy systems depends intimately on the properties of their coating materials. Hence, for further enhancement in the efficiency of various energy systems, researchers have focused on manipulating the microstructure of coating materials. This phenomenon leads to the utilization of nanostructured materials for coating purpose. Nanomaterials have unique characteristics like higher specific surface area [19], crystallinity, high porosity, low density [20], quantum confinement with discrete energy levels [21], ballistic electronic transport [22], and depressed melting temperature [23]. As a result of these characteristics, nanomaterials show enhanced photo catalytic activity [24], better paramagnetic properties [25], enhanced charged separation and migration properties [26], improved diffusion rate and reactivity [23], and superior mechanical properties [27] compared to its bulk counterpart.

These interesting behaviors of nanomaterials had attracted the researchers throughout the globe to employ these materials for various energy applications. Liang et al. had used cone-shaped nanostructures (nanocones) over GaAs solar cell which leads to an efficiency of 17 %, the highest energy conversion rate till date [28]. In another study by Chai et al., mesoporous lanthanum phosphate nanostructures had been used as a promising electrolyte for polymer electrolyte membrane fuel cells [29]. The lanthanum phosphate nanostructures had exhibited a maximum power density of 84.0 mW cm^{-2}, while the conventional Nafion 115 membrane shows a power density of only 79.2 mW cm^{-2}. For energy storage applications like batteries and supercapacitors, Deng et al. had reported for titanium-di-oxide (TiO_2) and tin (Sn)-based nanomaterials as lithium-ion battery electrodes. These nanomaterials are reported to possess enhanced cycle life, safety, and high efficiency than conventional lithium cobalt oxide ($LiCoO_2$) and graphite electrodes. Thus, incorporating nanomaterials have developed the energy efficiencies of different system and they are playing a crucial role in reinventing an efficient source [30].

In geothermal-based energy systems, Cheng et al. had reported for carbon nanotube (CNT)/polytetrafluoroethylene (PTFE) composite coatings on heat exchanger which acts as an efficient super hydrophobic coating [31]. In addition to that, it enhances the thermal conductivity, corrosion resistance, and abrasion resistance of heat exchangers [31]. Keshri et al. had studied CNT-reinforced aluminium oxide coatings for marine and high temperature turbine coating applications and observed a 30 % reduced wear loss due to the reinforcement of 8 wt% of CNT's [32].

Though utilizing such efficient nanomaterials has imposed several advantages on enhancing the efficiency of an energy system, it has few limitations in their synthesis and characterization. Notably, nanomaterials are quite difficult to disperse uniformly in a system, Finnegan et al. had reported for the dispersion issue of titania nanoparticles in aqueous solution [33]. Secondly, Yacaman et al. had recognized that most of the nanoparticles did not have unique stable entities due to its quantum confinement effect [34]. In addition to these, synthesis of nanoparticles also possesses lots of challenges. Nanoparticles exhibit a variety of size, shape, and crystal structure-dependent physical and chemical properties; hence a superior control in the reaction of nanomaterials was highly emphasized by Song et al. [35]. Apart from this, Baer et al. had reported for the challenges in the characterization of nanoparticles using UV–Vis, XPS, XRD, and TEM techniques [36]. Hence, in order to minimize these issues, a lot of care should be taken on fabricating nanostructured coatings. Since the applications of these nanocoatings are varied as thin films and thick coatings, fabrication techniques and processing parameters also get differed, which are detailed in the subsequent topic.

2.3 Thin Film and Thick Coating Fabrication Methods

Coating fabrication techniques can be broadly classified into two types: (1) thin film fabrication and (2) thick coating fabrication techniques. As represented in Fig. 2.3, thin film techniques are subdivided into physical vapor deposition (PVD) and chemical vapor deposition (CVD) techniques. Thermal evaporation is one of the PVD techniques, which provides highly uniform films for mass production. Yet, this evaporator method is limited to low melting point metals and dielectric materials, poor adhesion, and poor density, which imposes a serious drawback on its choice [37]. In order to create the films of higher melting point materials like Ce, Zr, W, etc., electrons are used to dissociate atoms from the target material; this method of deposition is known as electron beam (E-beam) deposition. This method

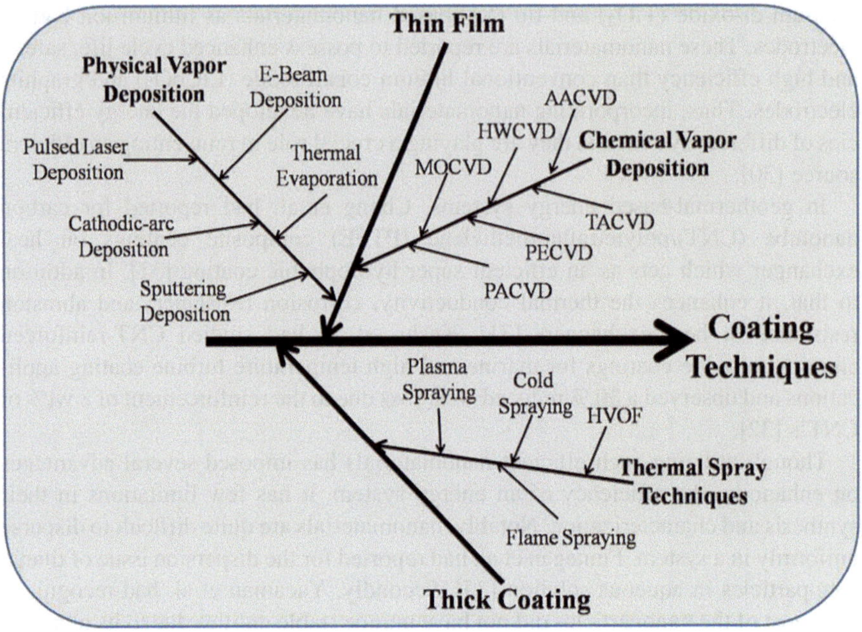

Fig. 2.3 Various techniques for the fabrication of thin films and thick coatings

provides high adhesion and high density of target material on the substrate, but the deposition rate of this technique is very low, which limits its potential for large-scale industrial applications.

In addition to this, for depositing various hard coatings, methods like cathodic arc deposition and sputtering techniques are used. Among them, arc deposition process employs higher energy input than the sputtering process. Therefore, the amounts of ionized target atoms are more for cathodic arc deposition technique, i.e., 50–100 %. However, this arc deposition has its major limitations of forming macro particles while depositing low melting point materials and forming excess atoms that are not completely ionized. On the other hand, sputtering technique had overridden these disadvantages and they provide high adhesive and high density materials even for large area deposition of targeted materials. The films produced by this method exhibit uniform deposition and high reproducibility with an easy control over the growth rate [38]. Pulsed laser deposition (PLD) is another PVD technique which also uses thermal energy to dissociate atoms, ions, and clusters of target material. The thermal energy was produced by high power-pulsed laser beams, hence this technique is considered to be an appropriate method to deposit complex oxides and to maintain the target composition in thin film. In general, all PVD methods are restricted to line of sight deposition, therefore these techniques were not considered to be advantageous to coat complex structures. Moreover, all these techniques are vaccum-based technologically challenging process. In order to outweigh these shortcomings of PVD techniques, chemical vapor deposition can be considered.

Likewise, through CVD method, deposition of amorphous, single-crystalline, polycrystalline thin films and coatings are being made for a wide range of applications [39]. Most importantly, materials like Si, Ge, III-V group semiconductor, II-VI semiconductors were being coated for energy conversion devices (e.g., solar cells). The distinctive advantages of CVD systems are: production of highly thick and pure coating, its capability to produce uniform films with good reproducibility and adhesion at higher deposition rates, its ability to coat complex structures since it is not a line of sight technique [39]. Apart from this, it possesses flexibility of using wide range of precursors such as halides, hydrides, and organometallics. Conventionally, CVD method is called as thermally assisted because thermal energy is used to activate chemical reactions. Apart from heat, chemical reactions can also be activated using various energies like plasma, light, etc.; hence, the other sub categories of CVD like plasma-enhanced CVD (PECVD), photo-assisted CVD (PACVD), aerosol-assisted CVD (AACVD), hot wire CVD (HWCVD), and metal organic CVD (MOCVD) [30] had emerged as shown in Fig. 2.3. These thin film coating techniques are extensively applied for energy systems like solar, fuel cells, batteries, and super capacitors.

Other energy production systems like nuclear reactors, thermal power plants, wind and tidal power plants which operate at harsh environments do not accommodate these thin films. Normally, thermal sprayed thick coating techniques like flame spraying, high velocity oxy fuel technique (HVOF), plasma spraying, and cold spraying are employed to coat these devices [40]. These techniques are suitable for long range deposition and the principle behind these is to melt, accelerate, and deposit the material feedstock on a substrate, while the melting source varies for each type of spraying. In flame spraying technique thermal energy is used to melt the material, while in HVOF spraying the acceleration of the thermally molten material is increased. In the case of plasma spraying, plasma is utilized for melting the feedstock; this method is highly suitable for coating materials of high melting point. This plasma-based system possesses high densification than thermal-sprayed and HVOF-sprayed coatings [40]. In cold spray technique, the feedstock material is not melted but they are accelerated to a highest velocity by supersonic de-laval nozzle. Hence, this technique can lead to dense and ultra-thick coatings which can be applied for energy devices that operate at tough environment. The application of these thin film and thick coating techniques in various energy systems has been discussed extensively in the subsequent topics.

2.4 Coatings for Solar Cells

Photovoltaic technology directly converts clean and renewable solar energy into electrical power and it is considered to be an effective solution for the existing world energy crisis and environmental pollution. In recent years, considerable efforts have been taken to study the feasibility of high efficiency and low-cost solar cells. The first generation of solar cells are silicon (Si) wafer-based, which involves challenges in materials processing like wafer cleaning, high silicon consumption, chemical etching, and poor outdoor performance [41]. Thereby, this technique has imposed high cost on its solar module.

In second-generation Si-based thin film solar cells, the manufacturing cost was greatly reduced because of the utilization of thin films through various coating techniques. Table 2.1 shows various materials that have been coated by diverse coating techniques for different thin film solar cells in the past 4 years. Figure 2.4 describes about the various layers in a thin film Si solar module; it has both n doped layer (abundant of electrons) and p doped layer (abundant of holes) sandwiched with a concentric i layer which acts as a bridging layer. But the efficiency of this module drastically reduced compared to the wafer-based first-generation module. The lowered efficiency was attributed to excessive light reflection at the front interface of solar cells, which leads up to ~30 % loss in incident light [53]. Hence, there is a need of anti-reflection layer which can lower down these

Table 2.1 Various efficiencies obtained by different coating techniques for various thin film solar cells

S. no.	Year	Type of solar cell	Material coated	Coating technique	Reported efficiency	Reference
1.	2014	Organic solar cells	Graphene	TECVD	2.07 % with superior stability	[42]
2.	2014	CdTe thin film solar cells	CdTe	RF magnetron sputtering	12.78 %	[43]
3.	2014	Dye-sensitized solar cell (DSSC)	Metal sulfide/graphene	CVD	5.25 %	[44]
4.	2014	DSSC	Graphene	PECVD	6.01 %	[45]
5.	2014	Quantum dot-sensitized solar cells (QDSSC)	CdS on carbon nanotube	PECVD	PCE = 7.5 %	[46]
6.	2013	a-Si solar cells	a-Si:H thin films	PECVD	11.9 %	[47]
7.	2013	CIGS solar cells	Boron doped ZnO	MOCVD	18.3 %	[48]
8.	2013	Si-based single/double junction solar cells	Amorphous and micro crystalline Si	PECVD	Stabilized efficiency of 11.5 %	[49]
9.	2013	Metal-insulator-semiconductor (MIS) transistor-based solar cells	ITO (Indium tin oxide)	Sputtering	10.85 %	[50]
10.	2012	DSSC	Pt	ITO	10.71 %	[51]
11.	2011	CIGS thin film solar cells	Cu, In, Ga, Se	Sputtering	20.3 %	[52]

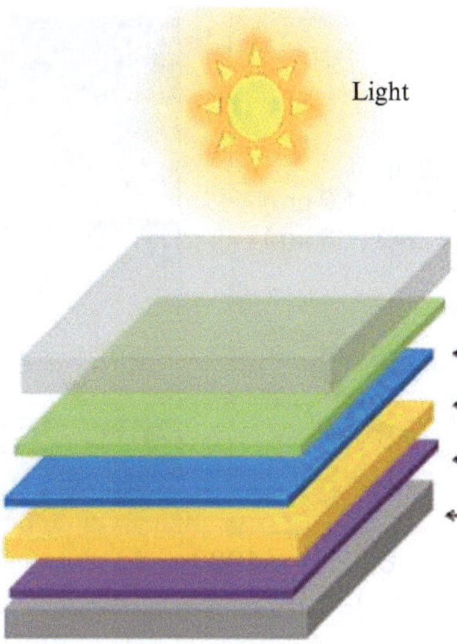

Fig. 2.4 Explanation of various layers that are involved in a thin film Si solar cell to increase its efficiency

reflection losses as shown in Fig. 2.4. Zhang et al. had reviewed various anti-reflection coating materials and found a significant reduction in the reflection losses from 30 % to 10 % [53].

Recently, Lee et al. [50] had investigated the feasibility of nano-textured surfaces with different morphologies as anti-reflective layer for solar cells. This group has investigated and compared the reflectance of ZnO nanowire coatings prepared by hydrothermal method, with the reflectance of MgO-ZnO core shell-type nanowire coatings and branch-type nanostructures prepared by means of RF magnetron sputtering and e-beam evaporation, respectively. Figure 2.5 shows the changes in the reflectance due to different morphologies [50]. The SEM images reveal that ZnO nanowires are of the length of 1.8–1.9 μm and have tapered shapes of approximated 200 nm in length and 60° in angle from the top. When magnesium oxide (MgO) films were deposited on such ZnO nanowires by e-beam deposition technique, the tapered shapes of ZnO nanowires get reduced to 120 nm and form a branch-like structure with diameter less than 20 nm. On the contrary, the MgO deposition on ZnO by RF magnetron sputtering produced the core-shell nanowires with ununiformed MgO shells at the sidewalls and the top. These changes in morphology due to different deposition techniques had shown an adverse effect in reflectance phenomena. Among them, ZnO coated by MgO which is a low refractive index material has shown reduced reflectance compared with uncoated ZnO. Apart from this, the porosity of MgO nanobranches is about 75 % and this porosity plays a major role in reducing the reflectance of these MgO-coated ZnO anti-reflective layers for Si-based solar cells.

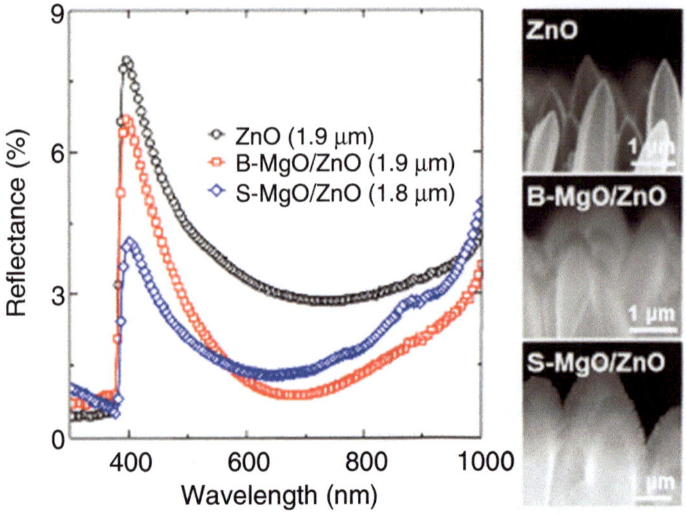

Fig. 2.5 Representation of the reflectance spectra of ZnO nanowires, MgO/ZnO branched nanostructures (B-MgO/ZnO), and MgO/ZnO core shell nanostructures (S-MgO/ZnO). Their respective SEM images are also shown (Reprinted with permission from [50], Lee et al., *ACS Appl. Mater. Interfaces*, 6 (2014). © 2014, ACS Publications)

Apart from Si-based solar cells, other semiconductor materials with high optical absorption coefficient were also being considered. Li et al. had reported for cadmium telluride (CdTe) coatings processed by RF magnetron sputtering, followed by chlorine (Cl) treatment and enhanced its efficiency from 8 % to 12.78 % [43]. This enhancement was mainly attributed to the reduction in the band gap of CdTe as well as changes in defect structure due to processing techniques followed by Cl treatment. However, cadmium is an extremely toxic metal; researchers are always on the lookout for other eco-friendly materials. Hence, solar cells made of Cu(In, Ga)Se_2, dye-sensitized solar cell (DSSC), and organic materials had gained attraction among modern researchers.

Recently, Wu et al. had reported for enhanced efficiency of CIGS solar cells due to novel vapor-based selenization coating of selenium (Se). They had reported the enhancement in the efficiency from 7.1 % to 10.8 % for this particular type of solar cells [9]. Enhanced efficiency is contributed to the decrease in buffer layer (Cd)/CIGS the interface recombination and the enhanced separation of electron and hole carriers by the back surface field on the surface of the CIGS film [9]. Other than CIGS, DSSC is also one of the most promising photovoltaic devices because of its low cost and high efficiency. Figure 2.6 explains about the components of a DSSC which typically includes a working electrode (i.e., anode) for light harvesting and charge separation, a counter electrode (i.e., cathode) to transport the electrons received from an external circuit, and an electrolyte filled between the two electrodes that provides redox coupling called dye regeneration. Nanostructures of titanium dioxide (TiO_2) and platinum (Pt) are most commonly used DSSC working

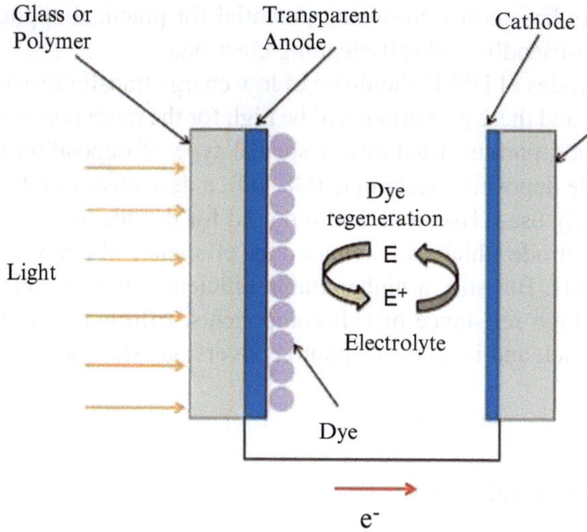

Fig. 2.6 Working principle of dye-sensitized solar cell (DSSC)

electrode. These materials improve the DSSC's performance using electron-hole pair multiplication mechanism known as "multiple exciton generation". In this mechanism, absorption of a single photon produces two electron-hole pairs which lead to enhancement in photo conversion efficiency (PCE) of DSSCs.

Apart from this nanostructure of TiO_2 and Pt, one dimensional nanostructure can also improve the electron transport in DSSC. Generally, Pt metal or Pt nanoparticles are used as conventional coating material for counter electrodes due to their enhanced electro catalytic performance. As a substitution for that rare Pt metal, metal sulfide like nickel sulphide (NiS) with graphene was proposed as counter electrode by Hui-bi et al. [44]. They achieved a photo conversion efficiency of ~5.25 % close to Pt coatings, by coating the NiS/graphene composite through chemical vapor deposition method [44]. Since graphene has large surface area and high electron mobility, it is considered as a potential alternative material for the Pt electrodes. However, it shows a lower electrocatalytic activity compared to Pt in DSSC due to lesser oxygen reduction reaction (ORR) sites. Meanwhile, NiS nanoparticles can show higher electrocatalytic activity due to large active sites. Hence, the composite films of NiS/graphene show high electrocatalytic activity due to two important reasons: (1) uniform deposition of relatively small-sized NiS nanoparticles on graphene in the range of 5–20 nm and (2) good contact and well intercalation of NiS nanoparticles into graphene sheets.

In addition to that, Yang et al. had reported for counter electrodes with uniformly erected vertical walls of graphene grown by plasma-enhanced chemical vapor deposition (PECVD) [45]. This vertical arrangement has two significant advantages: (1) they had enhanced the photovoltaic efficiency of DSSC from 5.25 % to 6.01 % by producing less dense and porous structures and (2) these structures had provided super-hydrophobic property with contact angle greater than 140° [45].

Super-hydrophobicity has shown promising potential for practical applications as excellent, cheap, eco-friendly, and self-cleaning electrode.

The counter electrodes of DSSC should be of low charge transfer resistance with high current density, and these properties will be high for the more porous structure. In order to attain such porous structures, a special type of deposition technique named glancing angle deposition technique (GLAD), a derivative of e-beam deposition, was deliberately used. Hsu et al. had reported for this technique for coating porous Pt counter electrode which had increased the efficiency of DSSC to a greater extent of 10.71 % [51]. But still, a higher stable efficiency of solar cells was not obtained due to the high resistance of cell components. Efforts are continuing to overcome this drawback and increase its photo conversion efficiency.

2.5 Coatings for Fuel Cells

Fuel cells are highly efficient electrochemical energy convertors, which generate electricity from its chemical energy. It consists of an electrolyte sandwiched between porous anode and cathode as represented in Fig. 2.7. Oxygen supplied to the cathode reacts with the incoming electrons from the external circuit to become oxygen ions, which are transported through the conducting electrolyte. At anode electro-oxidation of the fuel (i.e., hydrogen to water) is achieved by the transported oxygen ions, while electrons are liberated to the external circuit with the generation of heat. The efficiency of any fuel cell is defined by its operating temperature. Lower the operating temperature, higher the efficiency of the fuel cell. Electrolytes and its thickness play a vital role in determining the operating temperature of a fuel cell. Lower film thickness of electrolyte has more effect in ionic conductivity, because these thin film electrolyte conducts oxygen ions at lower temperature

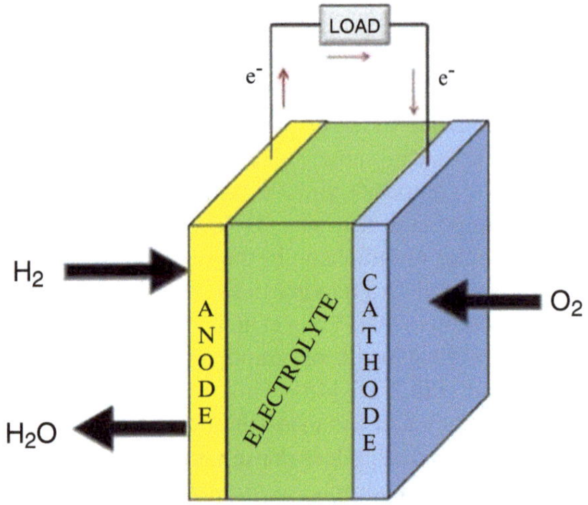

Fig. 2.7 Elucidation of the schematic and operation of a solid oxide fuel cell

(~400 °C), which is useful for fuel cell commercialization. The main advantages of depositing thin film electrolytes are: (1) decreased film thickness, being 100–1,000 nm, lowers the ohmic losses across the fuel cell, (2) this processing allows sintering temperatures below 1,000 °C or even no sintering at all, and (3) the possibility exists to manipulate the electrical conductivity by controlling the nano-crystalline microstructure of a thin film [54]. And the following relation proves that ionic conductivity is dependent on film thickness:

$$J = \frac{\sigma_x \mathrm{OCV}}{L}, \quad (2.1)$$

where, J is the oxygen diffusion permeability, σ is the ionic conductivity, OCV is the open circuit voltage, and L is the thickness of the electrolyte. From the above relation, the oxygen ion permeability increases with decrease in thickness of the electrolyte film. Therefore, thin films are being extensively used in all types of fuel cells; Table 2.2 shows such thin films coated by various coating techniques with their obtained efficiency. Kim et al. [62] had studied gadolinium (Gd)-doped ceria

Table 2.2 Thin films coated by different coating techniques for various types of fuel cells and their obtained efficiency

S. no.	Year	Fuel cell type and component	Material coated	Fabrication method	Obtained efficiency	Reference
1.	2014	All fuel cells, electrolyte	Au nanoparticle dispersed within the $EtSO_4$	Sputtering	Ionic conductivity 5 mS cm^{-1}	[55]
2.	2013	PEFC, electrode	Pt-Ni alloy nano rods vertically aligned	Sputtering GLAD	ORR is 3.5 times higher than the normal Pt	[56]
3.	2013	PEFC, electrode	$TiSi_2Ox$-coated nitrogen-doped carbon nanotubes (NCNTs)	Magnetron sputtering and CVD	ORR is 1.7 times higher	[57]
4.	2013	Alkaline fuel cells, anode	Ni-based anode	E-beam deposition	Specific energy of 7,485 Jg^{-1}	[58]
5.	2012	Direct alcohol fuel cells, Electrode	Pt catalyst supported on carbon nanowall (CNW)	PECVD	79.5 %	[59]
6.	2011	SOFC, electrolyte	Scandia-stabilized zirconia (ScSZ)	E-beam deposition	9.46×10^{-3} S cm^{-1} at 700 °C	[60]
7.	2010	SOFC, electrolyte	Scandium and cerium-stabilized zirconium	E-beam deposition	4.91×10^{-2} S m^{-1} at 873 K	[61]

thin films of thickness 0.33 μm by RF magnetron sputtering and they had decreased the operating temperature to the range of 300–500 °C with high ionic conductivity.

Apart from electrolyte of SOFCs, Wu et al. [63] had focused on evaluating the power density and total resistance of cathodes produced by GLAD method for solid oxide fuel cells (SOFC's) application. This group had evaluated the properties for cathodes consisting of nanocolumns of $La_{1-x}Sr_xCo_{1-y}Fe_yO_{3-\delta}$(LSCF) over samaria-doped ceria (SDC) with a glanced angle of 10°. The values were compared with the results obtained by normal non-glancing method of electron beam deposition. Results indicate that the thin film (electrode) prepared with a glancing angle shows a higher power density of 5.7 mW cm^{-2}. While, power density of the same samples prepared by non-glancing technique was reported as 3.5 mW cm^{-2}. It is because the SDC/LSCF boundary of non-glancing type coating is larger than that of glancing type coating [63]. Hence, the denser film (non-glancing type) had retarded the more oxygen flow than the porous (glancing type) film. In addition to that, the total resistance of the electrode fabricated by former technique was reported to be lower than that of the later technique. This lower resistance and higher power density had been reported to enhance the performance of a SOFC at an operating temperature of 600 °C [63].

In the case of electrode materials, supported catalysts like platinum (Pt) were used, to have high porosity and high conductivity. But the issue in using this Pt catalyst is that, it delivers a low kinetics of oxygen reduction reaction (ORR) and the cost of catalyst will account for 30–45 % of the total cost, so it is highly recommended by researchers to change Pt-based catalyst.

Recently, Ma et al. [58] had demonstrated a thin coating of very low amount palladium (Pd) catalyst loaded on conventional Ni-based electrode using e-beam deposition technique. This electrode membrane had been tested in a boro hydride fuel cell and found to show a specific energy of 4,728 Jg^{-1} equivalent to conventional Pt-loaded Ni electrode [58]. Thereby, thin film technique had served effectively in twofolds by reducing the material usage and by enhancing the material property.

Likewise, Zhang et al. had studied the performance of carbon nanowires supported by Pt for direct alcohol fuel cells (DAFC). These carbon nanowires were vertically aligned by inductively coupled PECVD method and it had contributed to an efficiency of 79.5 %, which is greater than conventional catalyst by 25 % [59]. This increase in efficiency was attributed to higher electrochemical surface area of carbon-based nanowires on comparison with conventional Pt electrode [59].

Further research conducted by Banis et al. [57] had decreased the percentage of Pt by studying the composite nanostructures of $TiSi_2O_x$ and annealed $TiSi_2O_x$ (An-$TiSi_2O_x$)-coated nitrogen-doped carbon nanotubes (NCNT), as supporters for Pt catalysts. These composite structures were prepared by a combined method of magnetron sputtering and CVD. However, these CNT-based nanostructures have shown 1.7 times larger oxygen reduction reaction (ORR) activity in proton exchange membrane fuel cells (PEMFCs) for 50 % lower Pt load [57]. Figure 2.8 shows the electrocatalytic activities for ORR of these composite nanostructures. Among them Pt/An-$TiSi_2O_x$-NCNT coating has shown a higher peak potential than

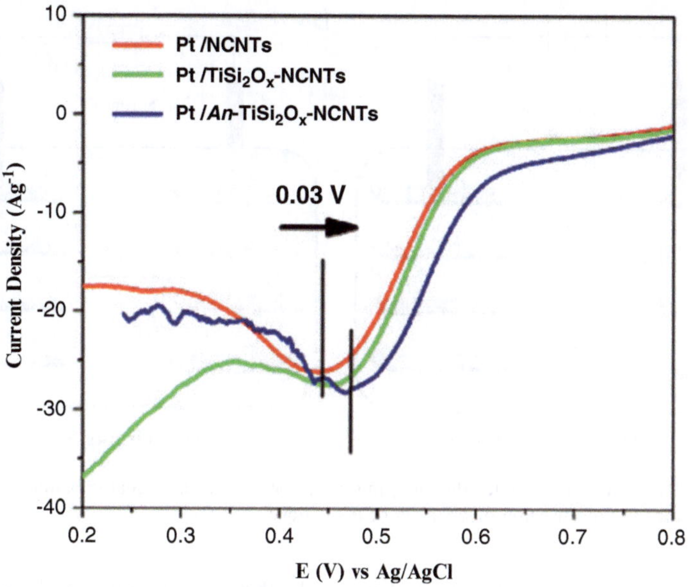

Fig. 2.8 Representation of the oxygen reduction curves of Pt-deposited composite nanostructures compared with Pt-deposited nitrogen-doped carbon nanotubes (Reprinted with permission from [57], N. Banis et al., *The Journal of Physical Chemistry C* 117, (2013). © 2013, ACS Publications)

the Pt/$TiSi_2O_x$-NCNT and Pt/NCNT. The peak potential of Pt/An-$TiSi_2O_x$-NCNT is 0.47 V, which is a 30 mV positive shift in peak potential compared to the other two composite nanostructures. This enhancement of catalytic activity was attributed to high porous nature of those nanostructures. Upon increasing the porosity, Kariuki et al. had obtained 3.5 times higher ORR activity by using (Pt-Ni) alloy nanorod arrays. They achieved this higher porosity by magnetron sputtering glancing angle deposition (GLAD) technique [56]. Thus, the above discussion clearly indicates the impact of coating on the efficiency of fuel cells, and further researches are undergoing to implement commercial low temperature fuel cell.

2.6 Coatings for Batteries

Effective energy storage strategies are the key for exploiting alternative energy and for the replacement of fossil fuels and traditional energy sources. Rechargeable batteries play an important role in this regard and they currently dominate the market of smart phones, laptops, and also playing a significant role in industrial, transportation, and grid scale energy storage system applications. A rechargeable battery comprises of one or more electrochemical cells and each cell consists of a cathode (positive electrode), an anode (negative electrode), and an electrolyte

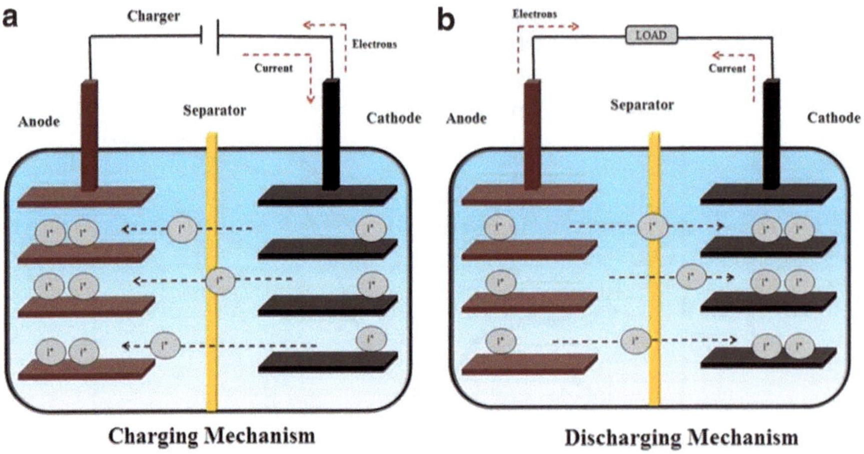

Fig. 2.9 Panel (**a**) demonstrates the charging mechanism in a rechargeable battery, while panel (**b**) represents the discharging mechanism of the battery

(separator). Energy is stored in these cells through charging and discharging process. On charging, the cathode gets oxidized and produces ions and electrons. Ions move from cathode to anode through the electrolyte, while the electrons flow through an external circuit and constitute the current flow. During discharging process, the ions which were already in anode get back to the cathode upon reduction. This whole charging and discharging process is termed as a cycle which is illustrated in Fig. 2.9.

Depending upon the material systems in batteries, they were classified as: lead acid battery, nickel-cadmium battery (Ni-Cd), lithium-ion battery (Li-ion), nickel-metal hydride battery, etc. Among all, Li-ion batteries were more significant due to their high volumetric energy, high power density, long cycle life, and low self-discharge properties. Commercial Li-ion batteries consist of lithium manganese oxide ($LiMn_2O_4$) and graphite as the cathode and anode materials, respectively, and they deliver a capacitance of ~150 Wh kg^{-1} [64]. However, the usage of these Li-ion batteries in hybrid electric vehicles (HEV), plug in hybrid electric vehicles (PHEV) and pure electric vehicles (PEV) requires five times more energy density than its present density. Hence, to increase the energy density of Li-ion batteries, development of new cathode and anode materials with extended cycle life is highly essential.

Upon selection of suitable anode materials that provide high capacity and ease diffusion of Li-ions, provide good cycling life, and are of high safety concerns, Fig. 2.10 illustrates the next generation of anode materials [65]. Most notably, Sun et al. had reported for coatings of ZnO quantum dots as Li-ion battery anode material. ZnO quantum dots of various sizes from 2 to 7 nm were homogeneously coated along the curvature of graphene sheets through atomic layer deposition technique (one of the PVD method) [66]. A high reversible specific capacity of 960 mAh g^{-1} at a current density of 100 mA g^{-1} is achieved for 2 nm ZnO quantum

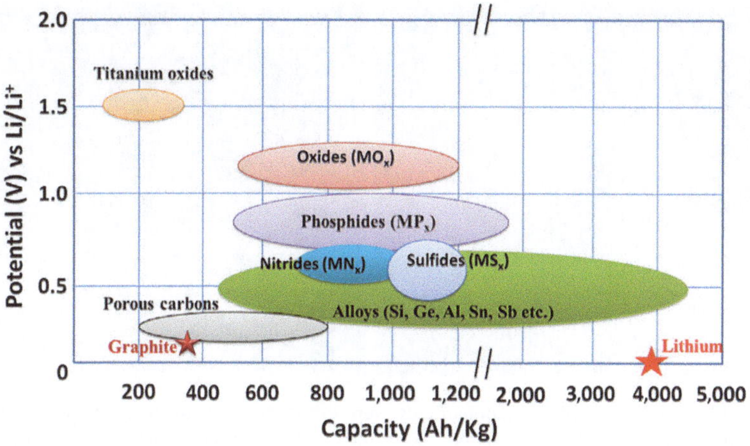

Fig. 2.10 Next-generation anode materials for the lithium ion (Li-ion) batteries (Reprinted with permission from [65], S. Goriparti et al., *Journal of Power Sources*, 257 (2014). © 2013, Elsevier Publications)

dots; it is because smaller particles exhibit better rate performance. In addition to that ZnO nanostructures have experienced large volume expansion or contraction during charging and discharging process, which is around ~228 %.

Recently, Zhang et al. had reported for molybdenum disulfide [MoS_2] nanospheres as another effective anode material for Li-ion batteries [67]. These MoS_2 nanospheres had shown highest specific capacity of 1,355.1 mAh g^{-1} for about 50 cycles [67]. This high specific capacity was attributed to the four electron transfer reaction of MoS_2 nanospheres, high specific surface area of nanospheres, and higher electrolyte/MoS_2 contact areas. In addition to that, Qin et al. had reported for the fabrication of 3D porous graphene network anchored with Sn nanoparticles encapsulated with graphene shells (Sn@G-PGNWs) as a superior lithium ion battery anode. This 3D structure with Sn nanoparticles of 5–30 nm encapsulated with thin graphene shells of ~1 nm is very tightly pinned on the surface of thin walls of porous graphene networks by a single step chemical vapor deposition method [68]. Figure 2.11 represents the SEM images of these 3D structures, which have unique interconnected pores, and the walls of these interconnected 3D porous networks indicate a clear curved profile. Subsequently, in high magnification SEM image shows Sn@ Graphene nanoparticles are homogeneously anchored on the walls of porous graphene network. These walls also possess high mechanical flexibility, which suppress the volume expansion and aggregation of Sn nanoparticles by preserving the structural and interfacial stabilization of Sn nanoparticles. Thus, the core shell structure of Sn-encapsulated graphene leads to enhanced electrical conductivity and structural integrity of overall electrode. Qin et al. group had achieved a very high rate performance of 1,022 mAh g^{-1} and extremely long cycling stability of 1,000 cycles with an efficiency of 96.3 % [68]. While, Ren et al. had showed highest energy storage

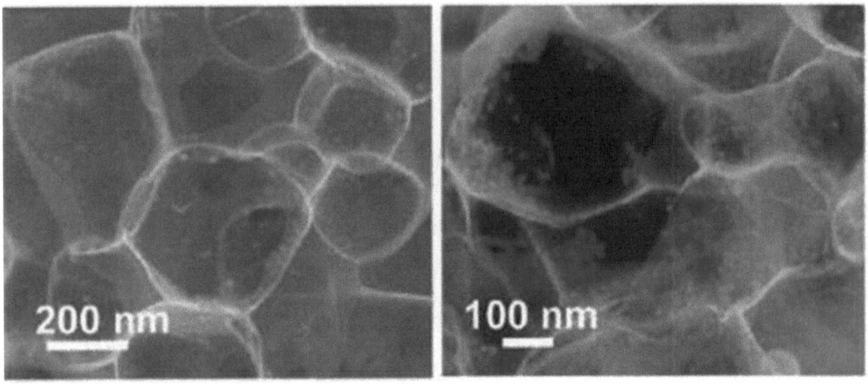

Fig. 2.11 SEM images of Sn nanoparticles were encapsulated with graphene shells (Sn@G-PGNWs) in two different magnifications (Reprinted with permission from [68], Qin et al., ACS Nano, 8, 2 (2014). © 2014, ACS Publications)

capability of lithium ion batteries with a coating of novel hierarchical Si nanowire-reduced graphene oxide composite fabricated using a solvothermal method followed by a chemical vapor deposition [69]. This group had grown well-dispersed Si nanowires on the reduced graphene oxide surface and in between graphene oxide sheets. They achieved the structural integrity over 100 cycles at a high lithium storage capacity of 2,300 mAh g^{-1} [69].

Apart from anode, the cathode part is also considered to be significant for attaining batteries with high energy density, high rate capability, and long cycle life. Lithium Iron Phosphate ($LiFePO_4$) [LFP] and Lithium Vanadium Phosphate ($Li_3V_2(PO_4)_3$) [LVP] were considered as the most potential coating candidates for the cathode of Li-ion batteries. However, these two materials have the drawback of poor rate capability which is due to slow lithium ion diffusion kinetics and poor electronic conductivity. Jin et al. had reviewed various $LiFePO_4$ and $Li_3V_2(PO_4)_3$-based materials like $LiFePO_4/C$, $Li_3V_2(PO_4)_3/C$, $Li_{1-x}Al_{2+x}FePO_4$, $Li_{3-x}Na_xV_2(PO_4)_3/C$, and $LiFePO_4$-$Li_3V_2(PO_4)_3$-based composites for coating cathodes [70]. Among them, Jin et al. group had concluded that coatings based on LFP-LVP composite materials had shown better electrochemical properties with an excellent capacity retention rate of 73.3 % and increased cycling stability [70]. It is because these composite materials have large number of defects which would greatly increase the electrochemical property of the battery.

In order to improve the safety and stability of batteries in terms of incorporating non-leakage and non-reactive properties in them and to have different battery shapes, their liquid electrolytes were being altered by solid polymer electrolytes like polyethylene oxide (PEO). However, solid electrolytes possess the issue of having lower ionic conductivity than liquid electrolyte. Recently, Tang et al. had increased the ionic conductivity of polyethylene oxide (PEO) coatings, by incorporating CNT as fillers to it [71]. The addition of bare CNT for electrical conductivity purpose has the risk of battery shorting. Therefore, Tang and group had packed these nanotubes within the insulating clay layers to form effective 3D nanofillers. Hence, upon coating this CNT-based 3D nanofillers, the Li salt dissociation had increased from 82.2 % to 95.7 % [71]. The dissociation of contact-ion pair has become so easy, due to the weakening of their interaction by clay-CNT 3D hybrids. Along with this, the mechanical property like tensile strength had also increased by 160 % due to the addition of these nanofillers [71].

2.7 Coatings for Supercapacitors

Electrochemical capacitors or supercapacitors have incited lots of interest in portable devices and electric vehicles where there has been an ever-increasing demand for high energy and power density energy storage devices. These devices have the ability to release the stored energy smoothly even at high energy density. Supercapacitors store energy by the electrical double layer capacitance (EDLC) mechanism, where the capacitance arises from the pure electrostatic charges accumulated at the electrode–electrolyte interface. Hence, it is evident that EDLC is a surface phenomenon, therefore large surface area nanomaterials are desired for high capacitance. The electrochemical performance of a supercapacitor is mainly determined by the electrode material; hence, various carbon-based nanostructures like CNTs, activated carbon, graphene, and templated carbon were coated as effective electrode material.

Lota et al. had reported for CNTs and its composites for electrochemical applications, where the CNTs are evaluated as electrodes with high energy density because of its unique tubular porous structures and superior electrical properties. This favor fast ion and electron transport in CNTs [72]; however, the major limitation is their small specific area and high practical applications. Therefore, modification of CNT had been reported to overcome these drawbacks. Recently, Ghaffari et al. had modified vertically aligned CNT by conformal coating of poly (3, 4-ethylenedioxythiophene), also called as PEDOT, which is a conducting polymer [73]. This PEDOT/aligned CNT composite was synthesized by oxidative chemical vapor deposition (OCVD) method and it shows a specific volumetric capacitance of 84.0 F cm^{-3}, while the non-coated CNT shows a specific volumetric capacitance of 3.9 F cm^{-3} [73]. Such increase in capacitance has been attributed to charge storage mechanism of electrode. Non-coated CNT undergoes electrical double-layer capacitance (EDLC) charge storage mechanism, while PEDOT-coated CNT stores energy through pseudocapacitive mechanism. Apart from this PEDOT coating has significantly increased the capacitance value and when these coatings combine with highly dense CNT, such very high volumetric capacitance is achieved.

Likewise, Warren et al. had demonstrated ruthenium oxide (RuO_2)-coated CNT as anode material for supercapacitors by atomic layer deposition technique [74]. This electrode has shown an ultra-high specific capacitance of 100 mF cm^{-3}; this very high specific capacitance was also contributed to the high surface area and density of CNT. Apart from this, Wei et al. had reported for CNT-graphene sandwich structures with CNT pillars grown in between graphene sheet [75]. This structure was synthesized by co-catalyst CVD method; because of this novel structure the specific capacitance had improved significantly to 385 F g^{-1} [75]. This enhancement was attributed to the sandwich layers of graphene and CNT. Here, CNT acts as a buffer layer and also avoids the aggregation of graphene. Along with this, the interconnected 3D network transports electrons more easily and hence the internal resistance of electrode decreases.

Similar 3D structures were reported by Chen et al., where nanostructures of manganese-di-oxide (MnO_2)-CNT-sponge hybrid electrode were fabricated and used as a novel electrode in supercapacitors [76]. This group had coated CNT on commercial sponges with pore sizes between 100–500 μm by simple dip-coating method. Then, they electrodeposited MnO_2 over these CNT-sponge materials. Figure 2.12a, b show the SEM images of MnO_2-CNT-sponge nanostructures, which illustrate the three-dimensional hierarchical flower like open-pore structures. These images also confirm the uniform deposition of MnO_2 nanoparticles onto the CNT-sponge skeleton. The most interesting point in these structures is that the backbone of sponge is free of junctions and this highly porous nanostructure remains the same even after the deposition of MnO_2 [76]. Since this MnO_2-CNT-sponge hybrid electrode is an integrated binder-free structure with highly porous and conductive channels, these structures can provide full access of electrolyte ions to the active electrode material in supercapacitors. Hence, the electrochemical performance of such electrodes outperforms the electrochemical activity of standard MnO_2-Pt electrode. Figure 2.13 shows the specific capacitance comparison between MnO_2-CNT-sponge hybrid electrode and MnO_2-Pt electrode, where the former shows a very high specific capacitance of 1,230 F g^{-1} which was achieved at

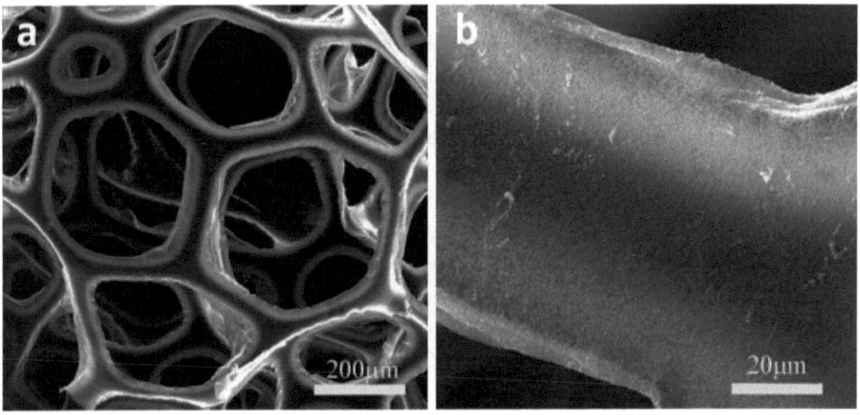

Fig. 2.12 Illustration of the SEM images of nano-structured MnO_2-CNT-sponge hybrid electrode (**a**) reveals the 3D hierarchical MnO_2-CNT-sponge electrode without any junctions, and (**b**) explains the uniform deposition of MnO_2 over the CNT-sponge skeleton (Reprinted with permission from [76]. Chen et al., *Nano Letters*, 11 (2011). © 2011, ACS Publications)

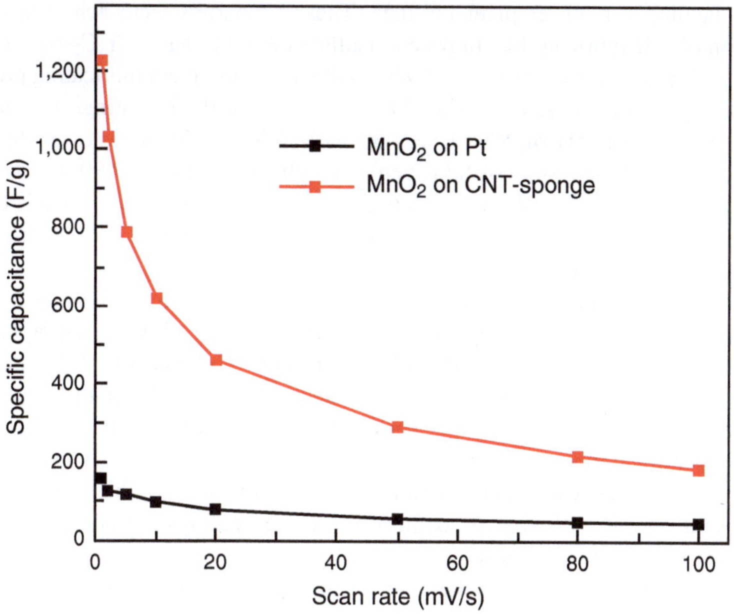

Fig. 2.13 Specific capacitance of MnO_2-CNT-sponge electrode is higher than the conventional MnO_2 on Platinum (Pt) electrode (Reprinted with permission from [76]. Chen et al., *Nano Letters*, 11 (2011). © 2011, ACS Publications)

a rate of 1 mV s^{-1}, while the later shows a specific capacitance of just 155 F g^{-1} for the same scan rate. Apart from this such MnO_2-coated CNT-sponge supercapacitor shows only 4 % degradation after 10,000 cycles, making it chemically more stable, and hence makes such novel capacitors as the future of energy storage systems.

2.8 Coatings for Fossil Fuels-Based Energy

Fossil fuels are still a predominant fuel material for the power suppliers. Generally, these fossil fuel materials like coal are of low grade type and it contains impurities like sodium, vanadium, and sulfur. These impurities form compounds such as sodium sulfate (Na_2SO_4), vanadiumpentoxide (V_2O_5), etc., which react adversely with the power generation equipment and hence lower down the performance and efficiency of the power plant. Coatings are most frequently applied on gas turbine components in order to restrict the surface degradation and to obtain accurate lifetime expectancies. Several thick coatings were being employed on the power plant components for resisting wear, hot corrosion, and erosion and also for enduring the hot environment of thermal power plant.

Yang et al. had reported for NiCr-Cr_3C_2-based anti-wear, anti-corrosion and anti-oxidation coatings by high velocity oxy fuel (HVOF) spraying followed by laser cladding for power plant turbines. These properties were achieved by the addition of self-lubricant like tungsten disulfide (WS_2) to NiCr-Cr_3C_2-based material [77]. The interesting property of WS_2 is that it can maintain lubricating property even at high temperature (539 °C). The wear results of these coatings confirm that the friction coefficient of NiCr-Cr_3C_2-30 wt% WS_2 coating was lower than that without WS_2. These results sustain even at a high temperature of 600 °C where the NiCr-Cr_3C_2 coating has shown a friction coefficient of ~0.43 while NiCr-Cr_3C_2-30 wt% WS_2 coating has ~0.28. Such self-lubricating property of WS_2 coatings was attained due to the lamellar growth of WS_2.

Recently, Jagadeeswaran et al. had deposited Stellite 6 alloy (Co-28.8Cr-2.6Ni-4.5W-2.5Fe-1.2C-1.3Si) using high velocity oxy-fuel (HVOF) spraying for resisting hot corrosion of fossil fuel power plant components. This Stellite-6 coating has shown superior performance than conventional cobalt-based, iron-based, and titanium-based alloys. They exhibit microhardness of 550 H_v while the conventional cobalt-based alloys show only 305 H_v [78]. This increased microhardness was resulted due to alloying different materials which had produced dense and pore-free Stellite 6 coatings and these coatings are stable even after 50 cycles of hot corrosion test.

In last one decade, researchers were putting much effort in enhancing the operating temperature of "thermal barrier coating." The primary function of TBCs is to insulate the underlying metal of gas turbine component from high temperature gases. At present, nano-structured yttria-stabilized zirconia (YSZ) materials have received significant interest in TBCs due to their extraordinary properties like low thermal conductivity, high coefficient of thermal expansion, and excellent mechanical properties. Recently, Zhou et al. had studied thermal diffusivity performance between nano-structured and traditional zirconia coatings, which were fabricated by plasma-sprayed technique [79]. The thermal diffusivity of nano-structured ZrO_2 coating was lower than the traditional ZrO_2 coating, which is attributed to micro pores and the decreased grain size of nano-structured zirconia coatings [79].

In addition to that, Lima et al. [80] had studied the diffusivity for heat-treated nano-structured and traditional YSZ coating. This group had quantified that the diffusivity of conventional YSZ coating was 57 % higher than that of the nano-structured coating. The thermal diffusivity value of nano-structured coating stabilizes after 1 h of heat treatment, while the thermal diffusivity of conventional coating continues to increase with heating time [80]. This property was mainly attributed to microstructural characteristics of the nano-structured YSZ coating. Nano-structured TBC have either fully molten region or semi-molten region, while the conventional YSZ coating has lamellar structure with inter-splat pores [80]. After heat treatment at 1,400 °C the nano-structured coating shows coarse pores, while the conventional YSZ coating does not show any difference from them as sprayed microstructure. Hence, this experiment clearly indicates the importance of nanomaterials in enhancing the performance and life of TBC.

However, these TBCs are also affected by fused sodium and vanadium-based salts, and it causes phase transformation in yttria-based TBC material. This phase transformation was accompanied by 3–5 % volume expansion, leading to cracking and spallation of TBC. Afrasiabi et al. group had reported for Al_2O_3 top coat on these YSZ barrier layer and performed hot corrosion test on this composite layer. This group did not found any spallation on YSZ/Al_2O_3 system. The presence of dense Al_2O_3 top layer on YSZ reduced the infiltration of molten salt and resulted in the further resistance of the coating against hot corrosion with increased hardness from 3 to 5 GPa [81]. As an upgradation to YSZ-based TBCs yttrium monosilicate (Y_2SiO_5) was observed to possess interesting properties like low thermal expansion coefficient and stability in water vapor atmospheres, which makes it as a promising protective layer intended for the hottest parts in the gas turbines [82]. Garcia et al. had coated such Y_2SiO_5 material as environmental barrier coatings and achieved a thermal conductivity of 0.8 W mk^{-1} with hardness of 6.0 GPa [82].

Apart from ceramic-based TBCs, cermet coatings are also used in energy applications mainly because of their high wear resistance and oxidation-resistant functionality. Cermet coatings of Ni-Al_2O_3 were cold sprayed by Sevillano et al. as wear-resistant sealants in power plant turbines [83]. They reported that depending upon the composition, fragmentation of alumina occurs during cold spraying, which leads to homogeneous dispersion of ceramic in metal matrix. Figure 2.14 reveals the SEM images of Al-Ni system with two extreme compositions of Al, one

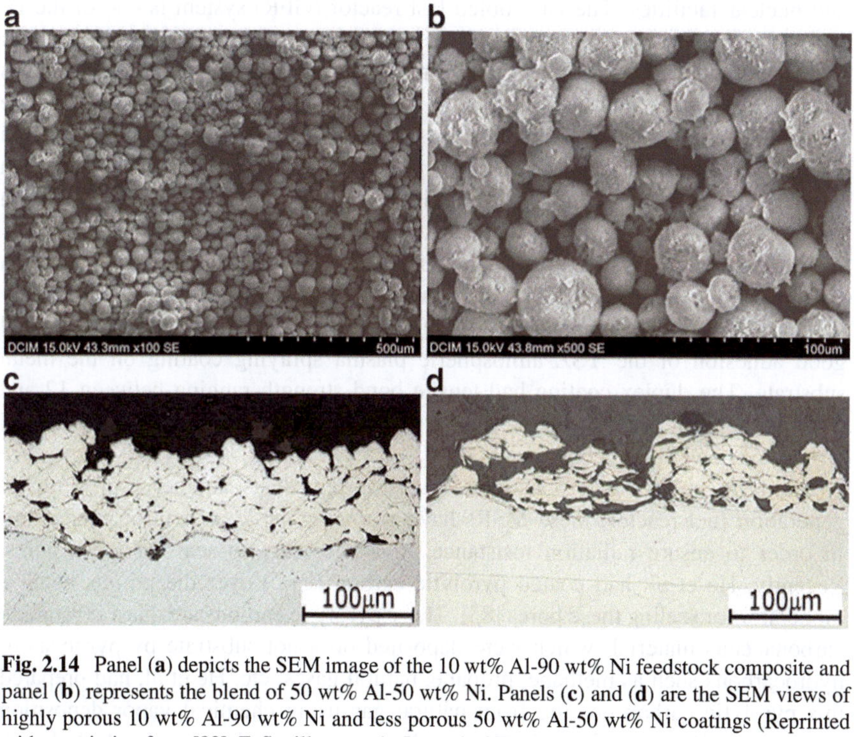

Fig. 2.14 Panel (**a**) depicts the SEM image of the 10 wt% Al-90 wt% Ni feedstock composite and panel (**b**) represents the blend of 50 wt% Al-50 wt% Ni. Panels (**c**) and (**d**) are the SEM views of highly porous 10 wt% Al-90 wt% Ni and less porous 50 wt% Al-50 wt% Ni coatings (Reprinted with permission from [83], F. Sevillano et al., Journal of Thermal Spray Technology 22, 5 (2013). © 2013, ASM International)

is 10 wt% Al-90 wt% Ni and the other is 50 wt% Al-50 wt% Ni. In Fig. 2.14a, b is the feedstock composition of these two blends before depositing on a substrate. While Fig. 2.14c, d represent the SEM images after cold spraying these feedstock of alumina-nickel blend, which reveals the change in the growth behavior of coating due to the difference in Al content. It is also been reported by this group that 50 wt% alumina feedstock has shown a very low porosity of 3 %; due to this negligible porosity, the hardness of pure Ni was increased from 175 to 338 H_v for Ni-Al_2O_3 composite, therefore promoting greater works hardness [83]. Hence, such cermet coating was used for the reduction of wear debris between the rotating and stationary components in gas path, thereby increasing the efficiency of gas turbine.

2.9 Coatings for Nuclear Energy

Nuclear energy technology offers significant advances in the area of power generation with the goals of providing sustainable energy generation with extension of nuclear fuel supply, minimization and management of nuclear waste, safe and reliable operation with improved accident management, competitive costs and financial risks of nuclear energy systems, control and security of nuclear material, and nuclear facilities. The gas-cooled fast reactor (GFR) system is one of the IV generation nuclear energy systems. This is a helium-cooled reactor, which operates with an outlet temperature of 850 °C. Protective coatings are used to protect various parts of the system and also protect the system in extreme conditions where the functional temperature can increase up to 1,250 °C with depressurization from 70 bars to atmospheric pressure. Such protective coatings must withstand high temperature, depressurization, and specific conditions of wear linked to erosion by high speed helium gas flow which is about 280 ms^{-1}.

For this purpose, Vert et al. had reported for thick yttria-stabilized zirconia coating (about 0.5 mm thick) on a thin metal substrate (1 mm thin) produced by suspension plasma spraying (SPS). The nano-structured SPS layer had showed good adhesion of the YSZ atmospheric plasma spraying coating on the metal substrate. The duplex coating had tensile bond strength ranging between 12 and 24 MPa; hence it was considered to be a potential candidate for resisting wear and erosion of power plant components [84].

Similarly, molten salt reactor (MSR) also represents another promising IV generation fuel reactor; these MSRs have porous graphite as neutron moderator. In order to ensure radiation resistance, it is necessary to seal off these pores. Recently, He et al. had coated pyrolytic carbon (PyC) over the porous neutron moderator for sealing these pores [85]. These pyrolytic carbons are quasi crystalline carbonaceous material, which were deposited on a hot substrate by pyrolysis of hydrocarbons such as methane, propane, natural gases, etc. He et al. had prepared this pyrolytic carbon coating from natural gas using chemical vapor deposition method at 1,300 °C. After this, the pyrolytic carbon coating was implanted with Xenon ($^{129}Xe^{26+}$) ions in order to understand the applicability of incorporated pyrolytic carbon coatings in nuclear energy field [85]. With increasing the dose of $^{129}Xe^{26+}$ ion, a wrinkled surface with elastic morphology appears. This formation of defects had resulted in increasing the average nanohardness of the coating by 95.5 %. In addition to that, the average normalized elastic modulus of this material had also increased from 2.63 to 3.35, thus quantifying the application of pyrolytic carbon coatings in next-generation nuclear reactors.

2.10 Coatings for Wind Energy

Turbines play a vital role in the generation of electricity through wind energy. These wind turbines are employed to operate at both low humidity and high humidity environments. Under high humidity environment, the icing of wind turbines is more prominent to occur. Icing causes mechanical failures and decreases the safety of the turbine; about 20 % of all wind turbines are located at sites where icing events are likely to occur during winter. Hence, there is an enormous demand for powerful and efficient methods to keep wind turbines free from ice. There are two main coatings that are used to achieve ice-free turbines, they are: (1) anti-icing coating and (2) de-icing coating.

Anti-icing systems try to avoid ice accretion, while de-icing systems are applied when ice has already built up. But, Sussoff et al. had reported for permanent ice phobic coatings which decreases the adhesion of ice to the surface. By this way, the accreted ice falls off from the rotor blades due to accreted mass, combined with centrifugal and vibrational forces [86]. This group had investigated various coatings with different ice phobic properties, coatings like hydrophilic coating, hydrophobic coating, and sol–gel-based coating containing fluorinated compounds and also viscoelastic rubber. The ice phobic properties and ice adhesion strength between coating and ice were investigated in this study [86]. Among all, viscoelastic rubber coating prepared from polydimethylsiloxanes (PDMS) has shown lower adhesion values. These materials are applied for ice phobic materials because of their combined effect of their low surface energy and outstanding elasticity property [86]. These coatings were based on the principle of lotus-leaf-inspired super hydrophobic effect and they have the significant drawbacks of poor mechanical property and are not resistive against erosion.

But under high humidity conditions, this super hydrophobic effect fails to perform as an effective ice phobic coating and it allows water condensation, frost formation, and increased ice adhesion on the turbine blades. And the reasons for this failure were large surface area and increased nucleation site density to condense droplets. In order to resolve this, Kim et al. had reported for extremely smooth surfaces with low wettability, using nanostructures which could delay frost formation and also reduces ice adhesion [87]. This group had reported for the coatings of slippery surfaces on bare aluminium, inspired from pitcher plant, and these surfaces are called as slippery, liquid-infused porous surfaces (SLIPS). Kim and his group had prepared this SLIPS surface using electro

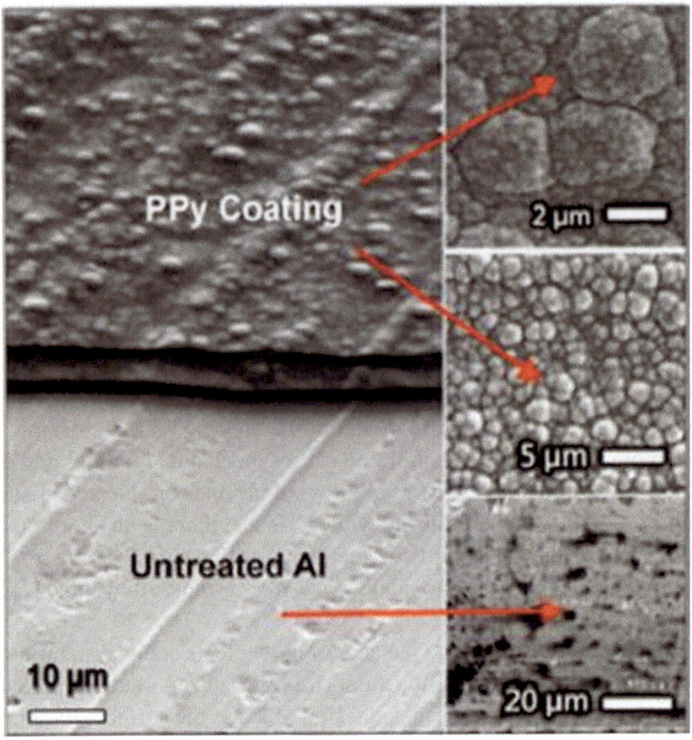

Fig. 2.15 SEM images that compare topography of polypyrrole (PPy)-treated Al substrate and untreated Al. *Insets* show higher magnification SEM images for the two areas (Reprinted with permission from [87], Philseok Kim et al., ACS Nano 6, 8 (2012). © 2012 ACS Publications)

deposition of nano-structured polypyrrole (PPy) on bare aluminium (Al) [87]. The nano-structured PPy coating is a good anti-corrosive coating material and has been created as a lubricating film that overcoats the solid support (i.e. bare Al). Figure 2.15 compares the surface features of PPy-coated Al surface and uncoated Al surface. These scanning electron microscope (SEM) images have shown a rough and globular morphology of PPy-treated layer which is not seen in the untreated layer. Further, higher magnification of these PPy-treated layer has revealed that each large globule consists of smaller nanoscale bumps, which had provided a surface that can infiltrate and retain the lubricant. As a result of this lubricant property, water droplets that condense on this interface have minimal exposure to the underlying solid defect sites and are capable of sliding of the surface at very low tilt angle [87]. Hence, such lubricating film on wind turbine blades tends to avoid icing and enhances the efficiency.

Likewise, during operation at low humidity condition, the compressor blades undergo intense abrasive action of sand, ash, and dust particles present in the air flow. As a result, the blade surface undergoes erosion and pitting corrosion. Such degradation on turbine blades will affect the aerodynamic and fatigue property of the blade material. Hence, special purpose corrosion and erosion-resistant coatings were highly essential under such environmental condition. Muboyadzhyan et al. had reviewed various high hardness metal compounds for this purpose and carried out abrasive wear test [88]. The results indicate that single layer coatings made up of carbides and nitrides of pure metals were shown to exhibit the highest erosion resistance. In terms of decreasing erosion resistance, the coatings were ordered as vanadium nitride (VN), vanadium carbide (VC), chromium carbide (Cr_3C_2), zirconium nitride (ZrN), titanium carbide (TiC), and titanium nitride (TiN) [88].

Apart from environment, the idle, start-up, or shut-down conditions due to wind cut off can breakdown the tribofilm (a protective coating) in rotating and lubricating parts of the turbine resulting in high friction and wear. Smooth and hard surface finishing together with proper lubrication systems are the essential requirements to overcome the challenges. Novel engineering surface treatment or deposition of hard coatings are necessary to enhance both wear and corrosion resistance.

Recently, Mubarok et al. had coated carbide-based materials such as SiC which has superior tribological and wear-resistant properties in corrosive and abrasive environments. High-frequency pulse detonation (HFPD) thermal spraying technique has been utilized by this group to deposit SiC. But the limitation of this SiC is that it decomposes or sublimates very easily at 2,500–2,600 °C, under atmospheric thermal spraying conditions. Therefore, SiC material was modified with high temperature materials. Hence, YAG-phase-coated SiC particles were used as feedstock material to spraying unit. Tribological tests recorded for these materials had resulted in the lowest coefficient of friction (COF) of 0.15 in polyalfaolefin (PAO) lubrication, which was the standard synthetic oil for offshore wind turbine gear and bearing systems [89]. Along with this, Bunker et al. had investigated ceramic thermal barrier coatings that are used to protect gas turbine hot section components. The TBC was air plasma-sprayed yttria-stabilized zirconia-oxide. For various degree of polishing from 10 to 1 μm the roughness value for TBC also varies. Results by this group had indicated significant effect of polishing degree on the surface heat-treated coefficients [90].

2.11 Coatings for Hydro Energy

Hydro turbine components undergo high erosion due to the impact of water droplets. This high energy particle impact wear (HEPIW) is mainly due to excessive silt content with particle size more than 90 μm. During monsoon season, it becomes impossible to control the silt content passing through the turbine. It increases from 250 to 10,000 ppm and the particle size exceeds 1,000 μm. The concentration of quartz in silt is very high ~90 %, therefore it causes rapid erosion to hydro power plant components. Components like runner blades, cheek plates, and profile portion of guide vanes, stay rings, and labyrinth seal suffer from this erosion. These parts are made up of corrosion-resistant martensitic, manganese, and austenitic steels. All these steels are considerably less resistant to erosion and require suitable protective coating.

To provide protection to these components, Mann et al. had reported of HVOF-sprayed tungsten carbide-cobalt-chromium (WC-Co-Cr)-based protective coatings. This selected powder of WC-10Co-4Cr had showed improved coating properties such as microhardness, fracture toughness, and high energy particle impact wear (HEPIW) resistance due to surface treatment approach by high power diode laser (HPDL) method [91]. The microhardness of "as-sprayed" HVOF coatings ranges from 1,450 to 1,590 H_v; after laser treatment this range had significantly increased to 1,800–1,890 H_v. In addition to that, HEPIW laboratory test resulted that HPDL-treated WC-10Co-4Cr HVOF-sprayed coating is 26 times better than conventional AISI 410 steel and eight times better than "as-sprayed" HVOF coating [91].

Similarly, CF8M and CA6NM were also extensively used grades of ASTM A743 steels in hydraulic turbine components. These materials were considerably used because of their excellent mechanical properties and resistance to corrosion. However, during longer hours of exposure, these steels are affected by slurry erosion because of their lesser hardness in comparison with very hard erodent particles of hardness of about 1,100 H_v. Recently, Deepak et al. had reported for Cr_2O_3 coatings deposited on CF8M and CA6NM turbine steels by high velocity oxy fuel (HVOF) spray process [92]. These coatings have micropores and voids between the coating and substrate. Cr_2O_3 coatings had resulted in higher hardness on comparing with bare CF8M and CA6NM steels. Uncoated CF8M steel and CA6NM steel has the average microhardness of 220 H_v and 340 H_v respectively, while Cr_2O_3-coated CF8M steel and CA6NM steel shows the microhardness of 1,305 and 1,320 H_v respectively [92]. The slurry erosion resistance of Cr_2O_3-coated CA6NM steel is much better in comparison with CF8M steel; it is because of the higher hardness, higher yield strength, low porosity, fine grain or splat size, the absence of cracks, and good inter-splat bonding.

2.12 Coatings for Geothermal Energy

Geothermal energy has been claimed to be one of the most environmental friendly energies mainly because of its lower emission of greenhouse gases. However, geothermal power plants have been pinpointed as pollution sources mainly owing to the release of H_2S and other toxic elements like Cd, Hg, As, Mn, and B that are enriched in geothermal fluids. Along with this, geothermal water contains a lot of corrosion and fouling ions, including Cl^-, Mg^{2+}, Ca^{3+}, $HCo_3{}^-$, and $SO_4{}^{2-}$. Metal pipelines and equipments are being encountered by serious corrosion and fouling problems in the process of transporting geothermal water, which reduces the efficiency and service life of pipelines and increases the cost of geothermal energy exploitation.

In order to avoid these fouling problems, anti-fouling coatings were being employed over stainless steel pipelines. Generally, metal alloys are coated to avoid calcium carbonate ($CaCO_3$) fouling because metal ions can effectively restrain the crystal growth rate of $CaCO_3$ fouling and change $CaCO_3$ morphology on the pipeline surface. Recently, Wang et al. had reported for coating Ni-Cu-Al alloy powder composite and studied their performance in comparison with stainless steel, under epoxy-silicone resin in simulated geothermal water environment. Coatings were prepared by painting the slurry resin of Ni-Cu-Al alloy over the 304 grade stainless steels [93].

Wang group had immersed these specimens in simulated geothermal water for 72 h, and they observed large needle-like matters on the surface of stainless steel and epoxy-silicone resin coatings. The group revealed that these needle-like matters are the classical aragonite fouling in geothermal water. But they do not find such structures on Ni-Cu-Al composite coated specimen after immersion because metal ions of Ni^{2+}, Cu^{2+}, and Al^{3+} inhibit the induction period of crystal nucleation and crystal growth rate of $CaCO_3$ fouling on the surface of composite coating. Additionally, the same group had tested for extended immersion time of specimens in simulated geothermal water to 168 h [93]. Under this condition, the surface of composite coating has little needle-like $CaCO_3$ structures, while the surfaces of stainless steel and epoxy-silicone resin had exhibited bigger and longer needle-like $CaCO_3$ foulings. However, the amount of $CaCO_3$ fouling adhering to the surface of composite coating is much smaller than that adhering to stainless steel surface [93]. Further researches are being undergoing to induce antifouling properties for much extended hours to geothermal components, thereby increasing the efficiency of geothermal-based energy.

2.13 Future Trends in Coatings

Coatings are inevitable in all applications; the discussion in this chapter reflects the potential of coatings in various energy sectors. Even though both thin films and thick coatings were being applied for renewable and non-renewable energy sectors, their application will not compensate the energy demand that will keep on rising in future. Hence, to support the future of mankind notable attention has to be given to advanced coatings that consist of mesoporous structures, carbon allotropes, nanowires, alloys, nanotubes, chiral nanostructures, nanocomposites, superlattices, and hybrid nanostructures (i.e., like CNT + TiO_2 nanoparticles). In addition to that, prevailing energy challenges like carbon emission, energy harvesting, energy storing and energy saving have to be tackled in future with the combination of advanced coating materials and advanced coating techniques.

References

1. I. Ozturk, A. Acaravci, The causal relationship between energy consumption and GDP in Albania, Bulgaria, Hungary and Romania: evidence from ARDL bound testing approach. Appl. Energy **87**(6), 1938–1943 (2010)
2. G. Barat, Liquid biofuels potential and outlook in Iran. Renew. Sustain. Energy Rev. **16**, 4379–4384 (2012)

3. L.E. Chaar, L.A. Lamont, N.E. Zein, Review of photovoltaic technologies. Renew. Sustain. Energy Rev. **15**, 2165–2175 (2011)
4. A. Evans, A.B. Hütter, L.M. Rupp, L.J. Gauckler, Review on microfabricated micro-solid oxide fuel cell membranes. J. Power Sources **194**, 119–129 (2009)
5. T. Osaka, H. Nara, T. Momma, T. Yokoshima, New Si–O–C composite film anode materials for LIB by electro deposition. J. Mater. Chem. A **2**, 883–896 (2014)
6. S. Vepek, The search for novel and super hard materials. J. Vac. Sci. Technol. A **17**, 2401–2420 (1999)
7. Q. Chen, G. Hubbard, P.A. Shields, C. Liu, D.W.E. Allsopp, W.N. Wang, S. Abbott, Broadband moth-eye antireflection coatings fabricated by low-cost nanoimprinting. Appl. Phys. Lett. **94**(26), 263118 (2009)
8. H. Tan, R. Santbergen, H.M. Smets, M. Zeman, Plasmonic light trapping in thin-film silicon solar cells with improved self-assembled silver nanoparticles. Nano Lett. **12**, 4070–4076 (2012)
9. T.T. Wu, F. Hu, J.H. Huang, C. Chang, C.C. Lai, Y. Yen, H. Huang, H. Hong, Z.M. Wang, C. Shen, J. Shieh, Y. Chueh, Improved efficiency of a large-area Cu(In, Ga)Se2 solar cell by a nontoxic hydrogen-assisted solid Se vapor selenization process. ACS Appl. Mater. Interfaces **6**(7), 4842–4849 (2014)
10. Y. Lee, D.S. Ruby, D.W. Peters, B.B. McKenzie, W.P. Hsu, ZnO nanostructures as efficient antireflection layers in solar cells. Nano Lett. **8**(5), 1501–1505 (2008)
11. S.C. Singhal, Advances in solid oxide fuel cell technology. Solid State Ion. **135**(1–4), 305–313 (2000)
12. Z. Hu, K. Zhang, H. Gao, W. Duan, F. Cheng, J. Liang, J. Chen, Li_2MnSiO_4@C nanocomposite as a high-capacity cathode material for Li-ion batteries. J. Mater. Chem. A **1**, 12650 (2013)
13. Z. Huang, B. Zhang, S. Oh, Q. Zheng, X. Lin, N. Yousefi, J. Kim, Self-assembled reduced graphene oxide/carbon nanotube thin films as electrodes for supercapacitors. J. Mater. Chem. **22**, 3591–3599 (2012)
14. C. Ning, L.M. Yan, Z.W. Dong, Fouling and corrosion properties of SiO_2 coatings on copper in geothermal water. Ind. Eng. Chem. Res. **51**, 6001–6017 (2012)
15. F.P. Moreno, M.A. Jakab, N. Tailleart, M. Goldman, J.R. Scully, Corrosion-resistant metallic coatings. Mater. Today **11**(10), 14–23 (2008)
16. A. Salam, Y. Alsabagh, W. Tiu, Y. Xu, M.S. Virk, A review of the effects of ice accretion on the structural behaviour of wind turbines. Wind Eng. **37**(1), 59–70 (2013)
17. N. Dalili, A review of surface engineering issues critical to wind turbine performance. Renew. Sustain. Energy Rev. **13**, 428–438 (2009)
18. E. Korkut, M. Atlar, An experimental investigation of the effect of foul release coating application on performance, noise and cavitation characteristics of marine propellers. Ocean Eng. **41**, 1–12 (2012)
19. S. Zhang, P. Kang, T.J. Meyer, Nanostructured tin catalysts for selective electrochemical reduction of carbon dioxide to formate. J. Am. Chem. Soc. **136**, 1734–1737 (2014)
20. R. Urteaga, L.N. Acquaroli, R.R. Koropecki, A. Santos, M. Alba, J. Pallares, L.F. Marsal, A. Berli, Optofluidic characterization of nanoporous membranes. Langmuir **29**, 2784–2789 (2013)
21. D. Lee, Y. Kanai, Role of four-fold coordinated titanium and quantum confinement in CO_2 reduction at titania surface. J. Am. Chem. Soc. **134**, 20266–20269 (2012)
22. H. Sevincli, T. Lehmann, D.A. Ryndyk, G. Cuniberti, Comparison of electron and phonon transport in disordered semiconductor carbon nanotubes. J. Comput. Electron **12**, 685–691 (2013)
23. P. Chakraborty, M.R. Zachariah, Do nanoenergetic particles remain nano-sized during combustion. Combust. Flame **161**, 1408–1416 (2014)

24. J. Tao, M. Yang, J.W. Chai, J.S. Pan, Y.P. Feng, S.J. Wang, Atomic N modified rutile TiO_2 (110) surface layer with significant visible light photoactivity. J. Phys. Chem. C **118**, 994–1000 (2014)
25. B. Dobosz, R. Krzyminiewski, G. Schroeder, J. Kurczewska, Electron paramagnetic resonance as an effective method for a characterization of functionalized iron oxide. J. Phys. Chem. Solids **75**, 594–598 (2014)
26. T. Kushida, S. Tanaka, C. Morita, T. Tanji, Y. Ohshita, Mapping of minority carrier lifetime distributions in multicrystalline silicon using transient electron-beam-induced current. J. Electron Microsc. **61**(5), 293–298 (2012)
27. D. Golberg, Y. Bando, Y. Huang, T. Terao, M. Mitome, C. Tang, C. Zhi, Boron nitride nanotubes and nanosheets. ACS Nano **4**(6), 2979–2993 (2010)
28. D. Liang, Y. Kang, Y. Huo, Y. Chen, Y. Cui, J.S. Harris, High-efficiency nanostructured window GaAs solar cells. Nano Lett. **13**, 4850–4856 (2013)
29. D. Deng, M.G. Kim, J.Y. Lee, J. Cho, Green energy storage materials: nanostructured TiO_2 and Sn-based anodes for lithium-ion batteries. Energy Environ. Sci. **2**, 818–837 (2009)
30. A.K. Keshri, A. Agarwal, Wear behavior of plasma-sprayed carbon nanotube-reinforced aluminum oxide coating in marine and high-temperature environments. J. Therm. Spray Technol. **20**(6), 1217–1230 (2011)
31. K. Cheng, E. Yang, C.Y. Lee, Z. Ricks, V. Palmre, K. Kim, *Fine-Tuned Polymer Nano-Composite Coatings for Use in Geothermal Plants*, in ASME 2011 Conference on Smart Materials, Adaptive Structures and Intelligent Systems, vol. 1 (2011), pp. 713–719
32. Z. Chai, Q. Suo, H. Wang, X. Wang, Mesoporous lanthanum phosphate nanostructures containing H_3PO_4 as superior electrolyte for PEM fuel cells. RSC Adv. **3**, 21928 (2013)
33. M.P. Finnegan, H. Zhang, J.F. Banfield, Phase stability and transformation in titania nanoparticles in aqueous solutions dominated by surface energy. J. Phys. Chem. C **111**, 1962–1968 (2007)
34. M.J. Yacamán, J.A. Ascencio, H.B. Liu, J. Torresdey, Structure shape and stability of nanometric sized particles. J. Vac. Sci. Technol. B **19**, 1071–1023 (2001)
35. Y. Song, S.S.R. Kumar, J. Hormes, Synthesis of palladium nanoparticles using a continuous flow polymeric micro reactor. J. Nanosci. Nanotechnol. **4**(7), 788–793 (2004)
36. D.R. Baer, J.E. Amonette, M.H. Engelhard, D.J. Gaspar, A.S. Karakoti, S. Kuchibhatla, P. Nachimuthu, J.T. Nurmi, Y. Qiang, V. Sarathy, S. Seal, A. Sharma, P.G. Tratnyek, C.M. Wang, Characterization challenges for nanomaterials. Surf. Interface Anal. **40**, 529–537 (2008)
37. S. Kar, S. Chaudhuri, Shape selective growth of CdS one-dimensional nanostructures by a thermal evaporation process. J. Phys. Chem. B **110**(10), 4542–4547 (2006)
38. D. Lundina, K. Sarakinos, An introduction to thin film processing using high-power impulse magnetron sputtering. J. Mater. Res. **27**(5), 780–792 (2012)
39. K.L. Choy, Chemical vapour deposition of coatings. Prog. Mater. Sci. **48**, 57–170 (2003)
40. K. Niessen, M. Gindrat, Plasma spray-PVD: a new thermal spray process to deposit out of the vapor phase. J. Therm. Spray Technol. **20**(4), 736–743 (2011)
41. S. Olson, K. Hummler, B. Sapp, *Challenges in Thin Wafer Handling and Processing*, in IEEE (2013), pp. 62–65
42. K. Kim, Ultrathin organic solar cells with graphene doped by ferroelectric polarization. ACS Appl. Mater. Interfaces **6**, 3299–3304 (2014)
43. H. Li, X. Liu, B. Yang, P. Wang, Influence of substrate bias and post-deposition Cl treatment on CdTe film grown by RF magnetron sputtering for solar cells. RSC Adv. **4**, 5046 (2014)
44. H. Bi, W. Zhao, S. Sun, H. Cui, T. Lin, F. Huang, X. Xie, M. Jiang, Graphene films decorated with metal sulfide nanoparticles for use as counter electrodes of dye-sensitized solar cells. Carbon **61**, 116–123 (2013)
45. C. Yang, H. Bi, D. Wan, F. Huang, X. Xie, M. Jiang, Direct PECVD growth of vertically erected graphene walls on dielectric substrates as excellent multifunctional electrodes. J. Mater. Chem. A **1**, 770 (2013)

46. C. Li, J. Xia, Q. Wang, J. Chen, C. Li, W. Lei, X. Zhang, Photovoltaic property of a vertically aligned carbon nanotube hexagonal network assembled with CdS quantum dots. ACS Appl. Mater. Interfaces **5**, 7400–7404 (2013)
47. T. Matsui, M. Kondo, Advanced materials processing for high-efficiency thin-film silicon solar cells. Sol. Energy Mater. Sol. Cells **119**, 156–162 (2013)
48. T. Kobayashi, T. Kumazawa, Z.J. Kao, T. Nakada, Cu(In, Ga)Se2 thin film solar cells with a combined ALD-Zn(O, S) buffer and MOCVD-ZnO:B window layers. Sol. Energy Mater. Sol. Cells **119**, 129–133 (2013)
49. S.Y. Myong, L.S. Jeon, Improved light trapping in thin-film silicon solar cells via alternated n-type silicon oxide reflectors. Sol. Energy Mater. Sol. Cells **119**, 77–83 (2013)
50. J.W. Lee, B. Ye, D. Kim, J.K. Kim, J. Heo, H.Y. Jeong, M. Kim, W. Choi, J.M. Baik, ZnO nanowire-based antireflective coatings with double-nanotextured surfaces. ACS Appl. Mater. Interfaces **6**, 1375–1379 (2014)
51. S. Hsu, C. Tsai, C. Lu, Y. Tsai, T. Huang, Y. Jhang, Y. Chen, C. Wua, Y. Chen, Nanoporous platinum counter electrodes by glancing angle deposition for dye-sensitized solar cells. Org. Electron **13**, 856–863 (2012)
52. P. Jackson, D. Hariskos, E. Lotter, S. Paetel, R. Wuerz, R. Menner, W. Wischmann, M. Powalla, New world record efficiency for Cu(In, Ga)Se2 thin-film solar cells beyond 20%. Prog. Photovolt. Res. Appl. **19**, 894–897 (2011)
53. Q. Zhang, E. Uchaker, S.L. Candelaria, G. Cao, Nanomaterials for energy conversion and storage. Chem. Soc. Rev. **42**, 3127 (2013)
54. P. Arunkumar, M. Meena, K. Suresh Babu, A review on cerium oxide-based electrolytes for ITSOFC. Nanomater. Energy **1**, 288–305 (2012)
55. S.C. Hamm, S. Basuray, S. Mukherjee, S. Sengupta, J.C. Mathai, G.A. Baker, S. Gangopadhyay, Ionic conductivity enhancement of sputtered gold nanoparticle-in-ionic liquid electrolytes. J. Mater. Chem. A **2**, 792 (2014)
56. N.N. Kariuki, W.J. Khudhayer, T. Karabacak, D.J. Myers, GLAD Pt–Ni alloy nanorods for oxygen reduction reaction. ACS Catal. **3**, 3123–3132 (2013)
57. M.N. Banis, S. Sun, X. Meng, Y. Zhang, Z. Wang, R. Li, M. Cai, T.K. Sham, X. Sun, $TiSi_2O_x$ coated N-doped carbon nanotubes as Pt catalyst support for the oxygen reduction reaction in PEMFCs. J. Phys. Chem. C **117**, 15457–15467 (2013)
58. J. Ma, Y. Sahai, Effect of electrode fabrication method and substrate material on performance of alkaline fuel cells. Electrochem. Commun. **30**, 63–66 (2013)
59. C. Zhang, J. Hu, X. Wang, X. Zhang, H. Toyoda, M. Nagatsu, Y. Meng, High performance of carbon nanowall supported Pt catalyst for methanol electro-oxidation. Carbon **50**, 3731–3738 (2012)
60. Y.S. Hong, H.H. Yoon, Preparation of scandia-stabilized zirconia electrolyte thin films for intermediate temperature-solid oxide fuel cells by electron beam vapor deposition, Jpn. J. Appl. Phys. **50**, 01BE091–01BE095 (2011)
61. D. Virbukas, G. Laukaitis, J. Dudonis, D. Milčius, The properties of scandium and cerium stabilized zirconium thin films formed by e-beam technique. Solid State Ion. **188**, 46–49 (2011)
62. S. Kim, Y. Lee, G.M. Choi, Electrical conductivity of Gd-doped ceria film at low temperatures (300–500°C). Solid State Ion **262**, 411–415 (2014)
63. M.H. Wu, J.L. Huang, K.Z. Fung, D.F. Lii, Application of samaria doped sculptured ceria/ $La_{1-x}Sr_xCo_{1-y}Fe_yO_{3-\delta}$ in cathode of SOFCs at intermediate temperature. Vacuum **101**, 57–62 (2014)
64. B. Li, Y. Wang, H. Rong, Y. Wang, J. Liu, L. Xing, M. Xu, L. Weishan, A novel electrolyte with the ability to form a solid electrolyte interface on the anode and cathode of a $LiMn_2O_4$/graphite battery. J. Mater. Chem. A **1**, 12954–12961 (2013)
65. S. Goriparti, E. Miele, F. Angelis, E. Fabrizio, R.P. Zaccaria, C. Capiglia, Review on recent progress of nanostructured anode materials for Li-ion batteries. J. Power Sources **257**, 421–443 (2014)

66. X. Sun, C. Zhou, M. Xie, H. Sun, T. Hu, F. Lu, S.M. Scott, S.M. George, J. Lian, Synthesis of ZnO quantum dot/graphene nanocomposites by atomic layer deposition with high lithium storage capacity. J. Mater. Chem. A **2**, 7319 (2014)
67. X. Wang, Z. Zhang, Y. Chen, Y. Qu, Y. Lai, J. Li, Morphology-controlled synthesis of MoS_2 nanostructures with different lithium storage properties. J. Alloys Compd. **600**, 84–90 (2014)
68. J. Qin, C. He, N. Zhao, Z. Wang, C. Shi, E. Liu, J. Li, Graphene networks anchored with Sn@Graphene as lithium ion battery anode. ACS Nano **8**(2), 1728–1738 (2014)
69. J. Ren, C. Wang, Q. Wu, X. Liu, Y. Yang, L. He, W. Zhang, A silicon nanowire–reduced graphene oxide composite as a high-performance lithium ion battery anode material. Nanoscale **6**, 3353 (2014)
70. C. Jin, X. Zhang, W. He, Y. Wang, H. Li, Z. Wang, Z. Bi, Effect of ion doping on the electrochemical performances of $LiFePO_4$–$Li_3V_2(PO_4)_3$ composite cathode materials. RSC Adv. **4**, 15332 (2014)
71. C. Tang, K. Hackenberg, Q. Fu, P.M. Ajayan, H. Ardebili, High ion conducting polymer nanocomposite electrolytes using hybrid nanofillers. Nano Lett. **12**, 1152–1156 (2012)
72. G. Lota, F. Krzysztof, E. Frackowiak, Carbon nanotubes and their composites in electrochemical applications. Energy Environ. Sci. **4**, 1592–1605 (2011)
73. M. Ghaffari, S. Kosolwattana, Y. Zhou, N. Lachman, M. Lin, D. Bhattacharya, K.K. Gleason, B.L. Wardle, Q.M. Zhang, Hybrid supercapacitor materials from poly (3,4-ethylenedioxythiophene) conformally coated aligned carbon nanotubes. Electrochim. Acta **112**, 522–528 (2013)
74. R. Warren, F. Sammoura, A. Kozinda, L. Lin, *ALD Ruthenium Oxide-Carbon Nanotube Electrodes for Supercapacitor Applications*, in Proceedings of 27th IEEE Micro Electro Mechanical Systems Conference (2014), pp. 167–170
75. Z. Fan, J. Yan, L. Zhi, Q. Zhang, T. Wei, J. Feng, M. Zhang, W. Qian, F. Wei, A three-dimensional carbon nanotube/graphene sandwich and its application as electrode in supercapacitors. Adv. Mater. **22**, 3723–3728 (2010)
76. W. Chen, R.B. Rakhi, L. Hu, X. Xie, Y. Cui, H.N. Alshareef, High-performance nanostructured supercapacitors on a sponge. Nano Lett. **11**, 5165–5172 (2011)
77. M.S. Yang, X. Liu, J. Fan, X. He, S. Shi, G. Fu, M. Wang, S. Chen, Microstructure and wear behaviours of laser clad $NiCr/Cr_3C_2$–WS_2 high temperature self-lubricating wear-resistant composite coating. Appl. Surf. Sci. **258**, 3757–3762 (2012)
78. N. Jegadeeswaran, M.R. Ramesh, S. Prakrathi, K.U. Bhat, Hot corrosion behaviour of HVOF sprayed stellite-6 coatings on gas turbine alloys. Trans. Indian Inst. Metals **67**(1), 87–93 (2014)
79. C. Zhou, N. Wang, Z. Wang, S. Gong, H. Xu, Thermal cycling life and thermal diffusivity of a plasma-sprayed nanostructured thermal barrier coating. Scr. Mater. **51**, 945–948 (2004)
80. R.S. Lima, B.R. Marple, Nanostructured YSZ thermal barrier coatings engineered to counteract sintering effects. Mater. Sci. Eng. A **485**, 182 (2008)
81. A. Afrasiabi, A. Kobayashi, Hot corrosion control in plasma sprayed YSZ coating by alumina layer with evaluation of microstrcuture and nanoindentation data (H, E). Vacuum **88**, 103–107 (2013)
82. E. Garcia, P. Miranzo, M.I. Osendi, The prospect of Y_2SiO_5-based materials as protective layer in environmental barrier coatings. J. Therm. Spray Technol. **22**(5), 680 (2013)
83. F. Sevillano, P. Poza, C.J. Munez, S. Vezzu, S. Rech, A. Trentin, Cold-sprayed Ni-Al_2O_3 coatings for applications in power generation industry. J. Therm. Spray Technol. **22**(5), 772–782 (2013)
84. R. Vert, P. Carles, E. Laborde, G. Mariaux, E. Meillot, A. Vardelle, Adhesion of ceramic coating on thin and smooth metal substrate: a novel approach with a nanostructured ceramic interlayer. J. Therm. Spray Technol. **21**(6), 1128–1134 (2012)
85. X. He, J. Song, H. Xia, J. Tan, B. Zhang, Z. He, X. Zhou, Z. Zhu, M. Zhao, X. Liu, L. Xu, S. Bai, Direct characterization of ion implanted pyrolytic carbon coatings deposited from natural gas. Carbon **68**, 95–103 (2014)

86. M. Susoff, K. Siegmann, C. Pfaffenroth, M. Hirayama, Evaluation of icephobic coatings – screening of different coatings and influence of roughness. Appl. Surf. Sci. **282**, 870–879 (2013)
87. P. Kim, T.S. Wong, J. Alvarenga, M.J. Kreder, W.E. Martinez, J. Aizenberg, Liquid-infused nanostructured surfaces with extreme anti-ice and anti-frost performance. ACS Nano **6**(8), 6569–6577 (2012)
88. S.A. Muboyadzhyan, Erosion-resistant coatings for gas turbine compressor blades. Russ. Metall. **6**(3), 183–196 (2009)
89. F. Mubarok, S. Armada, I. Fagoaga, N. Espallarga, Thermally sprayed sic coatings for offshore wind turbine bearing applications. J. Therm. Spray Technol. **22**(8), 1303–1309 (2013)
90. R.S. Bunker, *The Effect of Thermal Barrier Coating Roughness Magnitude on Heat Transfer with and Without Flow Path Surface Steps*, in International Mechanical Engineering Congress and Exposition (2003), doi:10.1115/IMECE2003-41073
91. B.S. Mann, V. Arya, B.K. Pant, High-power diode laser surface treated HVOF coating to combat high energy particle impact wear. J. Mater. Eng. Perform. **22**(7), 1995–2004 (2013)
92. D.K. Goyal, H. Singh, H. Kumar, V. Sahni, Slurry erosive wear evaluation of HVOF-spray Cr_2O_3 coating on some turbine steels. J. Therm. Spray Technol. **21**(5), 838–851 (2012)
93. G.G. Wang, L.Q. Zhu, H.C. Liu, W.P. Li, Galvanic corrosion of Ni–Cu–Al composite coating and its anti-fouling property for metal pipeline in simulated geothermal water. Surf. Coat. Technol. **206**, 3728–3732 (2012)

Chapter 3
Ternary and Quaternary Semiconducting Compounds Thin Film Solar Cells

P. Malar

Abstract This chapter deals with solar cells that are based on thin films of ternary and quaternary semiconducting compounds. These ternary and quaternary compounds are derived by an ordered substitution of group IV elements of the periodic table by the elements belonging to nearby groups while maintaining the average number of valence electrons per atom to be four. In case of ternary compounds, there exist two sets of compounds, namely, I-III-VI_2 and II-IV-V_2. Among the I-III-VI_2 compounds, $CuInSe_2$ and Cu(In,Ga)Se_2 were considered as potential candidates for absorbers in solar cells. In 2003, CIGS-based solar cells with recorded efficiency of 19.2 % has been reported by NREL. Recently, The Centre for Solar Energy and Hydrogen Research Baden-Württemberg (ZSW) has achieved a record of 20.8 % conversion efficiency with a copper indium gallium diselenide (CIGS) thin-film PV cell.

Following the introduction on energy demands and history of solar cells, different generations of solar cells are described along with their working principle. After the discussion on ternary compounds for thin film solar cells, currently considered quaternary compounds I_2-II-IV-VI_4 for absorbers in thin film solar cells are described.

3.1 Introduction

Recent apprehensions due to the global warming have steered the already established research and developments on renewable energy technologies and energy storage systems. Renewable energy sources such as solar energy, wind energy, and thermal energy are considered because of their abundance and green nature. These alternate sources promise for easing of increasing energy demands that currently rely majorly on fossil fuels. Risk factors associated with nuclear power cause concern and many countries across the world concentrate on the development of alternate renewable energy technologies for their power requirements.

P. Malar (✉)
Research Institute, SRM University, Kattankulathur, Chennai 603 203, Tamil Nadu, India
e-mail: malar.p@res.srmuniv.ac.in

S. Babu Krishna Moorthy (ed.), *Thin Film Structures in Energy Applications*,
DOI 10.1007/978-3-319-14774-1_3

Photovoltaics deal with conversion of sunlight into electricity. The effect was discovered by Becquerel [1] in 1839. He observed development of photovoltage due to incident light on the electrode in an electrolyte. Following which, several people worked on different materials, namely, solid selenium and cuprous oxide [2–5]. After 40 years of the first report on photovoltaic effect, photovoltaic cells based on the above materials have been reported. The generation of charge carriers by incident photon absorption happens in materials that exhibit semiconducting nature. In 1950s, various research groups have reported photovoltaic cells based on different absorber materials systems such as Si [6], Cu_xS/CdS heterojunction [7], and GaAs [8]. At first, all these systems have exhibited photovoltaic conversion efficiencies of about 6 %. Apart from Si, the other semiconducting materials that are found suitable absorbers in solar cell applications are binary, ternary, and quaternary compounds. They belong to II-VI (Eg, CdTe), III-V (Eg, GaAs), I-III-VI_2 ($CuInSe_2$ and Cu(In,Ga)Se_2/CIGS), and recently I_2-II-IV-VI_4 groups of compounds with direct band gap and having strong absorption edge near the fundamental absorption edge.

3.2 Classifications into Different Generations of Solar Cells

Based on the materials and technology, the solar cells can be divided into three main generations. Recently, new class of "inorganics in organics solar cell" is gaining attention, which is grouped under fourth-generation solar cells. Illustration in Fig. 3.1 indicates the classification of solar cells according to bulk and thin film absorber materials.

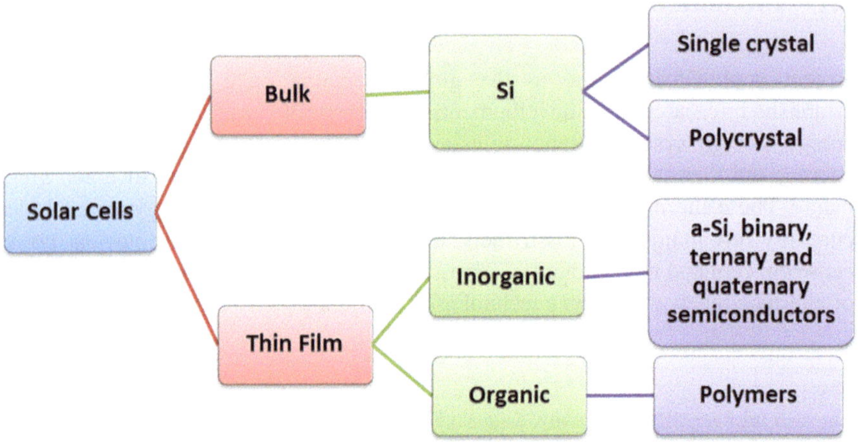

Fig. 3.1 Classification of solar cell according to bulk and thin film light absorbers

3.2.1 *Inorganic Solar Cells*

In the inorganic solar cells that belong to first as well as second generations (bulk Si and thin film solar cells), the solar energy conversion results from charge generation, charge separation, and charge transport. Charge generation or electron-hole pair creation happens because of absorption of light. Unless the electron-hole pairs created are separated and transported, the light-electricity conversion is incomplete. A pn junction (homojunction in case of bulk Si and heterojunction in thin film CdTe/CIGS solar cells) is used for charge separation and transported through external circuit. Figure 3.2 illustrates the functioning of a p-n junction-based solar cells. Usually, the n-type material is maintained thin compared to the p-type layer to facilitate light to pass through it as well as to enable charge separation.

Figure 3.2a indicates p and n-type semiconductor in isolation. When they are brought together (Fig. 3.2b), depletion region formation takes place. When light photons are incident on the p-n junction (Fig. 3.2c), electron-hole pairs are generated. The electrons travel through the external circuit to recombine with the holes in the p-region. This creates a flow of electric current (Fig. 3.2d). Detailed description of the functioning of thin film solar cells have been presented by various research groups across the world [9–13].

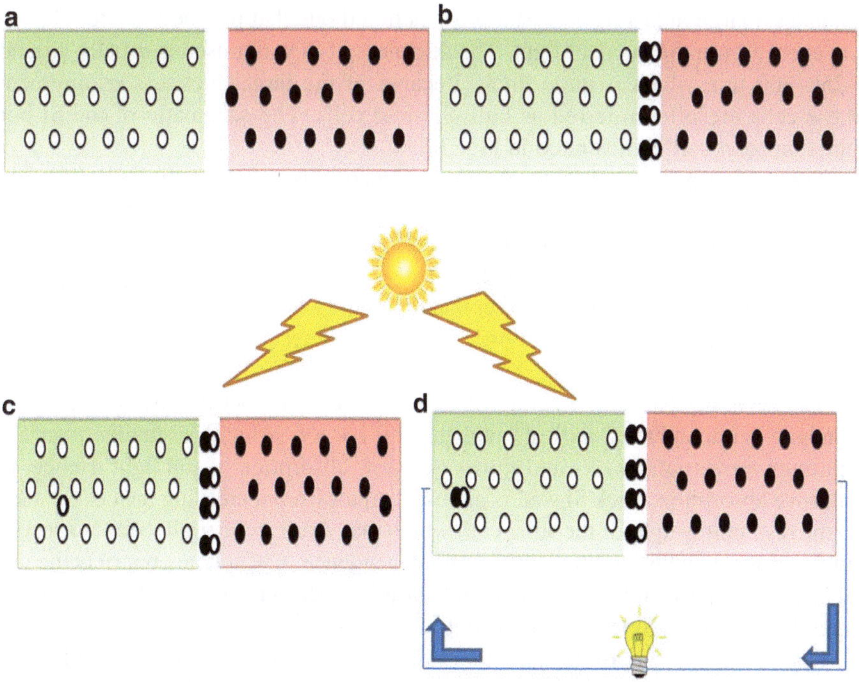

Fig. 3.2 Illustration of working of p–n junction-based solar cells. (**a**) p and n type semiconductors under isolation, (**b**) pn junction under equilibrium, (**c**) pn junction under illumination; generation and separation of electron-hole pair, (**d**) charge (electron) transport and recombination

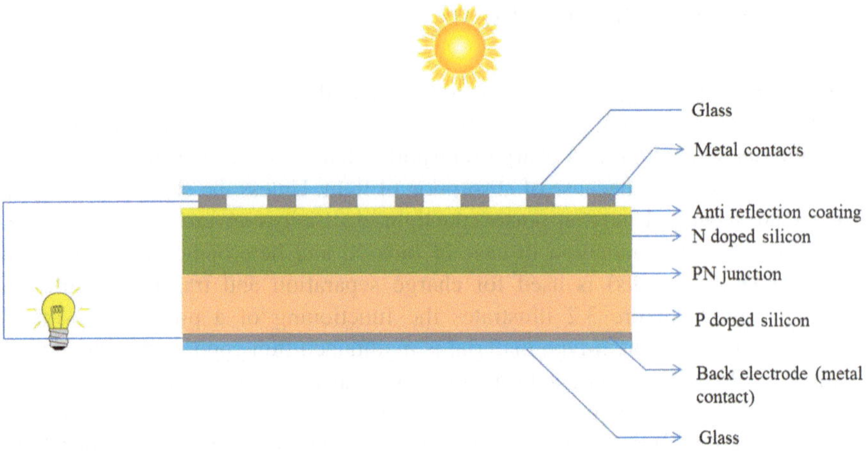

Fig. 3.3 Schematic of bulk Si solar cell

3.2.2 First-Generation Solar Cells

The first-generation solar cells are made of crystalline Si wafers. These cells are relatively expensive primarily due to the material cost. The design of these cells includes a large-area single-crystal p-n junction diode that produces useful electrical energy from the incident light. The thickness of Si wafer used is usually around ~250–300 μm, which is quite thick. Because of the large thickness requirement these cells are often referred as bulk Si solar cells. The schematic of the Si p-n junction solar cell is illustrated in Fig. 3.3.

The doping process of Si to obtain p-n junction is achieved through diffusion process. The maximum reported single cell efficiency for this type is about ~24.7 % and usually commercially available panel efficiencies vary between 13 % and 20 % average efficiencies [14]. Though Si solar cells are technologically matured and commercially available, the very fact that high purity Si requirement in bulk form is the major disadvantage leads to high initial cost investments compared to power output.

The necessity for bulk quantity of Si arises from the fact that Si is an indirect bandgap semiconductor with a band gap of 1.1 eV (which is smaller than the optimum terrestrial 1.4 eV) and shows very low absorption of sun light. Hence, to improve absorption, thick Si wafers are used. Apart from increasing the cost, it adds to the resultant weight of the solar cell, which may cause concern in certain specific applications. In addition, the temperature dependence of efficiency restricts the use of Si cells in space and under concentrated light conditions. Above all flexible solar module using bulk Si technology is out of choice.

Due to the above reasons, a number of alternate semiconducting materials have been developed and studied for photovoltaic applications. Electrically inferior compound semiconductors in place of expensive single crystalline Si are one option. The compound semiconductors could be binary, ternary, or quaternary compounds. They can be used in thin film form because of the direct band gap nature exhibiting high sunlight absorption near surface. The less and less pure material requirements facilitate initial cost reduction. In addition, low-temperature processing and higher throughput manufacturing have enabled these thin film technologies to be a cheaper technology compared to Si.

3.2.3 Second-Generation Thin Film Solar Cells

Solar cells technology based on thin film absorber materials is categorized under second-generation solar cells. The absorber materials used are mainly compound semiconductors, the most predominant being CdTe and CIGS apart from a-Si. The main selection criteria for absorbers in second-generation solar cells are low cost raw materials, lesser amount compared to bulk Si, band gap near optimum 1.4 eV, and high absorption coefficient near the fundamental absorption edge. An elaborate discussion on derivation and characteristics of these compound semiconductors is presented in Sect. 3.2.3.2.

3.2.3.1 Amorphous Silicon Cells

These solar cells are made with hydrogen-passivated amorphous silicon thin film absorbers [15]. The cells are built from bottom to top on a glass substrate with metal contact, n-type a-Si layer, intrinsic layer, p-type a- Si layer, and transparent contact layer. Hydrogenation of the a-Si layer is done to reduce the number of dangling bonds and hence the defect density. These cells exhibit typical efficiencies in the range of 9.5 %. But due to the light-induced degradation of the hydrogenated a-Si layers, these devices are prone to reduced lifetime issues. a-Si solar cells are commonly used in portable electronic devices such as calculators.

3.2.3.2 Binary and Ternary Compounds for Solar Cells

Derivation of Compound Semiconductors for Absorbers in Solar Cells

An ordered substitution of group IV elements of the periodic table by the elements belonging to other nearby groups while maintaining the average number of valence electrons per atom to be four gives rise to two families of binary and ternary compounds, namely, II-VI, III-V, I-III-VI_2, and II-IV-V_2 [16]. Figure 3.4 shows the schematic illustration of derivation of compound semiconductors through substitution. This illustration follows a depiction that is similar to the one presented in [17].

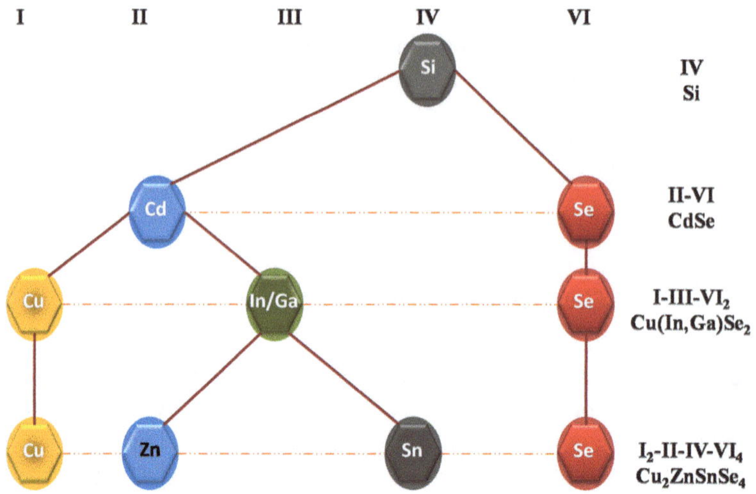

Fig. 3.4 Schematic illustration of derivation of binary, ternary, and quaternary compounds from the tetrahedrally bonded group IV elements

Cadmium Telluride

Cadmium telluride is a binary semiconducting compound that belongs to II-VI group of compounds. It is a direct band gap semiconductor with a band gap of 1.44 eV, which is closer to the optimum band gap required for solar cells. CdTe crystallizes in wurtzite structure. It can be doped p or n type. p-type CdTe has been studied much as an absorber material in solar cells. These cells can be constructed bottom up on a glass substrate with metal contact, or top down on a glass substrate with transparent conducting electrode. Both are made from a heterojunction with thin cadmium sulfide as emitter layer. The First Solar is the pioneer company for CdTe technology and took almost a decade to develop high-efficiency panels with the CdTe thin-film solar cells exhibiting 17.3 % and their solar panels showing efficiencies of 14.4 % [18].

Copper Indium Diselenide

$CuInSe_2$ belongs to the I-III-VI_2 family of ternary chalcopyrite semiconductors. These I-III-VI_2 ternary compounds crystallize with chalcopyrite structure, which belongs to the tetragonal crystal system. $CuFeS_2$ is a natural mineral example of the compounds, which have the chalcopyrite structure [19], and has the space group of $I\bar{4}2d$. Due to the lower symmetry of the chalcopyrite crystal structure when compared to diamond or zincblende, these compounds exhibit nonlinear optical properties. All of these ternaries are direct band gap semiconductors and show high absorption coefficient (~10^5 cm^{-1}) near the fundamental absorption edge (Fig. 3.5).

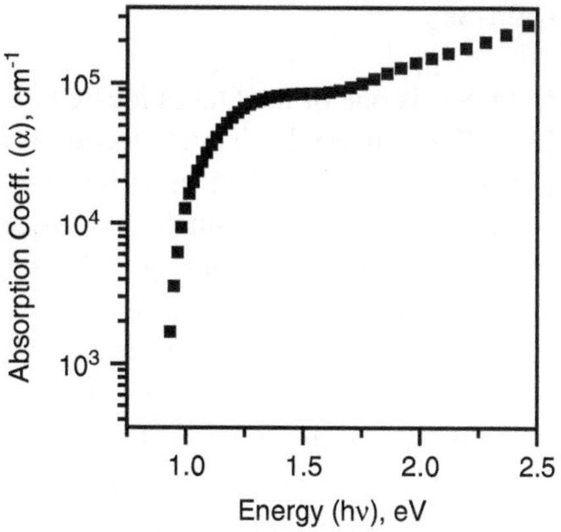

Fig. 3.5 Plot of α versus energy (hν) of $CuInSe_2$ thin films grown by physical vapor evaporation (Reprinted with permission from [20], Malar et al., *Vacuum*, **75** (2004). © Elsevier)

In the two families of ternary compounds, $CuInSe_2$ (CIS) and related solid solutions with Ga, belonging to I-III-VI_2 family of compounds, are well explored for their extreme suitability as an absorber material in thin film solar cells.

It is a widely studied compound both in bulk as well as in thin film form [21]. The band gap value lies in the range of 0.9–1.04 eV. It is the strongest absorber of sunlight among the various photovoltaic materials such as crystalline Si, amorphous hydrogenated Si, and CdTe. An intriguing nature of $CuInSe_2$ is that it exhibits unusual tolerance to stoichiometric deviations. In addition, unlike its II-VI binary analogue, $CuInSe_2$ can be made n or p type by introducing native defects, with the level of doping being decided by the stoichiometry. Another significant feature of $CuInSe_2$ in comparison to the binary and elemental semiconductors is the fact that the polycrystallinity does not affect its optoelectronic behavior. These exceptional and pertinent characteristics have made $CuInSe_2$ and related compounds $Cu(In,Ga)Se_2$ (CIGS) potential candidates for high-efficiency low-cost thin film heterojunction solar cells [22].

Ordered Defect Compounds

The ternary chalcopyrites exhibit large tolerance to off-stoichiometry when compared to the binary analogues. Compositions away from the ideal stoichiometry imply the introduction of native defects. This compositional flexibility allows the modulation of structural, optical, and electrical properties without any extrinsic impurities. At the extreme limits to structural tolerance, compounds that deviate from the chalcopyrite structure get stabilized. In Cu ternaries, Cu-poor tetrahedral phases of different compositions have been found to exist in the homogeneity region of (Cu-III-VI_2) – (III_2-VI_3) tie line. Existence of off-stoichiometric compounds has been explained based on ordering of the defect pair $(2V_I^- + III_I^{2+})$ [23]. Accordingly, these compounds are called as ODCs. The formation energy associated with this defect pair is unusually low and therefore the stabilization of ODCs is energetically favorable (Zhang et al. 1997).

$CuIn_3Se_5$

$CuIn_3Se_5$ is one of the ODCs located on the (Cu_2Se)- (In_2Se_3) quasi-binary tie-line [24]. The structural tolerance to off-stoichiometry of chalcopyrite $CuInSe_2$, at the extreme limits, from 47.5 to 55.0 mol% of In_2Se_3, gives compounds with different compositions such as $CuIn_5Se_8$, $CuIn_3Se_5$, and $Cu_3In_5Se_9$. XRD studies reveal that $CuIn_3Se_5$ belongs to the tetragonal system and has a chalcopyrite-related structure. The XRD pattern of $CuIn_3Se_5$ is similar to that of $CuInSe_2$ except for few additional reflections. These extra reflections due to (002), (110), (200), (202), and (114) planes are characteristic of ODC-$CuIn_3Se_5$ and do not satisfy the extinction conditions for the space group $I\bar{4}2d$ of the chalcopyrite structure. Honle et al. [25] have proposed a space group $P\bar{4}2c$, based on the refinement of crystallographic data of Cu-poor β-phase with a composition of $Cu_{0.39}In_{1.20}Se_{2.00}$. On the other hand, Hanada et al. [26] have proposed the space group $I\bar{4}2m$, based on the convergent beam electron diffraction studies of $CuIn_3Se_5$. In spite of the variance existing on the type of space group, there are several reports available on the growth and characterization of single crystals as well as thin films of $CuIn_3Se_5$ and ODCs related to other compounds of the I-III-VI_2 system.

Theoretical band structure calculations have shown that $CuIn_3Se_5$ is a direct band gap semiconductor and the fundamental absorption edge lies in the range of 1.23–1.31 eV. The option for $CuIn_3Se_5$ in solar cells based on $CuInSe_2$ thin films arose after it was reported that the segregation of n-type $CuIn_3Se_5$ over p-type $CuInSe_2$ results in a p-n heterojunction. This intrinsically formed thin film heterojunction was shown to enhance the efficiency of $CuInSe_2$-based thin film solar cells. Due to these beneficial characteristics, $CuIn_3Se_5$ has also received considerable attention in photovoltaic applications.

$Cu(In,Ga)Se_2$

Alloying of $CuInSe_2$ with Ga to form Cu $(In,Ga)Se_2$ (CIGS) seems to improve photovoltaic characteristics of $CuInSe_2$. With Ga alloying in $CuInSe_2$, CIGS band gap value increases towards optimum value and use of CIGS for absorbers in solar cells shows better electronic behavior with the rear contact. Several companies and R&D laboratories across the world (First Solar, Globalsolar, Nanosolar, Solopower, Muosolar, and Solexant) are keenly involved in developing CIGS-based thin-film solar cells and modules for the production of several gigawatt power per year. The schematic of CIGS-based solar cell is presented in Fig. 3.6. CdS seem to best buffer material and work well with all the absorber materials systems such as CdTe, $CuInSe_2$, CIGS, or CZTS.

The current record for highest laboratory efficiency on CIGS-based thin-film solar cells is about 20.8 % with an active area of 0.5 cm^2 fabricated by Center for Solar-Energy and Hydrogen Research (ZSW) Company on glass substrates. They have demonstrated 18.7 % on flexible substrates [27]. There are few more

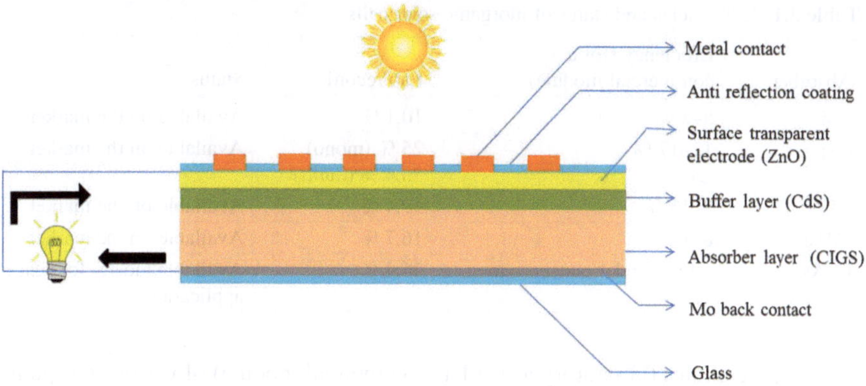

Fig. 3.6 Schematic of CIGS-based solar cell

companies like Ascent Solar Inc. and SoloPower who have developed CIGS cells on flexible substrates. The average efficiencies reported are around 11 %. Though CIGS is very promising and desired from application point of view, the In and Ga are getting scarce and consequently very expensive for mass production of large area CIGS and future deployment. Alternative absorber materials are being sought, which are earth-abundant and nontoxic.

CZTS and Related Compounds in Thin Film Solar Cells

Further substitution of group III element in I-III-VI_2 presents set of quaternary I_2-II-IV-VI_4($Cu_2ZnSn(S,Se)_4$/CZTS) compounds that can be used to mitigate the cost of the materials [11, 12]. The reason to choose CZTS as an absorber among other semiconductors is that all the elements in the absorber are abundant in the earth crust, nontoxic, and cost-effective. The carrier concentration, mobility, and resistivity of CZTS can be easily varied by changing intrinsic doping concentrations without employing extrinsic dopants, which are needed in the suitable range for thin-film solar cell applications. The lattice mismatch of tetragonal structured CZTS with hexagonal CdS may be acceptable to minimize interface states. These quaternary semiconductors have two different crystalline forms, stannite and kesterite. Kesterite CZTS has the space group $I\bar{4}$, whereas stannite structure is $I\bar{4}2m$. The quaternary p-CZTS has been found to be a suitable alternate in place of CIGS. CZTS presents a band gap of 1.5 eV that matches very well with the optimum bandgap for acquiring most of the photons from the solar radiation that reach earth. So far, the highest reported laboratory efficiency reported for CZTS (Se)/CdS/ZnO:Al thin-film solar cells on glass substrate is 10.1 % . However, this new group of quaternary materials are industrially novice and the basic studies on these materials are limited compared to CIGS or CdTe. The basic investigations on these new and less expensive quaternary materials are essential to improve efficiency of the cells before considering modules based on them.

Table 3.1 Efficiencies and status of inorganic solar cells

Absorber	Efficiency (for a commercial module)	Lab record	Status
a-si	5–8 %	10.1 %	Available on the market
c-si	12–17 %	25 % (mono) 20.4 % (multi)	Available on the market
CIS/CIGS	9–15 %	19.4 %	Available on the market
CdTe	8–12 %	16.7 %	Available on the market
GaAs	~25 % under concentrators	26.4 %	Available mostly for space applications

The efficiencies (laboratory as well as commercial module) of various inorganic solar cells are listed in Table 3.1 [28–30]. Among the listed ones, GaAs-based solar cells are mostly used for space applications due to its thermal stability of performance under concentrated light conditions. Rest of the solar cells based on Si, CdTe, and CIS/CIGS are commonly used for terrestrial applications.

3.2.4 Routes for Enhancement of Performance in Thin Film Solar Cells

Various factors have been considered for improving the efficiency of solar cells. One of them is light management, which involves concentration of the incident photon flux through light trapping structures [31, 32]. The photon flux available for electron-hole pair generation is equal to the incident light flux less the reflected and transmitted component. Reflection losses can be minimized through anti-reflection coating and transmission losses can be minimized by incorporating light trapping features in the device. Light trapping or increased path length of photon is done by incorporating reflective mirror at the back side of the cell and textured front and rear surfaces. Concentration of the incident light flux using reflective optics can reduce the required module area and subsequently the system cost.

3.2.5 Third- and Fourth-Generation Solar Cells

These solar cells are very different from the first two generations. The common factors between first two generations are the fact that they are based upon inorganic semiconductors for photon absorption and carrier generation. Secondly, with both first- and second-generation absorber materials the solar cell relies on p-n junction for charge carrier separation.

Third-generation devices include polymer-based solar cells, dye-sensitized solar cells and tandem cells, etc. Polymer solar cells or organic solar cells are treated separately in the following chapter. The very new fourth generation of devices [33] incorporates inorganics (quantum dots, carbon nanotubes, etc.) in organic solar cells.

3.3 Conclusion

Solar cells based on inorganic thin films, especially a-Si, CdTe, CIGS, and CZTS, are considered second-generation devices. These material systems are studied with the aim of replacing the first-generation bulk Si solar cell technology. The major concerns related to bulk Si technology being expensive and not suitable under concentrated light conditions (space applications). In addition, having a flexible solar cell is not possible due to the thickness requirement of the bulk Si solar cells. The second-generation solar cells involving the above inorganic materials reduce the material costs considerably and promise for flexible solar cell designs. The obtained module efficiencies in these thin film-based devices closer to bulk Si is indicative of the fact that the second-generation technology is reaching a matured level and factors such as light trapping and enhanced light absorption for improved efficiencies need consideration in near future.

Acknowledgements I thank my students Mr. Kunal J Tiwari and Ms. S. Sivasangari for their help during the preparation of this book chapter.

References

1. A.E. Becquerel, Recherches sur les effets de la radiation chimique de la lumiere solaire au moyen des courants electriques. C. R. Acad. Sci. **9**, 145–149 (1839)
2. W.G. Adams, R.E. Day, The action of light on selenium. Proc R. Soc. Lond. **A25**, 113 (1877)
3. B. Lange, Z. Phys. **31**, 139 (1930)
4. L.O. Grondhal, The copper-cuprous-oxide rectifier and photoelectric cell. Rev. Mod. Phys. **5**, 141 (1933)
5. W. Schottky, Uber den enstelhungsort der photoelektronen in kuper-kuperoxyydul-photozellen. Phys. Z. **31**, 913 (1930)
6. D.M. Chapin, C.S. Fuller, G.L. Pearson, A new silicon p-n junction photocell for converting solar radiation into electrical power. J. Appl. Phys. **25**(5), 676–677 (1954)
7. D.C. Reynolds, G. Leies, L.L. Antes, R.E. Marburger, Photovoltaic effect in cadmium sulfide. Phys. Rev. **96**, 533 (1954)
8. D.A. Jenny, J.J. Loferski, P. Rappaport, Photovoltaic effect in GaAs p-n junctions and solar energy conversion. Phys. Rev. B **101**, 1208–1209 (1956)
9. K.L. Chopra, *Thin Film Phenomena* (McGraw-Hill, New York, 1969)
10. K.L. Chopra, S.R. Das, *Thin Film Solar Cells* (Plenum, New York, 1983)
11. S.R. Kodigala (ed.), Cu(InGa)Se_2 based thin film solar cells, in *Thin Films and Nanostructures (Book Series)*, vol 35 (Academic press, UK, 2010), pp. 2–685
12. S.R. Kodigala, *Thin Film Solar Cells from Earth Abundant Materials Growth and Characterization of* $Cu_2ZnSn(S, Se)_4$ *Thin Films and their Solar Cells* (Elsevier, London, 2014)

13. J. Nelson, *The Physics of Solar Cells* (Imperial college, London, 2008)
14. M.A. Green, K. Emery, Y. Hishikawa, W. Warta, Solar cell efficiency tables (version 37). Prog. Photovolt. Res. Appl. **19**(1), 84–92 (2011)
15. X. Deng, E.A. Schiff, Amorphous silicon–based solar cells, in *Handbook of Photovoltaic Science and Engineering* (Wiley, Chichester, 2003). ISBN 0-471-49196-9
16. J.L. Shay, J.H. Wernick, *Ternary Chalcopyrite Semiconductors: Growth, Electronic Properties and Applications* (Pergamon press, Oxford, 1975)
17. A. Walsh, S. Chen, S.-H. Wei, X.-G. Gong, Kesterite thin-film solar cells: advances in materials modelling of Cu_2 $ZnSnS_4$, Adv. Energy Mater. **2**, 400–409 (2012)
18. NREL Continuum magazine, Issue 3 (online) www.nrel.gov/continuum/spectrum/photovoltaics.cfm
19. J. Buerger Martin, *Elementary Crystallography: An Introduction to the Fundamental Geometrical Features of Crystals* (Wiley, New York, 1963)
20. P. Malar, V.D. Das, S. Kasiviswanathan, Characterization of stepwise flash- evaporated $CuInSe_2$ films. Vacuum **75**(1), 39–49 (2004)
21. A. Rockett, R.W. Birkmire, $CuInSe_2$ for photovoltaic applications. J. Appl. Phys. **70**, R81–R97 (1991)
22. R. Kannan, M.A. Contreras, C.L. Perkins, S. Asher, F.S. Hasoon, J. Keane, D. Young, M. Romero, W. Metzger, R. Noufi, J. Ward, A. Duda, Properties of 19.2 % efficiency $ZnO/CdS/CuInGaSe_2$ thin-film solar cells. Prog. Photovolt. Res. Appl. **11**, 225–230 (2003)
23. S.B. Zhang, S.-H. wei, A. Zunger, Stabilization of ternary compounds via ordered arrays of defect pairs. Phys. Rev. Lett. **78**, 4059–4062 (1997)
24. U.C. Boehnke, G. Kuhn, Phase relations in the ternary system Cu-In-Se. J. Mater. Sci. **22**, 1635–1641 (1987)
25. W. Honle, G. Huhn, U.C. Boehnke, Crystal structures of two quenched Cu-In-Se phases. Cryst. Res. Technol. **23**, 1347–1354 (1988)
26. H. Takeshi, A. Yamana, Y. Nakamura, O. Nittono, T. Wada, Crystal structure of $CuIn_3Se_5$ semiconductor studied using electron and X-ray diffractions. Jpn. J. Appl. Phys. **36**, L1494–L1497 (1997)
27. P. Jackson, D. Hariskos, R. Wuerz, W. Wischmann, M. Powalla, Compositional investigation of potassium doped $Cu(In,Ga)Se_2$ solar cells with efficiencies up to 20.8 %. Phys. Status solidi RRL wiley **8**(3), 219–222 (2014)
28. Y. Liu, C.-C. Chen, Z. Hong, J. Gao, Y.(.M.). Yang, H. Zhou, L. Dou, G. Li, Y. Yang, Solution-processed small-molecule solar cells: breaking the 10 % power conversion efficiency. Nature **3**, 1–7 (2013)
29. D. Joly, L. Pellejà, S. Narbey, F. Oswald, J. Chiron, J.N. Clifford, E. Palomares, R. Demadrille, A robust organic dye for dye sensitized solar cells based on iodine/iodide electrolytes combining high efficiency and outstanding stability. Nature **4**, 1–7 (2014)
30. Solar Cell Comparison Chart – Mono-, Polycrystalline and Thin Film (online) http://energyinformative.org/solar-cell-comparison-chart-mono-polycrystalline-thin-film/. Last updated September 27, 2013
31. M. Peters, B. Blasi, S.W. Glunz, A.G. Aberle, J. Luther, C. Battaglia, Optical simulation of silicon thin film solar cells. Energy Procedia **15**, 212–219 (2012)
32. M. Zemen, O. Isabella, K. Jager, R. Santbergen, Advanced light management approaches for thin film silicon solar cells. Energy Procedia **15**, 189–199 (2012)
33. K.D.G. Imalka Jayawardena, L.J. Rozanski, C.A. Mills, M.J. Beliatis, N. Aamina Nismy, S.R.P. Silva, 'Inorganics-in-organics': recent developments and outlook for 4G polymer solar cells. Nanoscale **5**, 8411–8427 (2013)

Chapter 4
Organic Semiconductors: A New Future of Nanodevices and Applications

Anamika Dey, Ashish Singh, Dipjyoti Das, and Parameswar Krishnan Iyer

Abstract Organic materials with attractive electronic and optoelectronic properties are in high demand for functional electro-optical device applications. Research on synthetic polymer and organic small molecules has gained a great deal of attention due to their potential applications in low cost, ultra-thin, and flexible products and are expected to have a revolutionary role in modern day life. Basically, organic semiconductors are imperative for a range of optoelectronic applications including organic photovoltaic devices, light-emitting diodes, organic light-emitting transistor and organic field effect transistor-driven photonics since they possess photoluminescence, electroluminescence, and nonlinear optical properties in addition to liquid crystalline, structural patterning, printing and solution casting abilities. Electronic devices based on organic materials can be fabricated by simple processing techniques and are under intense investigation in academic and industrial laboratories because of their potential for mass production of flexible and cost-effective devices models. Organic devices also possess light-weight and transparent advantages compared to silicon-based devices. Significant progress has been achieved in the development of novel materials and new device engineering in the last decade. In this regard, arylene diimide families have shown high promise as useful building blocks for the fabrication of next-generation electronic and optoelectronics devices, providing deeper insight into their transistor characteristics at the molecular scale to practical applications ranging from medicine to high end security systems. These organic compounds are being developed for improved resistance to thermal and environmental stresses, which is one of the major challenges in the field. Several other classes of compounds based on fullerene (C_{60}) and chemically modified thiophenes are also being explored and developed for use in a plethora of devices encompassing interdisciplinary research areas. These devices, owing to their low-cost, ease of synthesis and processing, are expected to pave the way for next-

A. Dey • A. Singh • D. Das
Centre for Nanotechnology, Indian Institute of Technology Guwahati,
Guwahati 781 039, Assam, India

P.K. Iyer (✉)
Department of Chemistry and Centre for Nanotechnology,
Indian Institute of Technology Guwahati,
Guwahati 781 039, Assam, India
e-mail: pki@iitg.ernet.in

S. Babu Krishna Moorthy (ed.), *Thin Film Structures in Energy Applications*,
DOI 10.1007/978-3-319-14774-1_4

generation electronics with energy-efficient security systems, sensors, photonics, and spintronic memories. In this chapter we present a brief review on the different types of organic devices, their fabrication processes, and recent applications by citing examples of oligomers and polymers.

4.1 Introduction: Transformation from Inorganic to Organic

Polymers, more frequently identified as plastics, have become inseparable from our everyday life. Due to their exceptional properties, relatively low cost, and ease of manufacture, polymer materials have replaced many of the traditional materials, such as wood, leather, metal, glass, and ceramics. Since the discovery of Bakelite, polymers began to be used as electrically insulating materials in electrical switches as well as other specialized products. This led to a widespread view of polymers as exclusively insulating materials. With the development of newer synthetic methodology newer hydrocarbon-based materials came into existence and the opinion that polymers were insulators was transformed in 1976, when Alan J. Heeger, Alan G. MacDiarmid, and Hideki Shirakawa discovered that it was possible to modify the conductivity of organic polymer "polyacetylene" by several orders of magnitude, approaching that of metals, by exposing it to iodine vapor [1]. They were awarded the Nobel Prize in Chemistry 2000 "for the discovery and development of conductive polymers". The conducting polymers represented a "new class of materials" that have electronic and optical properties of semiconductors and metals, but also with the processing advantages and mechanical properties of plastics. These findings not only unlocked the door for massive library of materials but also futuristic applications based on organic semiconducting and conducting materials being realized as the active material in electronic applications. This discovery has also led to the creation of a new research field residing at the boundary between chemistry, chemical engineering, physics, electronics, and commonly called as organic electronics.

A material can be classified as being either organic or inorganic based on the components they possess. Many centuries ago, substances that originated from living matter were considered as organic materials and it was believed that they possessed an indefinable "living force". Today, an enormous variation of organic materials can be synthesized and organic compounds are therefore commonly defined as any molecular material that contains the element carbon (C) in combination with other atoms.

There has been a remarkable progress in the development of organic semiconductors since the discovery of conducting polymers in the past 30 years [2]. Many of these organic materials, especially the polymers, can be processed from solution as inks. This allows for high-volume and low-cost manufacturing of organic electronic devices on a wide range of flexible substrates, e.g., paper, cloth, and plastic foils, by the use of traditional printing techniques, such as screen printing, gravure, offset, flexography, and inkjet [3, 4]. This stands in stark contrast to the very expensive and complicated methods used in traditional inorganic semiconductor device fabrication. The manufacturing technology of inorganic electronics also involves the use of many hazardous materials and solvents. Another benefit with organic semiconductors is that their physical and chemical functionality can be tailored by modifying their chemical structure [5]. A wide range of organic semiconductor devices have been developed, exemplified by the light-emitting diodes [6], solar cells [7, 8], sensors [9], and thin film transistors [10, 11]. These devices, owing to their low-cost, ease of synthesis and processing, are expected to pave the way for next-generation electronics with energy-efficient security systems, sensors, photonics, and spintronic memories.

4.2 Organic Semiconductors

4.2.1 Atomic Orbitals

The basic unit of matter is an atom, consisting of a dense, positively charged nucleus and a surrounding cloud of negatively charged electrons. Such microscopic systems are described by quantum mechanics, in which each elementary particle is associated with a *wave function* $\Psi(\boldsymbol{r}, t)$. The square of the modulus of the wave function, $|\Psi(\boldsymbol{r}, t)|^2$, is a density function that represents the probability of finding the particle at the location $\boldsymbol{r}$ at time t. The electrons of an atom can only reside in certain quantum states. Only those wave functions that are solutions of the Schrödinger equation are allowed. These wave functions are called *atomic orbitals* and they are categorized by three quantum numbers that determine the shape and energy of the orbital: the principal quantum number n that describes the energy, the orbital angular momentum quantum number l that describes the amplitude of the angular momentum, and the magnetic quantum number m_l that describes the orientation of the angular momentum. Each orbital can contain maximum two electrons, one of each spin (up or down). For historical reasons, the shells (determined by n) and the subshells (specified by l and m_l) are also labeled $K, L, M, N, \ldots$ and $s, p, d, f, \ldots$, respectively. The two most interesting kind of atomic orbitals in organic electronics are the s orbitals, which are sphere-shaped and nonzero at the centre of the nucleus, and the p orbitals, which resemble dumbbells with their two ellipsoid-shaped lobes that are separated by a nodal plane at the nucleus (Fig. 4.1).

4.2.2 Molecular Orbitals and Bonds

The electrons in the outermost shell, the so-called *valence electrons*, determine the chemical, electrical, and optical properties of materials. They also participate in the formation of chemical bonds with other atoms. When two atoms are brought close

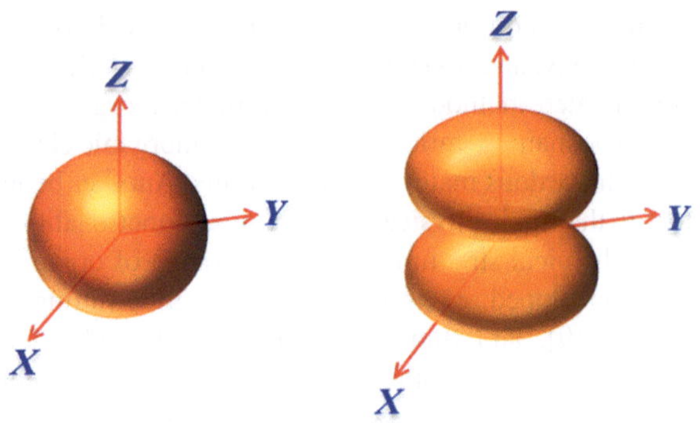

Fig. 4.1 Illustrations of an *s* orbital (*left*) and a *p* orbital (*right*)

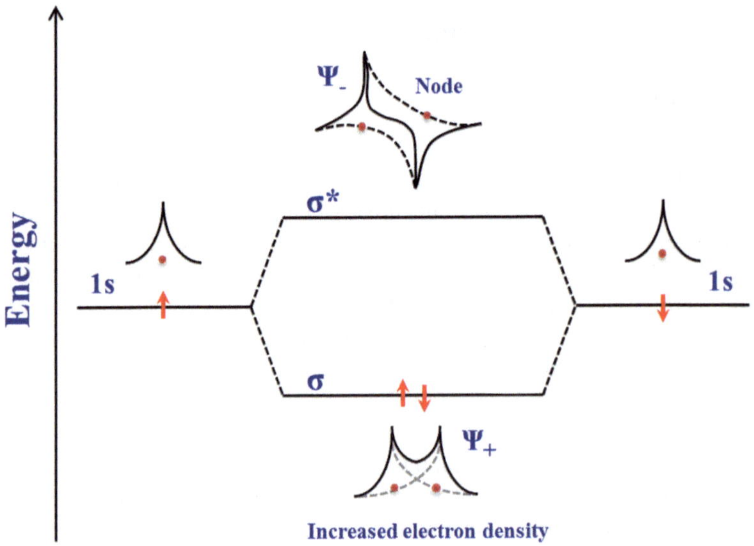

Fig. 4.2 The formation of bonding and antibonding molecular orbitals and the splitting of energy levels for a dihydrogen molecule

to one another, their atomic orbitals will start to overlap by valence electron interaction. These combined atomic orbitals form *molecular orbitals*, which can be represented by linear combinations of the atomic orbitals.

The interactions between the atomic orbitals are either constructive (Ψ_+) or destructive (Ψ_-), leading to an increase or a decrease of the electronic density between the nuclei, respectively (Fig. 4.2). The former is a *bonding* molecular orbital and the latter an *antibonding* molecular orbital. The bonding orbital stabilizes the molecule and has lower energy than the original atomic orbitals, while the antibonding orbital destabilizes the molecule and consequently has a higher energy.

Thus, there is a splitting of the original energy levels, where the separation of the energy levels indicates the strength of the atomic interactions. The electrons will fill the lower energy orbitals first, and if the total energy of the system is lower than that of the two isolated atoms, the atoms will form a stable bond.

The orbital having highest energy and which is occupied with electrons is called the *highest occupied molecular orbital* (*HOMO*), and the orbital with the lowest energy which is unoccupied is called the *lowest occupied molecular orbital* (*LUMO*). The bond is called an σ bond if it is symmetrical with respect to rotation about the bond axis, and a π bond if it is not. The orbitals in a π bond have a nodal plane passing through the nuclei. The corresponding bonds for the antibonding orbitals are called σ* and π*. The σ bonds are generally stronger than the π bonds due to the larger overlap of the atomic orbitals. Consequently, π orbitals have higher energy than σ orbitals.

The intramolecular bonds, in which the atoms share electrons, are called *covalent bonds*. The electrons in a bond may be shared unequally between the atoms, due to a difference in their electronegativity, which leads to the formation of an electric dipole moment along the bond axis. Such bonds are called *polar bonds*. A very large difference in electronegativity between the atoms involved in the bond can lead to explain that the electron pair gets located almost exclusively at the more electronegative atom. Such a bond is called an *ionic bond*.

There are two types of intermolecular bonds and both are much weaker than the intramolecular bonds. One type is the *van der Waals bond* that originates from interactions between permanent and/or induced dipoles. The second type is the *hydrogen bond*, which is an attractive interaction between hydrogen atoms and electronegative atoms carrying an electron lone pair, such as oxygen, nitrogen, etc. The hydrogen bonds are stronger than the van der Waals bonds.

4.2.3 Hybridization

All organic materials are based on molecules that include the element carbon in combination with other atoms. The carbon atom is very versatile since it is able to form single, double, and triple bonds. A carbon atom in its ground state ($1s^2\ 2s^2\ 2p_x^1\ 2p_y^1$) only has two unpaired valence shell electrons and should thus only be able to form two covalent bonds with other atoms. In order to explain the presence of methane, the virtual notion of "promotion" can be used. A carbon atom can form an excited state ($1s^2\ 2s^1\ 2p_x^1\ 2p_y^1\ 2p_z^1$) by promoting one of its $2s$ electrons to the empty $2p$ orbital so that there are four unpaired valence electrons available for bonding. The $2s$ and one, two or three of the $2p$ orbitals can be combined to form two sp, three sp^2, or four sp^3 *hybridized orbitals*, respectively. The respective hybrid orbitals have identical energies and shapes that resemble distorted p orbitals with unequal lobes, giving the orbitals a more directed orientation (see Fig. 4.3).

The carbon atoms in alkanes, e.g., methane (CH_4), form bonds to four other atoms exclusively via single bonds. These carbons are sp^3 hybridized and their

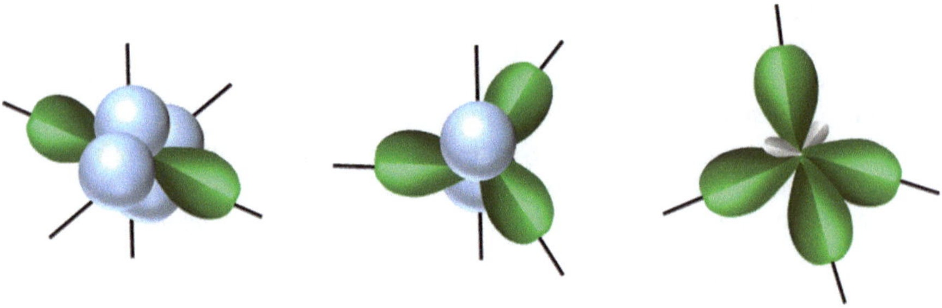

Fig. 4.3 Illustrations of sp (*left*), sp^2 (*centre*), and sp^3 (*right*) hybridized orbitals

orbitals have a tetrahedral arrangement with an angle of 109.5° between them. Carbon atoms that are involved in forming a double bond consist of one σ bond and one π bond, are sp^2 hybridized. These hybrid orbitals lie in one plane and are separated by an angle of 120°. The remaining unchanged $2p_z$ orbital, which is oriented perpendicular to the plane of sp^2 orbitals, participates in the double bond together with one of the sp^2 orbitals. Carbons that are sp hybridized form triple bonds. These hybrid orbitals point in opposite directions (180° angle) and are perpendicular to the two unchanged $2p^y$ and $2p^z$ orbitals.

4.2.4 Electronic Structure of Conjugated Materials

A conjugated molecule or polymer has a molecular framework that consists of alternating single and double carbon–carbon bonds. From a chemical structure point of view, the simplest example of a conjugated polymer is *transpolyacetylene*, which is composed of only carbon and hydrogen atoms. All the carbon atoms are sp^2 hybridized. The sp^2 orbitals form strongly localized σ bonds, which determine the geometrical structure of the molecule. The $2p_z$ orbitals, which are oriented perpendicular to the plane of the chain, overlap and form π orbitals that extend along the conjugated chain. The electrons in these π orbitals are not associated with any specific atom or bond and are therefore *delocalized*. The number of π and π* orbitals is proportional to the number of carbon atoms in the conjugated system. Hence, there is a splitting of the energy levels as the number of carbons is doubled. This is illustrated for a series of alkenes in Fig. 4.4.

For an infinitely long conjugated chain, e.g., *transpolyacetylene*, the energy difference between the energy levels becomes negligibly small, and the energies can then be described as continuous bands rather than discrete levels. The width of the band, W, depends on the coupling between the atomic orbitals. Strong coupling gives wide bands. Interestingly, the HOMO and the LUMO are degenerate if the bonds in the conjugated chain are equally long. The filled π band and the empty π* band will then coincide, resulting in a half-filled band. The polymer could thus be described as a quasi one-dimensional metal. However, according to Peierls' theorem, such a configuration is not energetically stable. Instead, the polymer will

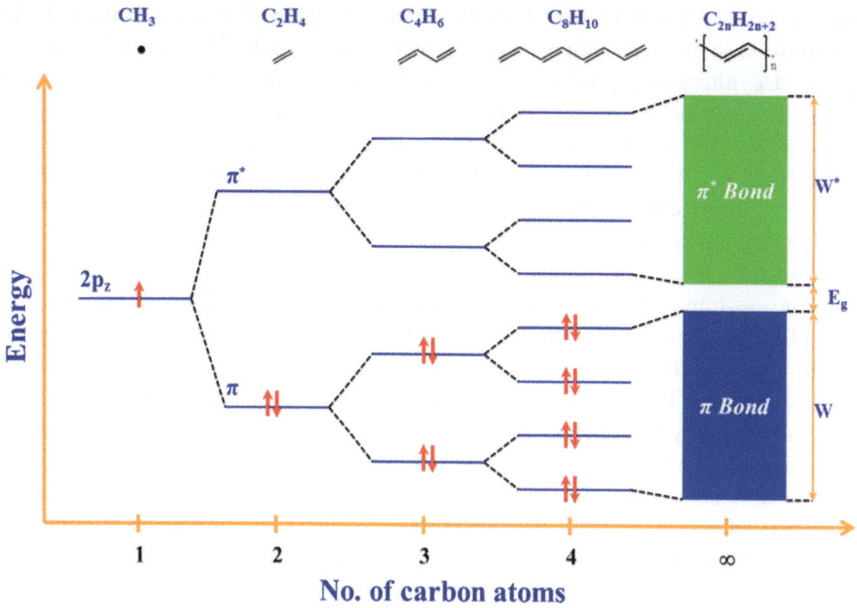

Fig. 4.4 Energy level splitting and band formation in conjugated molecules

dimerize and form alternating long single bonds (1.47 Å) and short double bonds (1.34 Å). This structural distortion, also known as the *Peierls distortion*, will stabilize the π band and destabilize the π* band, which produces a band gap, e.g., typically in the range of 1–4 eV [12]. The polymer is thus a semiconductor. The filled π band is commonly referred to as the *valence band* and the empty π* band as the *conduction band*.

4.2.5 *Molecules/Polymers*

There are basically two types of organic semiconductor: n-type and p-type. Again they are divided as conjugated polymers and conjugated small molecules. Conjugated polymers functionalized with flexible side chains are soluble and their thin films can be prepared by solution-based techniques, including spin-coating, flexography, gravure, and inkjet printing [13]. An example of such a material is the alkyl-substituted polythiophene, poly(3-hexylthiophene), etc. Regioregular head-to-tail P3HT self-organize into an ordered lamellar structure with a significant overlap between the frontiers π-orbitals of adjacent molecules (π–π stacking) [14]. This leads to a relatively high mobility (~0.1 $cm^2\ V^{-1}\ s^{-1}$) in the direction perpendicular to the lamellar plane [15]. Thus, the charge transport in ordered P3HT films is highly anisotropic and dependent on how the lamellae are oriented on the surface. A drawback with P3HT is that it is susceptible to oxidation [16]. In response to that, thienothiophene copolymers, such as poly(2,5-bis(2-thienyl)-3,6-dihexadecylthieno[3,2-β]thiophene), (P(T0T0TT16), have been developed that show better stability as compared to P3HT and higher carrier mobilities up to 1.1 $cm^2\ V^{-1}\ s^{-1}$ could be achieved [17–19]. In contrast to conjugated polymers, conjugated small-molecule materials typically have inferior film forming ability when cast from solution. Hence, these materials are usually deposited by thermal sublimation in vacuum or by organic vapor phase deposition. With a control over the substrate temperature, well-organized polycrystalline films can be obtained resulting in improved carrier mobility with these materials. Pentacene, which is one of the most studied small-molecule material, has shown mobilities as large as 6 $cm^2\ V^{-1}\ s^{-1}$ [20]. All the materials described above are mainly hole-transporting semiconductors. The observed electron mobility values in organic semiconductors are generally very low due to various reasons, including inefficient charge injection and more efficient trapping, e.g., in the presence of oxygen and water. However, several air-stable electron-transporting organic semiconductors with high electron affinity (>4 eV) have been developed, and one of those is hexadecafluorocopperphthalocyanine (F16CuPc; which has shown an electron mobility of 0.03 $cm^2\ V^{-1}\ s^{-1}$) [21]. Recently, conjugated polymers with high electron mobility have also been synthesized, and the most promising of those is the printable material poly{[*N*,*N*′-bis(2-octyldodecyl)-naphthalene-1,4,5,8-bis(dicarboximide)-2,6-diyl]-alt-5,5′-(2,2′-bithiophene)} (P(NDI2OD-T2), with mobilities as large as 0.85 $cm^2\ V^{-1}\ s^{-1}$.

4.3 Charge Transport in Organic Materials: Models and Theories

Inorganic semiconductor crystalline lattices (Si, Ge, etc.) are characterized by three-dimensional long range order and strongly coupled atoms. The charge transport therefore happens generally rather isotropically in delocalized energy bands separated by a forbidden energy gap where charge carriers (intrinsic or extrinsic) can move with a relatively large mean free path limited usually by scattering of the carriers at thermal lattice vibrations, i.e., phonons. Consequently the mobility of the charge carriers decreases with increasing temperature. In conjugated polymers the charge transport happens within a single chain (*intrachain*) and between different molecules (*interchain*) or even from a domain, commonly called grain, to another. Along the molecule, the HOMO and LUMO energy bands are generally narrow and can easily be modified by molecular disorder. The electric charge is delocalized along the π-conjugated segments of the polymer backbone for rather short distances (typically around 5 nm), separated by chemical defects, such as a non-conjugated sp^3 hybridized carbon atom on the polymer backbone, or by structural defects, such as chain kinks or twists out of coplanarity. A current conduction over longer distances implies therefore the possibility for the carriers to move from chain to chain. When the π-orbitals of different molecules overlap, this process becomes easier due to a degree of delocalization between different polymer chains. Oppositely, a reduced interchain charge transfer process turns out in strong reduction of the charge mobility in the material compared to the one associated to the intrachain transport. The efficiency of the interchain conduction process strongly depends on the molecular structure that determines the way chains arrange with respect to one another. Planar molecules show in principle an easier stacking and better electronic coupling via the π-orbitals. With respect to this issue, films made of solution-processed conjugated polymers and films made from sublimation of analogous oligomers may present very different transport characteristics, correlated to the interchain spacing and degree of order. Consequently, due to disorder, the semiconductor cannot be regarded simply as having two delocalized energy bands separated by an energy gap. Instead, the charge transporting sites, which are the segments of the main chain polymer, generally gives rise to a Gaussian Density of States, implying that all the states are localized.

The shape of the density of states (DOS) is suggested to be Gaussian, because of the observed Gaussian shape of the optical spectra. The shape of the DOS, which is a manifestation of the disorder of the system, is however important for the description of charge transport. Hence, the charge transport and semiconductor properties of polymeric semiconductors are rather sensitive to the morphology of the polymer chains and the local structural order within the film. Structural and energetic disorder in conjugated polymer systems are therefore of importance in the description of charge transport involving different mechanisms and electronic processes. Several transport models have been proposed for organic semiconductors over the past decades, but despite the fast progress in organic electronics a univocal model has not been developed yet.

4.3.1 The Polaron Model

This model relies on the interaction of phonon–electron of the semiconductor. A polaron can be described as a quasi-particle generated by the lattice deformation due to the charge of the carrier. The hopping rate of carriers from occupied i to unoccupied j localized donor state depends on the height of the energetic barrier E_j–E_i and can be expressed as

$$v_{i\rightarrow j} \infty \frac{1}{\sqrt{E,T}} \exp\left[-\frac{\left(E_j - E_i + E_r\right)^2}{4E_r kT}\right], \tag{4.1}$$

where, E_{r} is the intramolecular reorganization energy. The charges moving by thermal activation hops between adjacent sites and have a mobility which is field (F)- and temperature (T)-dependent:

$$\mu = \mu_0 \exp\left[-\frac{E_r}{4kT} - \frac{(aF)^2}{4E_r kT}\right] \frac{\sin h\left(\frac{aF}{2kT}\right)}{\frac{aF}{2kT}}, \tag{4.2}$$

where, μ_0 is slightly temperature-dependent. However, using Marcus theory the polaron contribution to the activation of the mobility is insignificant. The activation of the mobility using this model amounts to 25–75 meV, while using the disordered model the activation energy amounts to 420 meV.

4.3.2 Multiple Traps and Release (MTR) Model

Another model used to account for the low mobility in amorphous organic materials is the multiple trapping and release model. Traps are levels localized at lattice defects or impurities in which the charge carriers are immobilized. These traps can be deep traps, which are located near the center of the band gap, or shallow traps, which are located close to the conduction or valence band.

The effect of these states on charge transport depends on their energy. If the energy of a localized state is separated from the mobility edge by more than a few $k_B T$, the state acts as a deep trap: the charge, once trapped in a deep trap, cannot be released by thermal excitations. Developed for hydrogenated amorphous silicon (a-Si:H) by Le Comber and Spear, the multiple trapping and release model has been used more recently by Horowitz et al. to explain the transport in sexithiophene FETs. This model assumes an exponential distribution of gap states. The charges injected or the charges which are already present in the organic semiconductor are trapped into localized states with a probability close to one and then released through a thermally activated process. The drift mobility μ_D is given by:

$$\mu_D = \mu_0 \alpha \exp\left[-\frac{E_T}{k_B T}\right], \tag{4.3}$$

where, μ_0 is the mobility at the band edge, α is the ratio between the effective DOS at the transport band edge and the density of traps, and E_T is the energy of the trap state. It has been demonstrated that the transport of carriers depends on the energy level of the trap states, the temperature, and the voltage applied.

4.3.3 Variable Range Hopping Model

In the previous model, as stated above, the assumption made is that most of the charge carriers are trapped in localized states. Then the amount of released charge carriers to an extended-state transport level (a concept close to the valence band for p-type semiconductors) depends on the temperature, the gate voltage, and the energy level of the localized states. However, while extended-state transport may occur in highly ordered vacuum-evaporated molecular films as stated by Horowitz, we do not expect such behavior in amorphous organic films where the charge carriers are strongly localized. In 1998, Vissenberg and Matters reported the concept of variable-range hopping (VRH), i.e., a carrier may either hop over a small distance with a high activation energy or hop over a long distance with a low activation energy. The temperature dependence of the carrier transport in such a system is strongly dependent on the density of localized states. In the case of a field-effect transistor (this model was developed in order to explain transport in polymeric FET), an applied gate voltage gives rise to the accumulation of charge in the region of the semiconducting layer that is close to the insulator. As these accumulated charge carriers fill the lower-lying states of the organic semiconductor, any additional charges in the accumulation layer will occupy states at a relatively high energy. Consequently, these additional charges will on average require less activation energy to hop away to a neighbor site. This results in a higher mobility with increasing gate voltage.

4.4 Organic Thin Film Deposition Techniques

4.4.1 Self-Assembly

Self-assembly represents a promising strategy for surface functionalization as well as creating nanostructures with well-controlled, tailor-made properties and functionality. Molecular self-assembly is the spontaneous association of molecules under equilibrium conditions into steady, structurally well-defined aggregates adhered by non-covalent bonds. In solid phase the surfaces are governed by the subtle interplay between molecule–molecule and molecule–substrate interactions that can be tuned by varying molecular building blocks, surface chemistry, and structure as well as substrate temperature. The intention of the tools developed in this discipline is to construct desired functional nano-architectures by the synthesis of molecular building blocks with predetermined intermolecular binding properties. There are basically two types of self-assembly. These are intramolecular self-assembly and intermolecular self-assembly. Commonly, the term molecular self-assembly refers to intermolecular self-assembly, while the intramolecular analog is more commonly called folding.

4.4.2 Vacuum Evaporation

Vacuum thermal evaporation involves heating the organic semiconductor using a resistive heating source under a vacuum environment with a pressure in the range of

10^{-8} to 10^{-6} Torr. For organic small molecules and oligomers that are solution-insoluble or which do not form uniform films directly from solution, vacuum thermal evaporation is an ideal processing technique. A number of small molecule organic semiconductors have been deposited using this method. Vacuum thermal evaporation has the advantages of forming films with high film uniformity and good "run-to-run" reproducibility. Multilayer deposition and co-deposition of several organic semiconductors are possible without the concerns of delamination or dissolution of the previous layers during subsequent deposition steps. However, this method has a relatively high material consumption and also a high initial cost for equipment setup. Nevertheless, this method has already been used for the manufacturing of organic small molecule-based light-emitting displays commercially.

4.4.3 Spin-Coating

Spin-coating technique has been used for several decades for the application of thin films on desired surface. A typical spin-coating technique process involves depositing a fluid consisting of the polymer or organic material onto the center of a substrate and then spinning the substrate at high speed (few thousand rpm). The centripetal acceleration causes the polymer or organic material to spread on to, and eventually off, the edge of the substrate leaving a thin film of resin on the surface. Final film thickness and other properties will depend on the nature of the resin (viscosity, drying rate, percent solids, surface tension, etc.) and the parameters chosen for the spin process. Factors such as final rotational speed, concentration, acceleration, temperature, and fume exhaust contribute to how the properties of coated films are defined. One of the most important factors in spin coating is the reproducibility. Slight variations in the parameters that define the spin process can result in drastic variations in the coated film.

4.4.4 Layer-by-Layer Electrostatic Deposition

Layer-by-Layer (LbL) deposition is another thin film fabrication technique. The films are formed by depositing alternating layers of oppositely charged materials with intermediate wash steps.

A solid substrate with a positively charged planar surface is immersed in the solution containing the anionic polyelectrolyte and a monolayer of the polyanion is adsorbed. Since the adsorption is carried out at relatively high concentrations of the polyelectrolyte, most ionic groups remain exposed to the interface with the solution and the surface charge is reversed. After rinsing in pure water, the substrate is immersed in the solution containing the cationic polyelectrolyte. Again, a monolayer is adsorbed but now the original surface charge is restored, resulting in the formation of multilayer assemblies of both polymers. This method has been used to deposit several layers of partially doped polyaniline and a polyanion, sulphonated polystyrene, etc. Formation of multilayers of organic compounds occurs on a solid substrate by sequential immersion of the latter in solutions of positively and negatively charged polyelectrolytes.

4.4.5 Langmuir–Blodgett Technique

A Langmuir–Blodgett film contains one or more monolayers of an organic material, deposited from the surface of a liquid onto a solid by immersing (or emersing) the solid substrate into (or from) the liquid. A monolayer is adsorbed homogeneously with each immersion or emersion step, thus films with very accurate thickness can be formed. This thickness is accurate because the thickness of each monolayer is known and can therefore be added to find the total thickness of a Langmuir–Blodgett film. The monolayers are assembled vertically and are usually composed of amphiphilic molecules with a hydrophilic head and a hydrophobic tail. An alternative technique of creating single monolayers on surfaces is that of self-assembled monolayers.

The LB-technique is one of the most promising techniques for preparing such thin films as it enables (1) the precise control of the monolayer thickness, (2) homogeneous deposition of the monolayer over large areas, and (3) the possibility to make multilayer structures with varying layer composition. An additional advantage of the LB technique is that monolayers can be deposited on almost any kind of solid substrate.

4.4.6 Physical Vapor Deposition

Physical vapor deposition (PVD) is fundamentally a vaporization-coating technique, involving transfer of material on an atomic level. It is an alternative process to electroplating. The basic PVD processes fall into two general categories: sputtering and evaporation. The main areas of application for PVD processes are thin films used in optical, optoelectronic, magnetic, and microelectronic devices. Other applications may be found in the areas of tribology, corrosion protection, thermal insulation, decorative coatings among others.

4.4.7 Chemical Vapor Deposition

Chemical vapor deposition (CVD) is a chemical process used to produce high-purity, high-performance solid materials. The process is often used in the semiconductor industry to produce thin films. In a typical CVD process, the wafer (substrate) is exposed to one or more volatile precursors, which react and/or decompose on the substrate surface to produce the desired deposit. Frequently, volatile by-products are also produced, which are removed by gas flow through the reaction chamber.

4.5 Organic Thin Film Devices

4.5.1 Organic Thin Film Transistors

Organic thin-film transistor (OTFT) technology involves the use of organic semiconducting compounds in electronic components. Several factors have motivated engineers to conduct and continue research in organic semiconductor technology. One of these factors is cost. Organic displays are relatively cheap, but until recently, they have proven slow in terms of carrier mobility. Slow carrier mobility translates into sluggish response time, which limits the ability of a display to render motion such as is common in animated computer games, etc. Researchers at Lucent Technologies have, however, developed a process for growing organic crystals with carrier mobility rivaling that of traditional TFT materials [22]. Further improvements are possible with these organic crystals. Another factor that motivates research in OTFT technology is their application diversity. Organic substrates allow for displays to be fabricated on flexible surfaces, rather than on rigid materials as is necessary in traditional TFT displays. A piece of flexible plastic might be coated with OTFT material and made into a display that can be handled like a common paper document. Sets of such displays might be bundled or stacked, to produce novels, magazines, or newspapers whose page contents can be varied periodically and animated. This has far-reaching ramifications for the technology world. For example, comic book characters might move around the pages and speak audible words making them more lively and interactive. More likely, such displays will find use in portable computers and communications systems.

4.5.1.1 Working Principles and Equations of Organic FETs

A typical organic thin film transistor consists of three basic components: an active semiconductor layer, a dielectric layer, and conductive electrodes. In general, OTFTs with p-type organic semiconductor layer conduct positive charges in active semiconductor layer. When a negative voltage is applied to the gate electrode, gate-induced positive charge carriers are formed in the organic semiconductor layer close to the interface between dielectric and semiconductor layers. These positive charge carriers can flow through the semiconductor layer, forming a conducting channel near the dielectric-semiconductor interface when the source-drain voltage (V_D) is applied. Thus, a source-drain current (I_D) can be generated. The charge carrier density created in the semiconductor layer for conducting can be modulated by gate voltage (V_G), and the source-drain current (I_D) can be modulated by source-

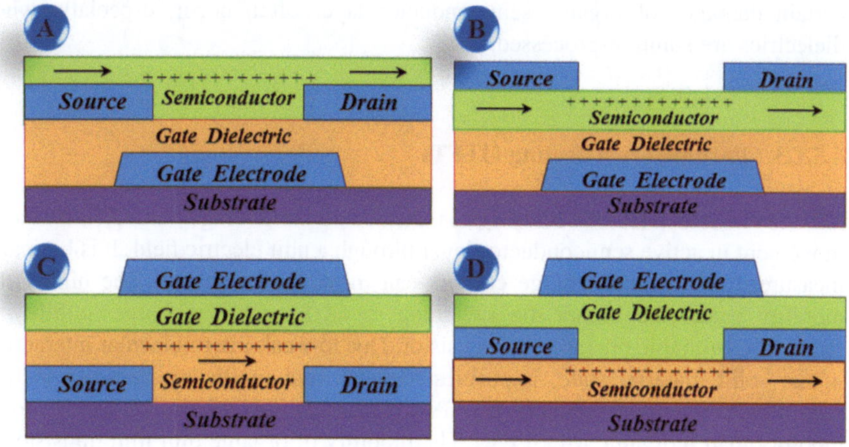

Fig. 4.5 Schematic illustrations of OTFT configurations: (**a**) Bottom gate with top contact; (**b**) Bottom gate with bottom contact, (**c**) Top gate with bottom contact; and (**d**) Top gate with top contact

drain voltage (V_D). Therefore, the source-drain current (I_D) is a function of both gate voltage (V_G) and source-train voltage (V_D).

4.5.1.2 Basic Structures

OTFTs have four basic device geometries as shown in Fig. 4.5.

As illustrated in Fig. 4.5, the source and drain electrodes are deposited on the top of semiconductor active layer for the top contact geometries (Fig. 4.5a, d), while the source and drain electrodes are deposited at the bottom of semiconductor active layer for the bottom contacted geometries (Fig. 4.5b, c). Among these four device geometries, the bottom gate with top contact geometry is most commonly used by researchers since high quality heavily *n*-doped silicon wafer substrates with a thin layer of thermally grown SiO_2 as gate dielectric layer are commercially available. Thus, organic semiconductors can be easily deposited on the top of thermally grown SiO_2 dielectrics by various deposition techniques including solution depositions for polymeric semiconductors and vacuum depositions for most small organic semiconductors.

OTFTs with top gate and bottom contact geometry (Fig. 4.5c) allow depositions of source and drain electrodes directly on any substrates such as glasses and flexible plastics. This geometry is particularly convenient for use of solution deposition techniques such as inkjet printing for the fabrication of source and drain electrodes with solution-processable conductive inks. OTFTs with top gate and top contact geometry (Fig. 4.5d) allow the deposition of a semiconductor layer on any substrates first. However, the deposition of a dielectric layer directly on the top of a semiconductor layer for top gate OTFTs usually presents a big challenge because certain damages of organic semiconductor layer often occur, especially when dielectrics are solution-processed.

4.5.1.3 Parameters Affecting OTFTs

Mobility: Field-effect mobility (μ) is defined as an average rate of the charge movement in active semiconductor layer through a unit electric field. It is basically measure of how easily charge carriers can move in the device. The higher the mobility value, the higher the switching speed for OTFTs. Large mobility is required for reliable operation of transistor. Due to weak intermolecular interaction in the solid material state, impurities, defects, and inefficient carrier injection capability at the metal contacts, OTFTs are characterized with much lower carrier mobility than inorganic MOSFETs. The mobility of organic thin film transistor is gate biased-dependent and tends to increase when gate bias increases. The field-effect mobility (μ) in the linear and saturated regions can be extracted from the following equations:

$$\begin{aligned} &\text{Linear regime } (V_\mathrm{D} \ll V_\mathrm{G}) : I_\mathrm{D} = \mu C_i (W/L)(V_\mathrm{G} - V_\mathrm{T})V_\mathrm{D} \\ &\text{Saturated regime } (V_\mathrm{D} > V_\mathrm{G}) : I_\mathrm{D} = \mu C_i (W/2L)(V_\mathrm{G} - V_\mathrm{T})^2, \end{aligned} \tag{4.4}$$

where W and L are channel width and length, respectively; V_G and V_D are gate voltage and drain voltage, respectively; C_i is capacitance per unit area of the gate dielectric layer; V_T is threshold voltage that can be determined from the relationship between the square root of I_D at the saturated regime and V_G of the device by extrapolating the measured data to $I_\mathrm{D} = 0$.

On/Off ($I_\mathrm{on}/I_\mathrm{off}$) Current Ratio: The ratio of current in the linear region over the current in the saturation region is called $I_\mathrm{on}/I_\mathrm{off}$. Current ratio depends upon various factors such as materials, channel length, and thickness of semiconductor. Very recently it is reported that there are two main causes of the off-current variations. One is the properties of the gate dielectric layer such as its roughness, thickness, etc. and the second one is the crystalline structure of the semiconducting layer. Short channel devices exhibit higher on/off current ratio over larger channel length. This ratio increases with decrease in the thickness of semiconducting layer.

Threshold Voltage: To know about impurity concentrations, interface states, and traps in an organic semiconductor, the threshold voltage and subthreshold current are measured during device analysis. The threshold voltage V_T of OTFTs varies with either the gate insulator capacitance or the thickness of the organic film. Lower V_T is useful in lowering device power consumption and in producing portable devices. Again the subthreshold current exponentially depends on the gate-bias as well as the drain source bias because below the threshold, the free carrier density exponentially depends on the local bias.

Contact Resistance: In a transistor, charge is injected from the source contact and removed at the drain contact, and a finite voltage is necessary to transport the current from the metal electrode into the accumulation layer at the active semiconductor-gate dielectric interface and vice versa. This leads to a contact resistance, namely, a potential drop at the contacts which prevents the full source-drain bias from being applied across the channel. Ideally the contact resistance should be ohmic and small in order to enable the whole voltage applied to the device and contribute to the transport current. It is an inherent property and independent of the measurement method. For top contact devices it strongly depends upon gate bias and sharply increases at low gate-source voltage, while contact resistance appears to be almost independent of the gate bias in bottom contact structures. In top contact structure the field effect mobility is higher and contact resistance is lower due to large injection area, whereas contact resistance is higher in bottom contact devices due to poor morphology. The total resistance R_{ON} of an OFET can be modeled as total contact resistance R_C (source and drain contacts) in series with the channel resistance R_{CH}.

Effect of Active Layer Thickness: Electrical parameters of OTFT do not solely depend upon gate capacitance; these can be modulated by film thickness and charge injection from the source electrode. There are trends which can be expressed as a function of the product of thickness of polymeric film and gate capacitance per unit area. It has been observed that with increasing permittivity of gate insulator and thickness of organic material, the mobility decreases in OTFTs.

4.5.1.4 Current Progress and Future Developments

Organic materials with improved carrier transport characteristics are in high demand for the fabrication of efficient electro-optical devices. Research on synthetic polymer and other small molecules has gained a great deal of attention in recent years due to their potential applications in developing low cost, ultra-thin and flexible products such as displays, solar energy conversion cells, smart cards, sensors, and security applications. Basically, conductive organic semiconductors are lighter, more flexible, and less expensive than inorganic conductors. This makes them a desirable alternative in many applications. It also creates the possibility of new applications that would be impossible using copper or silicon. Organic electronics not only include organic semiconductors, but also organic dielectrics, conductors, and light emitters. Such functional materials are expected to play revolutionary role in next-generation electronics.

The application of organic semiconductor in organic field effect transistors has been extensively investigated owing to their many potential advantages, including cost effectiveness, large-scale manufacture, wet processability, and mechanical compatibility with flexible displays. Recently, high performance p- or n-type organic FETs with high charge carrier mobility, high on-to-off current ratios, and high stability have been achieved due to the improvement of materials synthesis.

Various polycyclic aromatic compounds such as oligoarenes, oligoacenes, tetrathiafulvalenes, and disc-like polycyclic aromatic hydrocarbons are qualified as good candidates for FETs after accurate chemical tailoring. Researchers currently are exploring ways of "doping" organic semiconductors, like melanin, with relatively small amounts of conductive metals to boost conductivity. These organic compounds are being developed for resistance to thermal and environmental stresses, which is one of the major challenges in the field. Several other classes of compounds based on fullerene (C_{60}) and chemically modified thiophenes and arylene diimide series molecules are also being explored and developed in parallel for n-type organic semiconductors for use in a plethora of devices encompassing interdisciplinary research areas. These devices, owing to their low-cost, ease of synthesis and processing, are expected to pave the way for next-generation electronics with energy-efficient security systems, sensors, photonics, and spintronic memories.

4.5.2 Organic Solar Cells (OSCs)

The photovoltaic cells have been extensively studied since the 1950s when the first crystalline silicon solar cell, which had an efficiency of 6 %, was developed at Bell Laboratories. Since then, the efficiency has reached up to 24 % for crystalline Si solar cells, which is already close to the theoretical predicted upper limit of 30 %. But one of the major obstacles for the market implementation of photovoltaic (PV) cells is the large production costs for Si-based technology. Despite a great deal of effort to further reduce the price, large scale production of Si-based solar cells will be limited by the availability of raw materials, such as solar-grade Si. Therefore, to ensure a sustainable technology path for PV, the development of new materials and device structures is of utmost priority.

Organic photovoltaic (OPV) materials are promising alternative to their inorganic counterparts due to their low-cost, easy fabrication, and compatibility with large-area and flexible substrate. An organic solar cell or plastic solar cell is a type of solar cell that uses conductive organic polymers or small molecules for light absorption and charge transport to produce electricity from sunlight by the photovoltaic effect. Compared to the bulk, the optical and electronic properties of the molecules used in OPVs at an interface remain unexplored. In an organic device, the charge carriers have to pass through the organic/electrode interfaces. Therefore, the electronic structures of the molecules are important factors that affect the overall efficiency of the OPVs.

Solar Cell Parameters: There are primarily five basic electrical parameters for organic solar cell, namely,

- Short-circuit Current (I_{sc})
- Open-circuit Voltage (V_{oc})
- Fill Factor (FF)

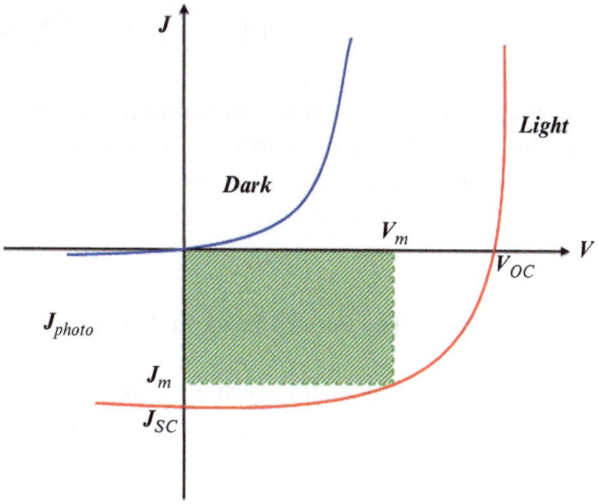

Fig. 4.6 Schematic of current–voltage (*I–V*) characteristics of a solar cell

- Efficiency (η)
- External Quantum Efficiency (EQE) or Incident Photon to Current Conversion Efficiency (IPCE).

Short-circuit Current (I_{sc}): The *short-circuit current* is the current through the solar cell when the voltage across the solar cell is zero; it is denoted by I_{sc}.

Open-circuit Voltage (V_{oc}): The *open-circuit voltage*, V_{oc}, is the maximum voltage available from a solar cell obtained at zero current. It measures the amount of forward bias on the solar cell. Hence, it is also a measure of the amount of recombination in the device.

Fill Factor (**FF**): The factor which defines the power extraction efficiency is known as the Fill Factor (FF) of the solar cell. It is obtained by the product of V_{oc} and I_{sc}. The FF is also obtained from the area of the largest rectangle which will fit in the *I–V* curve as shown in Fig. 4.6.

Efficiency (η): The overall *solar to electrical energy conversion efficiency*, η, for a solar cell is given by the photocurrent density measured at short-circuit (J_{sc}), the open-circuit photo voltage (V_{oc}), the fill factor of the cell (FF), and the intensity of the incident light (P_{in}). η can be expressed as:

$$\eta = \frac{J_{sc} \times V_{oc} \times \mathrm{FF}}{P_{in}}. \tag{4.5}$$

External Quantum Efficiency (**EQE**) **or Incident Photon to Current Conversion Efficiency** (**IPCE**): Another fundamental measurement of the performance of a solar cell is the "EQE," which is also called the *IPCE*. It is measured for the determination of a ratio of the number of collected charge carriers in the external circuit to the number of incident phonons. It can be expressed as:

$$\text{IPCE} = 1,240 \times \frac{J_{\text{sc}}(\lambda)}{\lambda \times P_{\text{in}}(\lambda)}, \tag{4.6}$$

where J_{sc} is the short-circuit photocurrent measured at different excitation wavelengths λ, P_{in} is the power density of incident light. Quantum efficiency is measured from 350 nm onwards as the power from the AM1.5 G contained in wavelengths lower than 350 nm is low.

4.5.2.1 Basic Processes in Organic Solar Cells

In general, for a successful organic photovoltaic cell six important processes have to be optimized to obtain a high conversion efficiency of solar energy into electrical energy:

1. Light absorption
2. Excitation
3. Exciton formation
4. Exciton diffusion
5. Exciton dissociation
6. Charge transport and collection

Organic materials are responsible for generating free charge carriers from sunlight in organic solar cells. The operational mechanism of OPVs is illustrated in (Fig. 4.7) and can be summarized as follows:

Light Absorption, Excitation, and Exciton Formation: An electron in the HOMO of an organic material absorbs a photon and is excited into the lowest unoccupied molecular orbital (LUMO), creating an exciton (bound electron hole pair).

For efficient collection of photons, the absorption spectrum of the photoactive organic layer should match the solar emission spectrum and the layer should be sufficiently thick to absorb the entire incident light. When the incident photon has an energy $h\nu \geq \text{Eg}$, where, Eg is the band gap of the semiconductor, an electron (e) in the HOMO of the donor would be excited to the LUMO, leaving a hole (h) in the HOMO level. This e–h pair is called singlet exciton having opposite spin. In an OSC, only a small region of the solar spectrum is covered. For example, a band gap of 1.1 eV is required to cover 77 % of the AM1.5 solar photon flux, whereas most solution processable semiconducting polymers (PPVs, P3HT) have band gaps larger than 1.9 eV, which covers only 30 % of the AM1.5 solar photon flux. In addition, because of the low charge-carrier mobilities of most polymers, the thickness of the active layer is limited to ~100 nm, which, in turn, results in absorption of only ≈60 % of the incident light at the absorption maximum. Thus, an efficient solar cell should have a wide absorption spectrum, so as to create as many e–h pairs as possible.

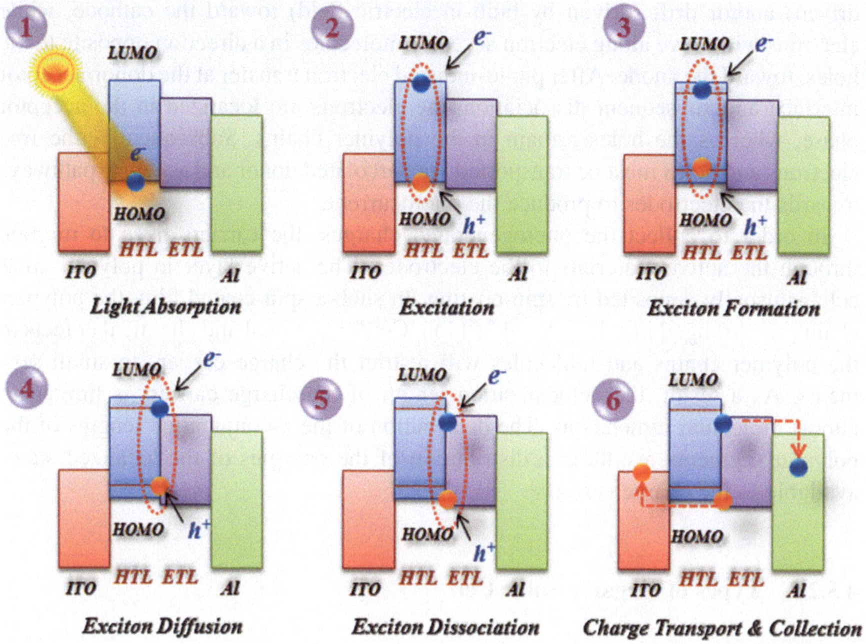

Fig. 4.7 Basic processes in organic solar cell

Electron Diffusion: The exciton diffuses to the interface of two materials (electron donor and electron acceptor). Because of the high exciton binding energy in the conjugated polymers, the thermal energy at room temperature is not sufficient to dissociate a photo-generated exciton into free charge carriers. Consequently, the configuration and operation principle of PV devices based on organic semiconductors differ significantly from those based on inorganic materials. Typically, in OSCs an efficient electron acceptor is used in order to dissociate the strongly bound exciton into free charge carriers.

Electron Dissociation: The exciton is dissociated or separated at the interface. Organic semiconductors are characterized by high excitonic binding energy of the order of 0.2–0.5 eV. As a result, photogenerated excitons dissociation occurs only when the potential drop at donor and acceptor interface is larger than the exciton-binding energy. After photo-excitation of an electron from the HOMO to the LUMO, the electron can jump from the LUMO of the donor to the LUMO of the acceptor. However, this process, which is called photo-induced charge transfer, can lead to free charges only if the hole remains on the donor due to its higher HOMO level. In contrast, if the HOMO of the acceptor is higher, the exciton transfers itself completely to the material of lower-band gap accompanied by energy loss.

Charge Transport and Collection: In this step, the newly separated charge carriers are transported and collected by the opposite electrodes of the cell. Positive holes will move along donor molecules due to diffusion (electrochemical gradient-driven) and/or drift (driven by built-in electric field) toward the cathode, while electrons will move along electron acceptor molecules in a direction opposite to the holes, toward the anode. After photo-induced electron transfer at the donor/acceptor interface and subsequent dissociation, the electrons are localized in the acceptor phase, whereas the holes remain in the polymer chains. Subsequently, the free electrons and holes must be transported via percolated donor and acceptor pathways towards the electrodes to produce the photocurrent.

In order to collect the photogenerated charges, the carriers have to migrate through the active materials to the electrodes. The active layer in polymer solar cells is usually deposited by spin-coating. In such a spin-coated film, the polymer chains are arranged in a disordered fashion. Conformational and chemical defects in the polymer chains and molecules will restrict the charge carriers to small segments. As a result, the delocalization length of the charge carriers is limited to almost molecular dimensions. The distribution of the π-conjugation lengths of the polymer segments results in a distribution of the energies of the localized states available to the charge carriers.

4.5.2.2 Types of Organic Solar Cell

In general, there are mainly five types of solar cells.

1. Single layer
2. Multilayer
3. Bulk Heterojunction
4. Dye sensitized
5. Tandem solar cell

Single Layer: Single layer organic photovoltaic cells are the simplest of the various forms of organic photovoltaic cells. These cells are made by sandwiching a layer of organic electronic materials between two metallic conductors, typically a layer of indium tin oxide (ITO) with high work function and a layer of low work function metal such as Al, Mg, or Ca. The basic structure of such a cell is illustrated in Fig. 4.8.

In 1994, this structure was created by Marks et al. using 50–320 nm thickness of poly (*p*-phenylene vinylene) (PPV) sandwiched between an ITO and a low work function cathode [23]. The reported quantum efficiencies for this device were around 0.1 % under 0.1 mW cm^{-2} intensity. This low quantum efficiency resulted from intrinsically low mobility of charges through semiconducting organics. The carrier mobility of semiconducting organics has shown an improving trend yet

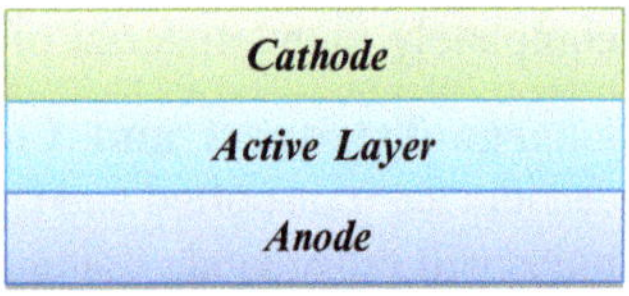

Fig. 4.8 Schematic of single layer organic solar cell

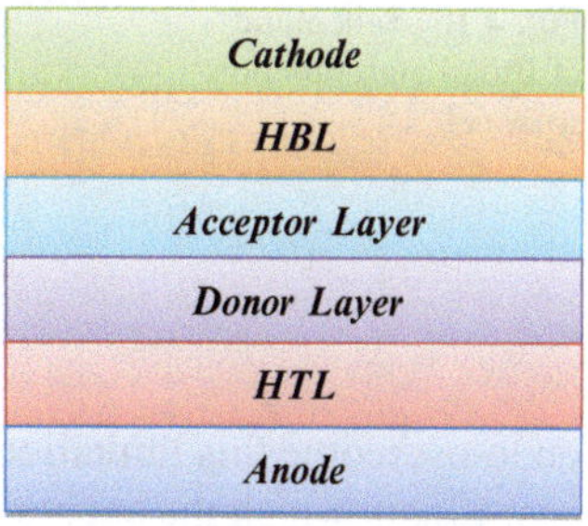

Fig. 4.9 Schematic of multilayer organic solar cell

around 10^{-3} cm^2 V^{-1} s^{-1} is very common, while the mobility of single crystalline silicon is about 10^3 cm^2 V^{-1} s^{-1} order. This indicates that the photogenerated charges in semiconducting organics require more time to be collected from electrodes. The slow charge transport itself decreases the efficiency of the OPV cell, but also increases the recombination chance of charges in the device. The other problem that causes the low PCE of OPV cells is exciton formation, which are strongly bound dipole charges of photo-excited semiconducting organics. Free electrons and holes are desired as an efficient charge carrier because the bound exciton requires an additional exciton dissociation step to make free carriers, which can decrease the carrier generation efficiency. In a single layer OPV cell, only one region to dissociate excitons into free carriers is the interface between semiconducting organics and a cathode. Later, it was recognized that the excitons are more efficiently dissociated at the interface between donor and acceptor, and a bilayer OPV cell was developed by inserting an acceptor layer between a donor semiconducting organic and a cathode.

Multilayer: This type of organic photovoltaic cell contains two different layers in between the conductive electrodes (Fig. 4.9). These two layers of materials differ in their electron affinity and ionization energy properties; therefore, electrostatic forces are generated at the interface between these two layers. The materials are carefully chosen to make the differences large enough, so that these local electric fields are strong, which may break up the excitons much more efficiently than the single layer photovoltaic cells achieve. The layer with higher electron affinity and ionization potential is the electron acceptor, and the other layer is the electron donor. This structure is also called a planar donor-acceptor heterojunction.

This structure was first realized by Tang in 1986 and the device structure was comprised of ITO/copper phthalocynines (CuPc)/perylene tetracarboxylic derivative (PV)/silver (Ag) [24]. The reported PCE was 1 % under simulated AM2 conditions. This tenfold PCE increase resulted from improving the exciton dissociation efficiency by adding electron transporting material that forms an offset energy band with hole transporting material. However, the reported PCE of bilayer OPV cells is still significantly lower than that of inorganic-based PV cells. One reason for this is the intrinsically short exciton diffusion length of excitons in organic semiconductors, which are typically in the range of 10–20 nm. Efforts are

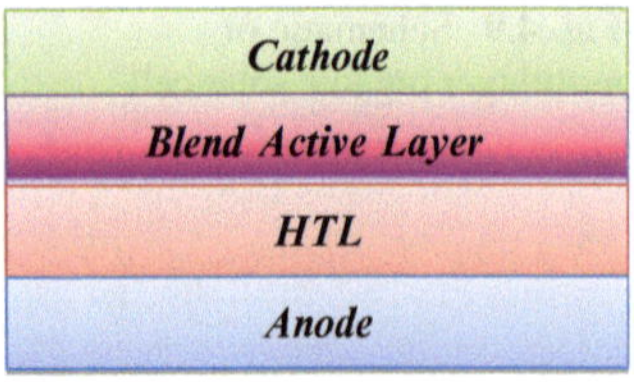

Fig. 4.10 Schematic of Bulk heterojunction solar cell

on to overcome this limitation in the bilayer OPV cell by using buckminsterfullerene, C_{60}, in which the exciton diffusion length is around 20 nm. Recently perylene tetracarboxylic derivatives were used instead of C_{60} as an acceptor in the device architecture which resulted in 3.5 % PCE [25]. This improvement of PCE is attributed to the longer travel distance of excitons in the triplet state of C_{60}. However, since donor and acceptor layers are ~40–60 nm thick, most of the absorbed photon energy is not efficiently converted to free carriers and dissipated by recombination, which indicates that photogenerated excitons are dissociated only near the interface between donor and acceptor. Other approaches to build bilayer OPV cells use conjugated polymers. Sariciftci et al. fabricated the first bilayer OPV cell that used a conjugated polymer where the hole transporting material was MEH-PPV and the electron transporting material was C_{60} [26]. In this structure, MEH-PPV plays a role in both absorbing photons and transporting holes to an anode, and C_{60} transports electrons to a cathode. In this OPV cell, the obtained PCE was 0.04 % under monochromatic incident light at 514.5 nm, and the performance was only slightly improved compared to single polymer layer PV cells. Halls et al. optimized the thickness of conjugated polymer and C_{60} layer and achieved 9 % quantum efficiency under 0.25 mW cm^{-2} intensity [27].

Bulk Heterojunction Solar Cell: In this type of photovoltaic cell, the electron donor and acceptor are mixed together, forming a polymer blend (Fig. 4.10). If the length scale of the blend is similar to the exciton diffusion length, most of the excitons generated in either material may reach the interface, where excitons break up efficiently. Electrons move to the acceptor domains then get carried through the device and collected by one electrode, and holes are pulled in the opposite direction and collected at the other side of the device.

Yu et al. investigated a phase-separated polymer blend composite made of poly [2-methoxy-5-(2′-ethyl-hexyloxy)-l, 4-phenylene vinylene], MEH-PPV, as a donor and cyano-PPV, CN-PPV, as an acceptor [28]. The photoluminescence and electroluminescence of both component polymers were quenched in the blend, which is indicative of rapid and efficient separation of photogenerated electron hole pairs with electrons on the acceptor and holes on the donor. Diodes made with such composite semiconducting polymers as the photosensitive medium show promising photovoltaic characteristics with energy conversion efficiency of 0.9 %, which is 20 times larger than that of diodes made with pure MEH-PPV.

The next significant improvement in PCE of blend OPV cells was achieved using poly(3-hexylthiophene) (P3HT) as a hole transporting polymer. Gang Li et al. fabricated a blend OPV cell using P3HT and PCBM, soluble C60 derivative [29]. The significant improvement in PCE results mainly from the crystallinity of P3HT. After casting P3HT and PCBM blend solutions on PEDOT: PSS/ITO, the blend film morphology was controlled by varying the evaporation rate. During the solvent annealing, P3HT and PCBM formed a well-mixed, interdigitated blend film where P3HT forms fibril-like crystalline morphology and PCBM aggregates are embedded. This phase-separated morphology of blend layers enhances the hole mobility of the conjugated polymer and also improves the absorption efficiency from the fibril-like P3HT aggregates. The best performance from the optimized blend OPV cell was 4.4 %.

Kim et al. also investigated a similar blend OPV structure focused on the regioregularity of P3HT [30]. They prepared various P3HT and PCBM blend solutions in which P3HT had different regioregularities ranging from 80 % to 96 %. The fabricated devices were thermally annealed to induce phase separation of P3HT and PCBM. The investigation revealed that a higher order of regioregularity of P3HT helps to form the crystallized fibril-like shape of P3HT, which improves charge transport as well as light absorption efficiency.

Tandem Solar Cell: Tandem solar cell or multi-junction solar cells are solar cells with multiple p–n junctions made of different semiconductor materials. The p–n junction of each material produces electric current in response to a different wavelength of light (Fig. 4.11). The use of multiple semiconducting materials allows the absorbance of a broader range of wavelengths, improving the cell's sunlight to electrical energy conversion efficiency.

Kim et al. successfully demonstrated improved performance of an OPV cell with 6.5 % PCE using tandem cell structure. In the blend layers, semiconducting polymers and fullerene derivatives consisted of photoactive layers [31].

A transparent titanium oxide (TiO_x) layer was used as an interfacial electrode that separates and connects the front cell and the back cell. The TiO_x layer served as an electron transport and collecting layer for the first cell and as a stable foundation that enabled the fabrication of the second cell to complete the tandem cell architecture.

Dye-Sensitized Solar Cell: A dye-sensitized solar cell is a low-cost solar cell belonging to the group of thin film solar cells. It is based on a semiconductor

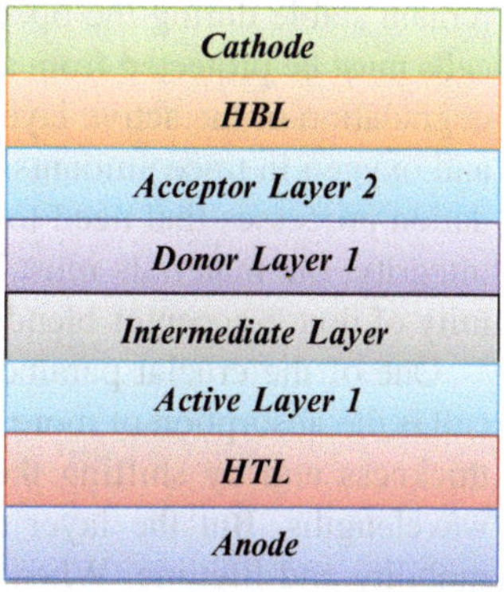

Fig. 4.11 Schematic of Tandem solar cell

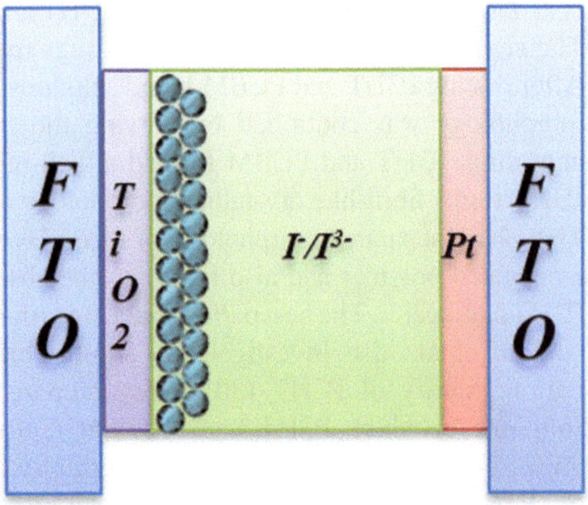

Fig. 4.12 Schematic of dye-sensitized solar cell

formed between a photo-sensitized anode and an electrolyte, a photo electrochemical system (Fig. 4.12). The modern version of a dye solar cell, also known as the Grätzel cell, was originally co-invented in 1988 by Brian O'Regan and Michael Grätzel at UC Berkeley.

The DSSC has a number of attractive features; it is simple to fabricate using conventional roll-printing techniques, is semi-flexible and semi-transparent which offers a variety of uses not applicable to glass-based systems, and most of the materials used are low-cost. In practice it has proven difficult to eliminate a number of expensive materials, notably platinum and ruthenium, and the liquid electrolyte presents a serious challenge to making a cell suitable for use in all weather. Although its conversion efficiency is less than the best thin-film cells, in theory its price/performance ratio should be good enough to allow them to compete with fossil fuel electrical generation by achieving grid parity.

4.5.2.3 Current Progress and Future Developments

Generally, any practical application of organic solar cells requires that the cells remain stable during the operation. The present-day organic, polymer-based solar cells must be protected from ambient conditions (humidity as well as air) to prevent degradation of the active layer and electrode materials by the effects of both water and oxygen in trace amounts. Even with proper protection there are several degradation processes that need to be eliminated to ensure stability. Apart from device integrity, the materials must be photochemically stable and the nanoscale uniformity of donor-acceptor blend in the active layer be preserved.

One of the crucial parameters for increasing the photocurrent of organic solar cell is the absorption of more photons. This can be achieved by increasing the layer thickness and by shifting the absorption spectrum of the active layer to longer wavelengths. But the layer thickness is presently limited by the charge carrier mobility and lifetime. When the mobility is too low or the layer is too thick, the transit time of photo-generated charges in the device becomes longer than the lifetime, resulting in charge recombination. The use of polymers such as P3HT that are known to have high charge carrier mobilities allows an increase in film thickness from the usual ~100 nm to well above 500 nm, without a loss of current.

Thus in future, there is a probability that modifying the organic semiconductor material structure, and varying their thickness in device preparation, can lead to further improvement in the performance of organic photovoltaic devices.

4.5.3 Organic Light-Emitting Diodes

Organic light-emitting diodes (OLEDs) are promising high-efficiency lighting sources that are presently being introduced in a wide variety of applications. An OLED consists of several semiconducting organic layers sandwiched between two electrodes. The device structure of a simple bi-layer OLED is shown in Fig. 4.13.

In an OLED, light is generally produced by the process of electroluminescence simply by injecting electron and holes in the emissive layer. For better electron injection, usually low work function material such as calcium, aluminum, and silver, or mixtures of Mg:Ag are used. Similarly, as anode, materials with high work function are used for better hole injection. However, since it is important that the light generated inside an OLED escapes from the device, one of the two electrodes used is transparent. Usually, ITO is the most widely used material as the anode material since it is transparent to visible light and has a high work function which promotes injection of holes into the emissive layer.

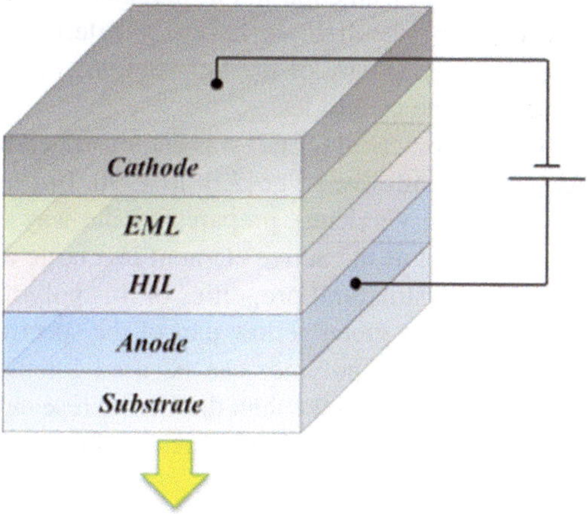

Fig. 4.13 Simplified device structure of an OLED

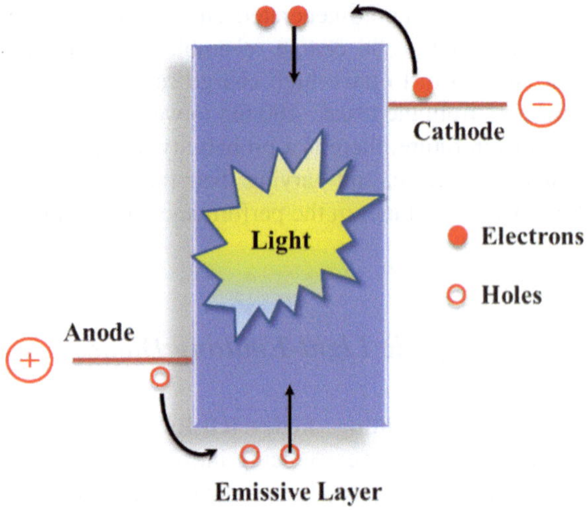

Fig. 4.14 The process of light generation in OLED

4.5.3.1 Mechanism

When a voltage is applied across the OLED such that the anode is positive with respect to the cathode, the current begins to flow and the electrons also start to flow through the device from cathode to anode (Fig. 4.14). Thus, the cathode gives electrons to the emissive layer and the anode withdraws electrons from the emissive layer or in other words, the anode gives holes to the emissive layer. Electrostatic forces bring the electrons and the holes towards each other and they tend to recombine. The recombination causes a drop in the energy levels of electrons, accompanied by an emission of radiation whose frequency is in the visible region. The color of the light is basically determined by the emissive material used, whereas the intensity of the light output can be controlled by the electric current applied by the electrodes.

The first practical OLED was demonstrated by Tang et al. in the year 1987 using Alq_3 as the emissive layer [32]. In 1990, Burroughes et al. demonstrated that poly (*p*-phenylene vinylene), prepared by the way of a solution processible precursor, could be used as the active element in a large-area light-emitting diode. Combination of good structural properties of this polymer, its ease of fabrication, and light emission in the green yellow part of the spectrum with reasonably high efficiency suggests that the polymer can be used for the development of large-area light-emitting displays. Since then the OLED research community has been divided into two groups, each favoring a specific fabrication method and class of materials resulting from that choice [6].

4.5.3.2 Small Molecule OLEDs

As the name suggests, small organic molecules are used as the active layer in this type of devices. SMOLEDs basically consist of three layers, namely the electron transport layer, emissive layer, and the hole-transporting layer (HTL). However to obtain better charge balance, highly efficient OLEDs usually are constructed of several organic layers, each especially suited for particular purpose. However, the need of these layers varies with different emissive material and one has to go for the best suitable device configuration for the emissive material used. A multilayer OLED is generally composed of anode, hole injection layer (HIL), hole-transporting layer (HTL), electron-blocking layer (EBL), emission layer (EML), hole-blocking layer (HBL), electron-transporting layer (ETL), electron injection layer (EIL), and finally a cathode. The device structure of a multilayer OLED with all possible optional layers has been depicted in Fig. 4.15.

The small molecule OLEDs, in general, are fabricated by a vacuum evaporation process under high vacuum conditions. Till date, thermal evaporation is practically the main process for manufacturing OLED fabrication and therefore all commercially available OLEDs are based on small molecule materials. This manufacturing process, however, suffers from the disadvantages of high costs, and by using this technique it is not possible to fabricate large-area devices. But compared to the wet processing techniques used for polymer-based OLEDs, it is easy to form well-controlled homogeneous films using thermal evaporation technique.

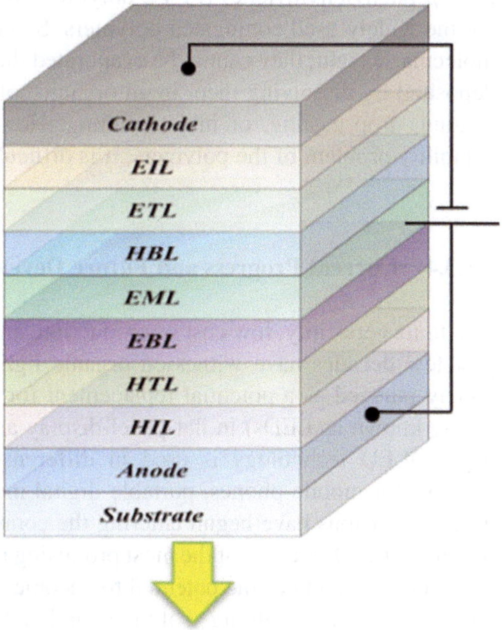

Fig. 4.15 Device structure a multilayer OLED with all possible optional layers

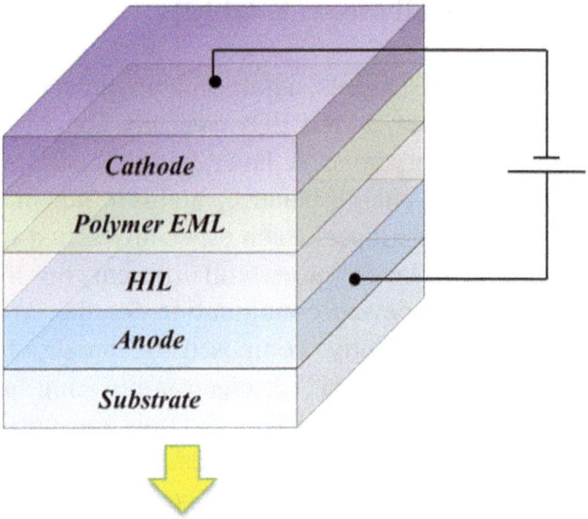

Fig. 4.16 Device structure a polymer light-emitting diode

4.5.3.3 Polymer Light-Emitting Diodes

In polymer light-emitting diodes, conjugated polymers are used as the active emissive layer. In general, a polymer OLED consists of an anode, a HIL, an emissive layer, and a cathode as shown in Fig. 4.16. Unlike the SMOLEDs, PLEDs have a comparatively simple structure as the functional materials required for better device performance can be introduced into polymer chains by co-polymerization technology.

In most of the PLEDs, PEDOT:PSS is used as the hole injection layer, whereas poly(*p*-phenylenvinylene) (PPV), polyfluorene (PF), and poly-*p*-phenylene (PPP) are the widely used conjugated polymers. Since the polymers possess a very high molecular weight, they cannot be evaporated thermally. Therefore, they are usually deposited by dissolving them in an organic solvent followed by spin coating, dip coating, drop casting, or inkjet printing. However, due to the common solvent solubility problem of the polymers, it is difficult to fabricate multilayer PLEDs.

4.5.3.4 Current Progress and Future Developments

Due to its versatility, low cost of production, and ease of large area fabrication, the past few decades have witnessed organic light-emitting devices (OLEDs) being widely pursued as a potential replacement for their inorganic counterparts (commonly known as LEDs) in flat-panel-display and light-emitting elements. Now-a-days, OLED technology is used in different commercial applications such as displays for mobile phones, portable digital media players, etc. Similarly OLED-based televisions have begun entering the commercial market area. In the field of lighting, OLEDs are one of the most promising new lighting technologies to emerge over the years and has the potential to become a highly energy-efficient, large area and eco-friendly lighting solution and alternative to existing technology.

The efficiency of the WOLEDs designed for solid state lighting has already surpassed that of the traditional incandescent lamp and compact fluorescent lamp and is at par with their inorganic counterparts. Recently, WOLEDs with efficiency as high as 156 lm W^{-1} has been claimed by NEC lighting which was co-developed with Yamagata University's research group led by Kido [33]. Although OLEDs have many advantages, the cost of the OLED-based products is still on the expensive side, and as the organic materials in general have a limited lifetime, the long-term stability of those devices is yet to be tested. However, considering the wide versatility of the organic semiconducting materials available, there is still huge scope for chemists, material scientists, and engineers for further improvement in the performance of organic light-emitting devices.

4.6 Applications of Organic Thin Film Devices

In recent times, functional "organic materials: molecular and polymers", being lightweight, low cost to fabricate and their ease of production, have made inroads into many areas of devices formerly occupied by "inorganic materials". Molecular monolayers deposited on Si exhibiting electronic functionalities are expected to push the scaling limits of Si microelectronics down to a few nanometers. Thin films of molecular semiconductors and conducting polymers are widely being used in various devices, such as, light-emitting diodes, optically pumped photodetectors, lasers, printed transistors, photorefractive applications, solar cells, biomedical application, gas sensors, etc. Recent advances show improved physical characteristics of molecular/polymer materials as active semiconductors, indicating a bright future for molecular and polymer devices. New concepts, innovative fabrication techniques, and material manipulation at the nanoscale are bringing about much improved performances, indicating that organic materials will herald a new technological base with high energy efficiency and inexpensive devices in the near future.

References

1. H. Shirakawa, E.J. Louis, A.G. MacDiarmid, C.K. Chiang, A.J. Heeger, J. Chem. Soc. Chem. Commun. 578 (1977)
2. A.J. Heeger, Chem. Soc. Rev. **39**, 2354 (2010)
3. A.C. Arias, J.D. MacKenzie, I. McCulloch, J. Rivnay, A. Salleo, Chem. Rev. **110**, 3 (2010)
4. D. Tobjörk, R. Österbacka, Adv. Mater. **23**, 1935 (2011)
5. J.E. Anthony, A. Facchetti, M. Heeney, S.R. Marder, X. Zhan, Adv. Mater. **22**, 3876 (2010)
6. J.H. Burroughes, D.D.C. Bradley, A.R. Brown, R.N. Marks, K. Mackay, R.H. Friend, P.L. Burns, A.B. Holmes, Nature **347**, 539 (1990)
7. T. Ameri, G. Dennler, C. Lungenschmied, C.J. Brabec, Energy. Environ. Sci. **2**, 347 (2009)
8. C.N. Hoth, P. Schilinsky, S.A. Choulis, C.J. Brabec, Nano Lett. **8**, 2806 (2008)

9. D.A. Bernards, R.M. Owens, G.G. Malliaras, *Organic Semiconductors in Sensor Applications* (Springer, New York, 2008)
10. H. Sirringhaus, Adv. Mater. **17**, 2411 (2005)
11. H. Klauk, Chem. Soc. Rev. **39**, 2643 (2010)
12. G. Tzamalis, V. Lemaur, F. Karlsson, P.O. Holtz, M. Andersson, X. Crispin, J. Cornil, M. Berggren, Chem. Phys. Lett. **489**, 92 (2010)
13. H. Yan, Z. Chen, Y. Zheng, C. Newman, J.R. Quinn, F. Dotz, M. Kastler, A. Facchetti, Nature **457**, 679 (2009)
14. H. Sirringhaus, P.J. Brown, R.H. Friend, M.M. Nielsen, K. Bechgaard, B.M.W. Langeveld-Voss, A.J.H. Spiering, R.A.J. Janssen, E.W. Meijer, P. Herwig, D.M. de Leeuw, Nature **401**, 685 (1999)
15. H. Sirringhaus, N. Tessler, R.H. Friend, Science **280**, 1741 (1998)
16. M.S.A. Abdou, F.P. Orfino, Y. Son, S. Holdcroft, J. Am. Chem. Soc. **119**, 4518 (1997)
17. I. McCulloch, M. Heeney, M.L. Chabinyc, D. DeLongchamp, R.J. Kline, M. Cölle, W. Duffy, D. Fischer, D. Gundlach, B. Hamadani, R. Hamilton, L. Richter, A. Salleo, M. Shkunov, D. Sparrowe, S. Tierney, W. Zhang, Adv. Mater. **21**, 1091 (2009)
18. I. McCulloch, M. Heeney, C. Bailey, K. Genevicius, I. MacDonald, M. Shkunov, D. Sparrowe, S. Tierney, R. Wagner, W. Zhang, M.L. Chabinyc, R.J. Kline, M.D. McGehee, M.F. Toney, Nat. Mater. **5**, 328 (2006)
19. H. Ohkita, S. Cook, Y. Astuti, W. Duffy, S. Tierney, W. Zhang, M. Heeney, I. McCulloch, J. Nelson, D.D.C. Bradley, J.R. Durrant, J. Am. Chem. Soc. **130**, 3030 (2008)
20. H.S. Tan, N. Mathews, T. Cahyadi, F.R. Zhu, S.G. Mhaisalkar, Appl. Phys. Lett. **94**, 263303 (2009)
21. Z. Bao, A.J. Lovinger, J. Brown, J. Am. Chem. Soc. **120**, 207 (1998)
22. A.L. Briseno, J. Aizenberg, Y.J. Han, R.A. Penkala, H. Moon, A.J. Lovinger, C. Kloc, Z. Bao, J. Am. Chem. Soc. **127**, 12164 (2005)
23. R.N. Marks, J.J.M. Halls, D.D.C. Bradley, R.H. Friend, A.B. Holmes, J. Phys. Condens. Matter **6**, 1379 (2014)
24. C.W. Tang, Appl. Phys. Lett. **48**, 183 (1986)
25. A. Yakimov, P. Peumans, S.R. Forrest, J. Appl. Phys. **93**, 3693 (2003)
26. N.D. Sacriciftci, Appl. Phys. Lett. **62**, 585 (1993)
27. J.M. Halls, Appl. Phys. Lett. **68**, 3120 (1996)
28. G. Yu, A.J. Heeger, J. Appl. Phys. **78**, 4510 (1995)
29. G. Li, Nat. Mater. **4**, 864 (2005)
30. Y. Kim, Nat. Mater. **5**, 197 (2006)
31. J.Y. Kim, K. Lee, N.E. Coates, D. Moses, T.-Q. Nguyen, M. Dante, A.J. Heeger, Science **317**, 222 (2007)
32. C.W. Tang, S.A. Vanslyke, Appl. Phys. Lett. **12**, 913 (1987)
33. T. Nozawa, NEC announces OLED device with world's highest efficiency, Nikkei Electronics (2013), http://techon.nikkeibp.co.jp/english/NEWS_EN/20130307/269942/. Accessed 7 Mar 2013

Chapter 5
Titania Nano-architectures for Energy

Tom Mathews and Rajini P. Antony

Abstract The chapter addresses the issues involved in identification of suitable nano-architectures of TiO_2 for selected applications pertaining to energy. The first part of the chapter provides a brief overview of the structure, properties, and various methods employed for modification of the electronic energy band structure of TiO_2. The subsequent section describes the synthesis of different nano-architectures of TiO_2 and their selected applications in the area of energy. In addition, role of titania one-dimensional nano-architectures in alternative energy generation processes is deliberated.

5.1 Introduction

Requirement of renewable materials for energy and environmental remediation is one of the biggest challenges in twenty-first century. In order to achieve the target, rapid advances in the field of materials development have been taking place, where researchers are in the process of developing novel, eco-friendly, and better performance materials in the field of photocatalysis, photovoltaics, photoelectrochemical cells, fuel cells, and environmental waste degradation. Development of novel better performance inorganic semiconducting materials for application in energy and environment has been the subject of intense research in the recent past. Within the class of inorganic semiconducting materials, oxide semiconductors have been paid much attention in the recent past as they display diverse range of properties because of the metal–oxygen bonding nature. Among semiconducting oxides, wide band gap oxides are of high interest as these exhibit wide ranges of electrical and optical

T. Mathews (✉)
Thin Films and Coatings Section, Surface and Nano-Science Division, Indira Gandhi Centre for Atomic Research, Kalpakkam 603 102, Tamil Nadu, India
e-mail: tom@igcar.gov.in

R.P. Antony
School of Materials Science and Engineering, Nanyang Technological University, 50 Nanyang Avenue, Singapore 639798, Singapore
e-mail: raji.anna@gmail.com

S. Babu Krishna Moorthy (ed.), *Thin Film Structures in Energy Applications*,
DOI 10.1007/978-3-319-14774-1_5

properties. Among the wide band gap semiconducting oxides, titanium (IV) oxide has attracted much attention due to its application in numerous fields ranging from paint to photocatalyst, hydrogen production, storage medium, solar cells, and biology [1, 2]. For example, due to its high refractive index, it is applied as antireflection coatings, optical wave guides, and narrow-band filters [3–5]. Because of its interesting photocatalytic properties, it finds use in various applications such as environmental remediation, hydrogen generation by photolysis of water, photovoltaic cells, electrochromic devices, photoelectrochemical anticorrosion coatings, self-cleaning, and antifogging coatings [6–14]. The two commonly available stable phases of TiO_2 are rutile and anatase. Rutile is preferred to anatase for optical and dielectric applications because of its high refractive index and dielectricity [15]. Whereas, anatase is preferred for all applications related to photocatalytic activity due to its higher charge carrier mobility [10, 16–18]. Several excellent reviews can be found in the literature which deals with the synthesis, properties, and applications of TiO_2 [19–23].

Ever since the discovery of size quantization effect in semiconductor nanomaterials, research in nano-structured materials has generated considerable interest [24, 25]. The properties of interest, viz. optical, electronic, and catalytic, are found to be related to size in the nanoscale regime and drastically differ from those of bulk counterparts. With respect to photo-activity, TiO_2 nanoparticles have specific advantages in enhancing light absorption due to the presence of large fraction of atoms in surfaces and interfaces. The primary mechanism of light absorption in semiconductors is the direct valance band to conduction band electronic transition, which is low in case of semiconducting materials like TiO_2 as direct transition is forbidden by crystal symmetry. However, when light absorption takes place at the crystal boundaries such as surfaces and interfaces, interspersed among crystallites, the momentum is not conserved and an indirect transition occurs leading to enhancement in light absorption. Hence, it can be expected that nanocrystals with high surface to volume ratio and large surface atom fraction will have enhanced light absorption. The particle size at which enhancement becomes significant is around 20 nm. When the particle size is confined, at least in one dimension, to a scale below the de Broglie wavelength, the wave function is spread over the particle and the photo-generated charge carriers need not diffuse to the surface to take part in surface/interface processes. This makes possible the timely utilization of photo-generated charge carriers. In addition, due to quantum confinement, the energy band structure becomes discrete thereby rendering the ensuing photophysical, photochemical, and surface properties different from those of bulk [26–28]. Since the properties of nano-structured TiO_2, relevant to the various applications, are significantly different from those of the bulk, recent research has been concentrated on controlling the size, shape, composition, and surface morphologies as well as surface properties for tailoring them to be utilized for specific applications.

Over the past decade, one-dimensional nano-architectures of TiO_2 such as nanotubes, nanorods, nanobelts, nanowires, and nanofibers have attracted widespread attention due to the unique physical and chemical properties associated with their size confinement in one dimension. Relatively simple preparation techniques used to achieve the requirements for various applications have made them quite attractive [29–33]. In nano-structured semiconductors the mobility of charge carriers is governed by quantum confinement and hence the electron and hole transport properties are influenced by their one-dimensional architecture [34]. The properties of one-dimensional nano-architectures of TiO_2 are comparable to that of nanoparticles but with high surface to volume ratio as well. The high specific surface area of one-dimensional nano-structured TiO_2 facilitates interaction between surfaces/interfaces and reactive species [35–37]. In addition, they have enhanced photoactivity and charge transfer at the interface due to better charge transport, higher surface defect density and aspect ratio. These properties along with the exotic surface geometry render the one-dimensional nano-structured TiO_2-based materials excellent candidates for application in photocatalysis, sensors, hydrogen generation by water splitting, photovoltaics, and biology.

5.2 One-Dimensional Architectures of TiO_2

Nano-architectures in different forms can be classified as zero-dimensional, one-dimensional, and two-dimensional. Among these, one-dimensional nanostructures have restricted their size in nanometer range in one dimension with larger dimension in the length scale which result in high aspect ratio (the ratio of length to diameter) for the one-dimensional nanostructures. With the invention of carbon nanotubes, great interest in investigation on properties and performances of one-dimensional nano-architectures has been triggered. Variety of nanostructures like nanotubes, nanowires, nanorods, nanobelts, and nanofiners can be found in the class of one-dimensional nanostructures which showed enhanced performance in the field of photovoltaics, field emission, and photocatalysis where high surface area, directional electron transport are required. Since semiconductors are the most common and diverse class of materials which have many application in various field due to its physical, chemical, and structural properties, achieving these classes of materials in one-dimensional forms in a controllable way by various synthesis techniques has been a major research area for materials scientist.

Among the one-dimensional nano-architectures, nanotubes are promising because of larger active surface area [38–41]. Nano-architectures like self-aligned nanotube arrays in thin film form provide enhancement in the performance of devices as well as expand the field of applications. For example, vertically aligned thin-walled TiO_2 nanotube arrays are expected to find application in photovoltaics and photocatalysis because of their high aspect ratio and large amount of surface states.

The present chapter focuses on synthesis, modification as well as photovoltaic and photocatalytic application of one-dimensional nano-architectures of TiO_2. The principle behind these processes will be discussed briefly. Emphasis is given to our recent work on one-dimensional TiO_2 coatings, self-aligned nanotube arrays, and nanotube powders of TiO_2. A brief review on structure, properties, and modification of electronic band structure by doping is given at the outset.

5.3 Structure

Titanium dioxide (TiO_2) exists in nature mainly in three forms, namely, anatase, rutile, and brookite [19, 42]. The most common polymorphs of TiO_2 are anatase and rutile and their structure can be expressed in terms of chains of octahedra in which Ti^{4+} ion is surrounded by an octahedron of six O^{2-} anions. The TiO_6 octahedra in both these phases are orthorhombically distorted. The distortion is significant in anatase, whereas it is marginal in rutile. Hence, anatase is of lower symmetry. Compared to rutile, the Ti–Ti distances are larger in anatase (3.79 and 3.04 Å vs. 3.57 and 2.96 Å), whereas the Ti–O distances are shorter (the Ti–O distance 1.934 and 1.98 Å vs. 1.949 and 1.98 Å in rutile). In rutile, each octahedron is in contact with ten other octahedra (two sharing edge oxygen pairs and eight sharing corner oxygen atoms), whereas in anatase, each octahedron is in contact with eight neighbors (four sharing edge and four sharing corner) [20, 43]. The rutile unit cell contains two Ti atoms and four oxygen atoms. The anatase unit cell contains four Ti atoms and eight oxygen atoms [44]. These differences in lattice arrangements cause difference in mass densities and electron band structures. The crystal faces with lowest energy are (110) and (100) for rutile [45, 46] and (101) and (001) for anatase [19]. The crystal structure data of both anatase and rutile phases are given in Table 5.1 [47]. Between these two phases, rutile is the thermodynamically stable one. However, in the nanosize samples occurrence of anatase phase is quite common [48, 49]. On heating, concomitant with coarsening, anatase is transformed to rutile in the temperature range 400–700 °C. Enthalpy of anatase to rutile phase transformation is low with values ranging from -1.3 to -6 ± 0.8 kJ/mol [50, 51]. The phase transition depends upon several factors such as strain, presence of impurities, particle size, applied pressure, and texture [52, 53]. It has been observed that the temperature of anatase–rutile transition increases with decrease in particle size. Lower the particle size, in the nanoregime, higher the stability of anatase phase. Due to these differences, anatase and rutile phases find different applications.

Table 5.1 Crystal structure data of anatase and rutile TiO_2

	Rutile	Anatase
Crystal structure	Tetragonal	Tetragonal
Lattice constants (Å)	$a = 4.5936$	$a = 3.784$
	$c = 2.9587$	$c = 9.515$
Space group	$P4_2/mnm$	$I4_1/amd$
Molecule/cell	2	4
Volume/molecule (Å^3)	31.2160	34.061
Density (g/cm^3)	4.13	3.79
Ti–O bond Length (Å)	1.949(4)	1.965(4)
	1.980(2)	1.965(2)
O–Ti–O bond angle	81.2°	77.7°
	90.0°	92.6°

5.4 Electrical and Electronic Properties

Titanium dioxide is an n-type semiconductor having a band gap of 3.2 and 3 eV, respectively, for anatase and rutile phases. Electronic structure of TiO_2 has been determined experimentally by various groups [54–57]. The valence bands of TiO_2 can be split into three main regions: the σ bonding in the lower energy region mainly due to O-p σ bonds; the π bonding in the middle energy region; and O-p π states in the higher energy region due to O-p π nonbonding states where the hybridization with d states of Ti is almost negligible. The conduction band is split into Ti e_g (>5 eV) and t_{2g} bands (<5 eV). The d_{xy} states of t_{2g} are dominantly located at the bottom of the conduction band. The rest of the t_{2g} bands are anti-bonding with p states of oxygen. The main peaks of the t_{2g} bands are identified to be constituted by mostly d_{yz} and d_{zx} states. The conductivity of TiO_2 is mainly dependent upon the presence of oxygen vacancies and Ti^{3+} cations, which makes TiO_2 an n-type semiconductor [58]. The oxygen vacancies formed can be represented using Kroger–Vink notation by the equation,

$$O_0^x(TiO_2) \rightarrow V_{\ddot{o}} + 2e^- + \frac{1}{2} O_2. \quad (5.1)$$

A color center at 3 eV due to the oxygen vacancy has been identified as the reason for the observed yellow color [59]. The nonstoichiometry of TiO_2 depends on the synthesis conditions and history of the samples. The anatase TiO_2 shows Arrhenius type thermally activated conduction, with higher carrier mobility; in comparison, conduction in rutile TiO_2 is low, stemming mainly from small polaron hopping. However, in both these crystal structures, high concentrations of donors lead to the formation of impurity bands [60]. The band gap of TiO_2 (3–3.2 eV corresponding to ~390 to 410 nm wavelength) implies excitation of electrons from VB to CB only under UV irradiation.

5.5 Synthesis of One-Dimensional TiO_2 Nano-architectured Coatings

Among the various nano-architectures of TiO_2, one-dimensional nano-architectures are preferred for various applications due to their high aspect ratio and enhanced electron transport properties compared to that of nanoparticles [29–31]. There exist various methods for the synthesis of one-dimensional TiO_2 nano-architectures such as tubes, rods, wires, and fibers in the form of coatings, powders, and suspensions, which include sol–gel, hydrothermal, chemical vapor deposition, electrospinning, electro deposition, and template-assisted growth [61–72]. Even though these techniques have several advantages, none of these can produce self-organized nanostructures on suitable substrates in a simple and cost-effective route, which

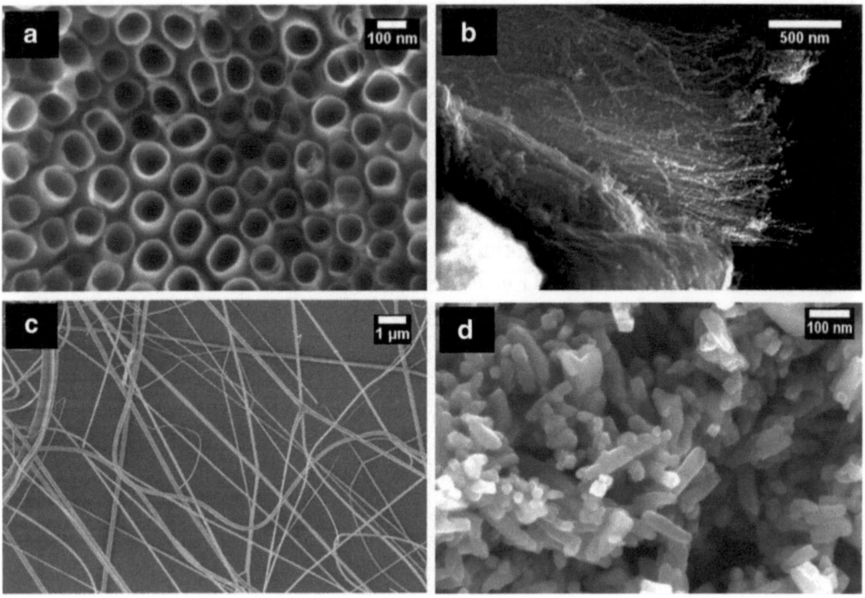

Fig. 5.1 FESEM images of the one-dimensional nanostructures fabricated (**a**) TiO_2 nanotube arrays by electrochemical anodization (**b**) TiO_2 nanotube powders by rapid breakdown Anodization (**c**) TiO_2 nanofibers by electrospinning, and (**d**) TiO_2 nanorods by hydrothermal method

can be used directly for device fabrication. The following section briefly discusses different techniques which produce disordered one-dimensional TiO_2 nano-architectures. This is followed by description of anodic oxidation technique which produces highly ordered self-organized TiO_2 nanotubular arrays [73]. Figure 5.1 shows the SEM pictures of one-dimensional TiO_2 nano-architectures such as tubes, rods, wires, and fibers synthesized in our laboratory using different techniques.

5.5.1 Sol–Gel-Assisted Synthesis

Sol–gel synthesis of TiO_2 is a wet chemical technique in which a colloidal suspension is formed from the acid-catalyzed hydrolysis of Ti alkoxide or Ti salts followed by gelation [74]. This technique employs gels derived from low to moderate molecular weight organic compounds for the synthesis of metal oxide nano-architectures [64, 75]. The driving forces behind such physical gelation involve hydrogen bonding, Vander Waals force, π–π interaction, and electrostatic interaction through cooperating noncovalent interactions of gelator molecules. The outstanding feature responsible for gelation is thought to rely on the formation of three-dimensional networks composed of highly intertwined fibers. The first stage of physical gelation is the self-assembly of gelator molecules, with the self-assembled bundle organogelator acting as a template in the sol–gel polymerization process. Oxide nanotubes are synthesized by coating these templates with oxide precursors followed by gel dissolution in a suitable solvent and subsequent calcination. By this reconstructive synthesis, hierarchical morphologies can be obtained, with the gelators acting as the structure-directing agents. Closely packed three-dimensional polymeric Ti–O–Ti chains formed by hydrolysis can also be converted to one-dimensional form by electrospinning the viscous precursor along with the co-solvent (generally high molecular weight polymer) or by electrophoretic deposition of TiO_2 suspensions into the pores of anodic alumina membrane and its subsequent removal.

Sol–gel-assisted synthesis has been adopted for the fabrication of TiO_2 [63, 76–84] nanorods and nanowires. The colloidal suspension (sol) is prepared by using metals salt or metal organic compound precursors. The precursors are mixed well and hydrolyzed followed by polymerization. The major precursors used for the synthesis of one-dimensional TiO_2 nano-architectures are titanium isopropoxide, titanium tert butoxide, and titanium tetra chloride [85–89]. For the fabrication of TiO_2 nanorods and wires growth directing surfactants such as oleic acid, oleamine are used [76, 90]. During the mixing of precursors in water under stirring the sols slowly convert to liquid gels, which upon optimized treatment followed by suitable heat treatment give nanorods or wires of TiO_2 of desirable crystal structure.

5.5.2 *Hydrothermal Synthesis*

It is generally conducted in Teflon-coated steel pressure vessels known as autoclaves and it assist in crystallizing the compound from high temperature aqueous solutions at high vapor pressures. TiO_2 nanoparticles, nanotubes, and nanorods are synthesized by hydrothermal treatment of Ti precursor with water. Hydrothermal treatment temperatures and conditions determine the extent of precursor conversions, which influences the geometrical features of the resulting nanostructures [61, 63]. Highly concentrated NaOH solutions and temperature in the range of 120–130 °C were found to be more appropriate for the formation of TiO_2 nanotubes by hydrothermal method. Formation of TiO_2 nanotubes by hydrothermal technique were proposed by several groups [63, 91, 92]. Wang et al. [91] proposed that during the reaction with NaOH, the Ti–O–Ti bonding existing between the basic building blocks of the anatase phase, the octahedron, was broken and a zigzag structure was formed where the free octahedra shared edges between the Ti ions with the formation of hydroxy bridges, leading to the growth along the [100] direction of the anatase phase. Two-dimensional crystalline sheets were formed from the lateral growth of the oxo bridges between the Ti centers (Ti–O–Ti bonds) in the [001] direction. These sheets roll up to saturate the dangling bonds emerging from the surface, subsequently lowering the total energy and resulting in the formation of TiO_2 nanotubes. Hydrothermal method has also been applied for the synthesis of

TiO_2 nanorods in powder as well as thin film forms. In this case of nanorods and nanowires the formation is assisted by using acetic acid, benzoyl alcohol, hydrochloric acid, ethylene glycol, etc. [93–95]. Alternative Ti precursors such as $Ti(OC_3H_7)_4$, $C_6H_{22}N_2O_8Ti$ were also used for the synthesis of TiO_2 nanowires and nanorods by hydrothermal route [96–98].

5.5.3 Electrospinning

The electrospinning technique is a simple, cost-effective approach for fabricating ultrafine fibers of polymeric and inorganic materials with diameters ranging from tens of nanometres to sub-micrometres according to the selection of the processing parameters. In this process, a polymer solution or melt is ejected from a small opening or a nozzle under the influence of a strong electrostatic field [99–102]. When the electrostatic force exceeds the surface tension of the precursor solution droplet formed at the tip of the nozzle, a charged fluid jet is ejected. During its flight to a target, the ejected, charged jet dries out, leaving ultrathin fibers on the target. For the synthesis of inorganic fibers the precursor solution is made by incorporating appropriate inorganic salts solutions into various polymer solutions by a sol–gel process. These deposited nanofibers are then subjected to heat treatment above the degradation temperature of the polymer in order to obtain the pure metal oxide nanofibers. For the synthesis of TiO_2 nanofibers a solution containing titanium isopropoxide and a high molecular weight poly vinyl pyrrolidone (PVP) is electrospun to give the titanium salt-embedded polymer fibers. These spun fibers are then calcined to burn off the polymer to get TiO_2 nanofiber network. The network both in the powder form as well as in thin film form has been applied in photocatalysis, dye-sensitized solar cells, and photo-electrochemical cells [102].

5.5.4 Electrochemical Oxidation

Among the one-dimensional nano-architectures, nanotubes are promising because of larger active surface area. In addition, self-aligned nanotube arrays in coating/thin film form provide enhancement in device performance as well as expand the field of applications due to their directionality. Hence, in this section synthesis of vertically aligned titania nanotube arrays by a simple electrochemical technique is detailed. Electrochemical anodization is a cost-effective technique, generally used for the corrosion protection of metals. Metals, generally valve metals (Al, Ta, Hf, Zr, and Ti), have a thin native oxide layer coating, measuring to few nanometers in thickness, on its surface due to interaction with atmospheric oxygen. Thickness of this native oxide layer can be further increased by thermal and anodic oxidation. Conversely, the oxide layers on the metal surface can be dissolved using a number of approaches, which includes surface etching and polishing. Taking the idea of both oxidation and etching, in 1995 Masuda and Fukuda have developed porous alumina coating by anodizing aluminum metal [103]. These porous structures possess high surface area and the pores have a narrow size distribution. The aspect ratio of the porous structures can be varied easily by varying the electrochemical parameters. By optimizing the electrochemical process parameters one can obtain ordered porous structures. The highly ordered porous anodic alumina has significant use as template for the synthesis of one-dimensional nano-architectures. This includes synthesis of one-dimensional structure-like nanofibrils as well as nanowire arrays of semiconductors, metals, and polymers, which find important applications in optical and magnetic devices, photonic crystal-based devices, and biocompatible equipments [104–107]. The idea has been extended to fabricate porous structures of TiO_2 by anodizing Ti metal. In 1999, Zwilling et al. successfully developed porous TiO_2 structure by anodizing Ti in fluoride-based acidic electrolytes [108]. Following this, extensive studies on anodization of Ti metal in various electrolytes have been done by several research groups and it has been widely accepted that fluoride ion containing electrolytes result in the formation of highly ordered self-organized nanostructures of TiO_2 [109, 110].

5.5.4.1 Mechanistic Aspect of Formation of TiO_2 Nanotube Arrays

Pourbiax diagram of Ti–H_2O system at 25 °C reveals that TiO_2 layer on Ti has excellent corrosion resistance [111, 112]. Valve metals possess wide passivation region in the range of few tens of volts [113]. When an external voltage is applied, this stable oxide layer thickness increases linearly, and the passivation region increases. The trend shows a deviation when the anodization is carried out in fluoride ion containing electrolytes. This is because the stable oxide layer is attacked by the fluoride ions yielding soluble fluoride complexes. The behavior of fluoride ions in the passivation of Ti metal has been extensively studied [114–116]. It is widely accepted that formation of oxide layer on Ti in the presence of fluoride ion is the result of competition between anodic oxidation and chemical dissolution of the oxide layer, by F^- ions due to the formation of $[TiF_6]^{2-}$ complexes in solution.

The key processes involved in the anodic formation of TiO_2 nanotubes, in a fluoride ion containing electrolyte, are initial oxidation at the metal electrolyte interface due to OH^- and O^{2-} ions, field-assisted oxidation of Ti metal (i.e., migration of oxide ion through oxide layer to the oxide-metal interface and oxidation of the metal resulting in increased oxide layer thickness), and field-assisted dissolution of metal oxide as well as metal, by fluoride ion at the metal oxide/ electrolyte interface and metal/metal oxide interface, respectively. The second process results in a thick oxide layer on Ti surface, whereas the third process results in the dissolution of the metal and metal oxide leading to the formation of nanopores and pits. At the initial stage, the pores formed by reaction with fluoride ions are of irregular shape and produce unequal local current densities. At a later stage the current densities will be shared equally by the pores and they start growing

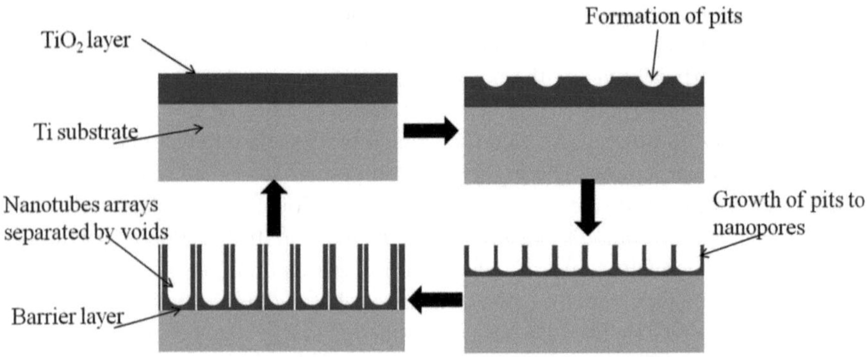

Fig. 5.2 Schematic of the formations steps of self-organized TiO_2 nanotubular structures

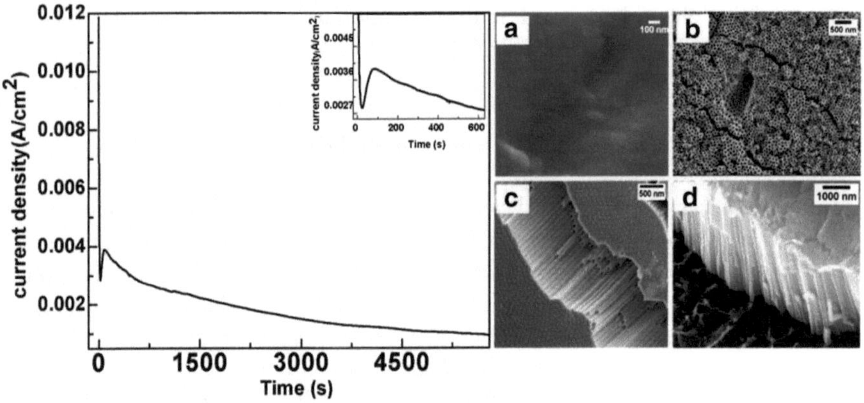

Fig. 5.3 Current density vs. Time curve recorded during electrochemical anodization which qualitatively reveals the formation of TiO_2 nanotube arrays. The FESEM micrographs reveal the (**a**) Ti foil before anodization, (**b**) Ti foil after 2 h of anodization (revealed the formation of TiO_2 nanotube arrays), (**c**) bottom view, and (**d**) the cross-sectional view of TiO_2 nanotube arrays. Figure adapted from authors' own work: Proceedings of International Conference on Nanoscience Engineering and Technology (ICONSET), Chennai, 2011, 326–330

uniformly. This indicates equilibrium between the field-assisted chemical oxidation and field-assisted chemical dissolution. Further, steady state equilibrium leads to growth of ordered porous nanostructures and finally to well-separated self-organized nanotubes. The schematic of the formations steps is depicted in Fig. 5.2. A qualitative picture regarding formation of TiO_2 nanotube arrays during anodization can be obtained from the current density transient recorded in situ. The current density transient and the formation of nanotube arrays are shown in Fig. 5.3a. The aspect ratio of the nanotubes can be varied by varying the process parameters such as applied voltage, time of anodization, electrolyte composition, and temperature [20, 109, 110, 117–119].

Correlating with the current density versus time graph, the mechanism of the nanotube formation can be explained as follows. Initially, when Ti metal is subjected to anodization, a passive oxide layer is formed, according to the reaction:

$$Ti + 2H_2O \rightarrow TiO_2 + 4H^+ + 4e^-. \quad (5.2)$$

The initial sharp decrease in the current density is due to the formation of this barrier oxide layer. By this reaction, H^+ ions are generated at the electrolyte/metal oxide interface. At the same time, in the presence of the H^+ ions, the fluoride ions nearer to the substrate will react with TiO_2 as well as Ti metal to form $[TiF_6]^{2-}$ complex according to the reactions:

$$TiO_2 + 6NH_4F + 4H^+ \rightarrow [TiF_6]^{2-} + 2H_2O + 6NH_4{}^+$$
$$Ti + 6NH_4F \rightarrow [TiF_6]^{2-} + 6NH_4{}^+.$$

Both H^+ and F^- ions enhance the rate of chemical dissolution at the interface resulting in increase in current density. The simultaneous chemical oxidation and chemical dissolution, assisted by the field ($E = V/d$, where V is the applied voltage and d the thickness of the oxide layer), occurring at the active area, will move the metal oxide/metal interface as well as the electrolyte/metal oxide more inside the metal substrate. After some time the current density approaches a constant value signifying steady state/equilibrium between the competitive oxidation and dissolution reactions. Since the anodization is carried out at constant potential, the field generated should be uniform throughout, i.e.; the active area should have equifield surface. The shape acquired by the active area should be hemispherical in order to satisfy the equifield concept. If the active area would have square or any other geometry, the field will not be uniform and will be maximum at the corners thereby increasing the chemical dissolution rate at the corners. This will prevent the continuous growth of nanotubes. The irregularity in the current density observed at the initial stage indicates that the active area is rough and therefore the field is not equivalent everywhere. At steady state the rough surface transforms into an equifield surface (hemispherical shape) because of attainment of equilibrium between the oxidation and dissolution reactions.

The history of TiO_2 nanotube fabrication by anodization technique can be classified into different generations depending upon the selection of process parameters. The first-generation nanotubes were synthesized in HF containing aqueous electrolytes [62, 73, 120]. In this case the lengths of the nanotubes were limited to few hundreds of nanometers due to high chemical dissolution rate [62, 120]. The synthesis of second-generation nanotubes of several micrometer lengths was achieved by replacing HF with fluoride salts such as KF, NaF, and NH_4F and varying the electrolyte pH [121]. But no nanotube formation occurred when the alkaline electrolyte medium was employed [121]. One of the limiting factors of second-generation nanotubes is the formation of unwanted precipitates. In the third-generation nanotubes, ammonium fluoride containing polar organic electrolytes such as formamide or dimethyl sulfoxide with less than 5 wt% of water was employed for anodizing Ti metal. The reduction in water content in the electrolyte decreases the rate of chemical oxidation and dissolution in fluoride containing electrolyte and therefore longer nanotube formation could be achieved [118, 122]. When fluoride ion containing ethylene glycol was used as an electrolyte, smooth-walled long TiO_2 nanotubes were obtained. It was reported that double side anodization of Ti foils resulted in nanotubes having 1 mm length, which is the highest achieved length [122]. Attempts were also made by various research groups, to anodize Ti thin films on different substrates like glass, silicon, quartz, and conducting glasses with aim of using the same in applications such as dye-sensitized solar cells (DSSC). In order to get high aspect ratio titania nanotube arrays (TNTA) on different substrates, the process parameters such as thin film deposition conditions, electrolyte composition, time of anodization, and temperature were varied and investigated [123–125].

These self-aligned nanostructures have limitations in photocatalytic applications such as organic degradation and hydrogen generation where large area exposure is needed for light absorption and reactant adsorption. Therefore, nanotubes in the powder form, which can be dispersed in the medium containing the reactants, are generally preferred. It is a common practice to synthesize self-aligned TiO_2 nanotubes by anodizing Ti in fluoride ion containing electrolytic medium. Extending this synthesis method in chloride ion containing medium, Hahn et al. synthesized nanotube powders of TiO_2 and WO_3 in sodium per chlorate and perchloric acid media [126]. In 2009, Fahim and Sekino synthesized nanotubes of TiO_2 in powder form by electrochemical anodization technique using Ti foils in perchloric acid and sodium chloride/sodium phosphate media and this electrochemical approach was termed as rapid breakdown anodization technique [127]. They obtained X-ray amorphous TiO_2 nanotube powders, which on heating crystallized into anatase phase in a temperature range 300–500 °C. Heating beyond 500 °C resulted in anatase to rutile phase transformation.

In the case of TiO_2 nanotube powders by anodization process the formation mechanism can be explained as follows. During initial stages of anodization, formation of a white layer of titanium dioxide was observed on the surface of the Ti metal, which started breaking down into the electrolytic solution as white powders. The rate of oxide formation is faster in the initial stage. The initial white TiO_2 layer formed on the metal surface started dissolving because of electrochemical etching by chlorine species giving rise to the formation of pits. The subsequent inward migration of oxide ions resulted in the formation of a thick porous layer of TiO_2 leading to reduction in current. In addition, the chlorine species etched the Ti metal causing dissolution of Ti as $[TiCl_6]^{2-}$. The electrochemical etching of Ti metal facilitated the easy breakdown of the oxide layer. Because of the high mechanical stress existing between the metal and metal oxide interface, originating from mismatch in crystal volume, a small disturbance is good enough to remove the metal oxide layer. This disturbance was provided by chemical reactions in the electrolyte, strong bubbling in the solution due to hydrogen evolution at the cathode, and local turbulence arising from magnetic stirring of the solution. The periodic release of oxide layer continued until the complete conversion of Ti metal to TiO_2.

5.6 Photocatalytic and Photovoltaic Applications of TiO_2 Nano-architectures

5.6.1 Photocatalytic Properties

Photocatalysis term is generally used when light energy is used to activate the material, which modifies the rate of a reaction without changing its chemical structure. The process can occur homogenously or heterogeneously. Heterogeneous photocatalysis is a rapidly emerging research field in Green chemistry. The process is termed as advanced oxidation processes (AOP) for the organic destruction, waste water treatment, generation of renewable energy sources, and organic synthesis [128].

5.6.1.1 Mechanism

Heterogeneous photocatalysis mainly involves an interface between the photocatalyst (semiconductors) and the liquid which contains the reactant. When the semiconductor is excited by photons of appropriate energy, electrons are excited from the VB to the CB creating holes and electrons in the VB and CB, respectively. These photogenerated charge carriers migrate to the TiO_2 crystallite surface, where they reduce and oxidize adsorbed electron acceptors and donors by interfacial charge transfer. The electron–hole pair has lifetime of the order of 10^{-9} s to react with the adsorbed species [129]. In TiO_2-based photocatalysis, usually the photogenerated electrons are trapped as Ti^{3+} at the surface and are transferred to the adsorbed molecular oxygen yielding peroxyl radical anions (O_2^{-}). In a suspension, depending on the pH value, these peroxyl radical anions can combine with H^+ ions to form superoxide radicals ($HO_2\cdot$). In the gas phase, the peroxyl radical anions directly attack the organic molecules. The holes participate in organic destruction either by directly oxidizing the organics (before being trapped either within the particle or at the particle surface) or by combining with hydroxyl anions to give hydroxyl radicals. The hydroxyl radicals then degrade the adsorbed organic species.

The probability and rate of the charge transfer processes for electrons and holes depend upon the respective positions of conduction and valence band edges and the redox potential levels of the adsorbate species. The schematic of the photoexcitation and its reaction pathway is shown in Fig. 5.4. For efficient reaction, the potential of the electron acceptor should be positioned below (more positive) the CB of the semiconductor and the potential of the electron donor should be above the VB of the semiconductor [130]. The former case applies to photocatalytic hydrogen generation and the latter applies to organic destruction. Therefore, assuming absence of any kinetic limitations, photo-induced interfacial redox reaction takes place with acceptors and donors whose respective reduction potentials are located between the CB and VB positions of the semiconductor. The photogenerated valence band holes are powerful oxidants (+1 to +3.5 V vs. NHE,

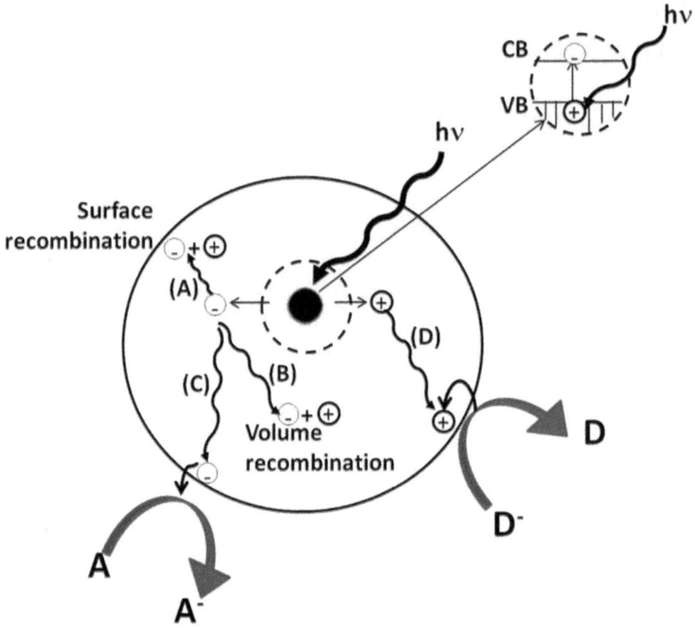

Fig. 5.4 Schematic of Photoexcitation process and its reaction pathway

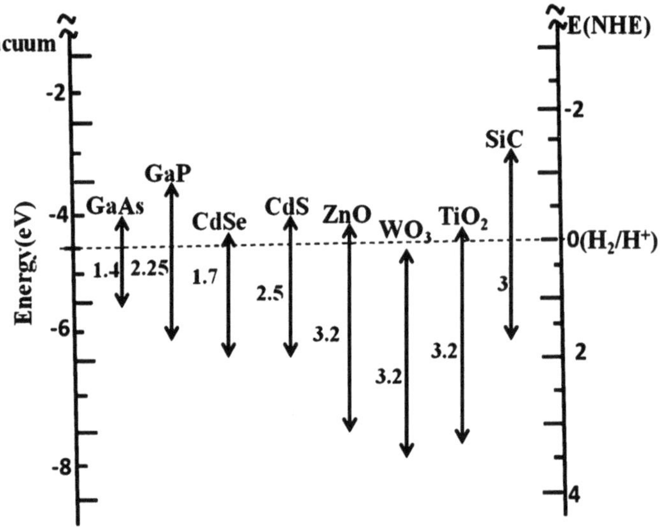

Fig. 5.5 Schematic of CB and VB positions of various semiconductors vs. NHE

depending on the semiconductor), whereas the CB electrons are strong reductants (0.5 to −1.5 V vs. NHE) [131–135]. The CB and VB positions of various semiconductors vs. NHE can be schematically represented as shown in Fig. 5.5.

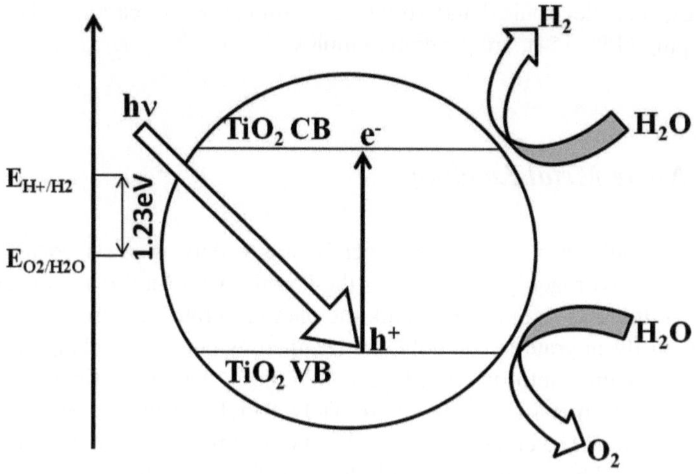

Fig. 5.6 Schematic of the TiO_2 photocatalytic process

Theoretically, all the semiconductors which satisfy the above-mentioned conditions exhibit photocatalytic activity. But the semiconductors such as CdS and SiC undergo photocorrosion and are not suitable for photocatalysis/water splitting applications. Limited numbers of photocatalysts are available which satisfies the criteria, among which TiO_2 is best suited. Ever since the publication of photo-assisted electrolysis of water using TiO_2 photo-electrode, by Honda and Fujishima [136] in 1972, lot of researches on photocatalytic action of TiO_2 have been going on which include water purification, organic destruction, and hydrogen generation. The schematic of the TiO_2 photocatalytic process is represented in Fig. 5.6.

5.6.1.2 Limitations

The photocatalytic efficiency of TiO_2 is limited due to the fast electron–hole recombination and relatively large band gap (3.2 eV), which need photons of energy greater than or equal to UV radiations (which constitutes only ~5 % of total solar energy) for VB to CB electronic excitation. The limitations can be surmounted by surface and electronic band structure modification of the semiconductor such that the charge carrier recombination is inhibited and wavelength response range is increased.

5.7 Modification of Pristine TiO_2

In order to increase the charge separation as well as to tune the band gap to make TiO_2 visible light active, surface modifications by noble metal loading [20–139], dye sensitization [140–143] coupling with lesser band gap semiconductors

[144–146], and electronic band structure modification by cation [147–150], or anion doping [151–154], are generally employed.

5.7.1 Noble Metal Loading

Transition metals, especially noble metals, are widely used as cocatalysts for photocatalytic hydrogen generation. As the Fermi level of noble metals are lower than that of the TiO_2 conduction band, the photogenerated electrons in the CB of TiO_2 can easily migrate to the noble metal, thereby enhancing charge separation and hence efficient photocatalysis. Figure 5.7a shows the schematic of the electron capture by noble metal from the CB of TiO_2, thereby achieving efficient charge carrier separation. The enhancement in the photo conversion of H_2O to H_2 and O_2 was first observed by loading TiO_2 with Pt [138]. Therefore, as of now, it is the most studied system. The metal (Pt) itself is important because of its specific catalytic activity towards hydrogen. Increase in hydrogen production due to platinum loading can be attributed to the trapping of photogenerated electrons on the metal sites yielding efficient charge carrier separation. It has been reported that there is an optimum size and loading for maximum generation of hydrogen [9]. Efficient hydrogen generations have also been reported for other metal-loaded TiO_2 such as Ag/TiO_2 and Au/TiO_2 [139]. In addition to the electron trapping ability of metal particles on TiO_2, noble metal particles, such as gold and silver, have optical properties dominated by their surface plasmon resonance (SPR). When light radiations corresponding to the resonant wave length of the noble metal nanoparticles are incident on the metal particles, the electromagnetic field of the photons couples with the oscillations of conduction electrons in the noble metal particles, resulting in strong-field enhancement of the local electromagnetic fields near the surface of the noble metal nanoparticles. When the semiconductor onto which the noble metal nanoparticles are loaded encounters the strong electric field, e^-/h^+ pairs are generated in the semiconductor in the near surface region. The charge carriers formed

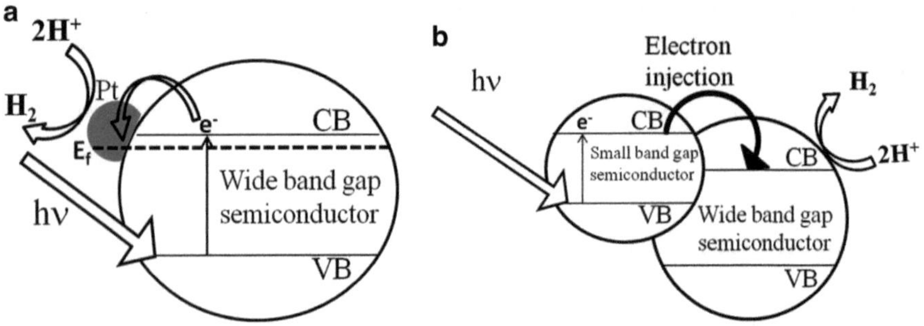

Fig. 5.7 Schematic of (**a**) the electron capture by noble metal from the CB of TiO_2 (**b**) Electron injection and prevention of electron–hole recombination in semiconductor combination

then migrate to the active surface sites more quickly than those generated in the bulk, thereby reducing the recombination rate and enhancing the photocatalytic activity [155, 156].

5.7.2 Semiconductor Coupling

Visible light utilization for the purpose of photocatalysis can be achieved by coupling TiO_2 with other semiconductors of lower band gap with conduction band placed at more negative level compared to that of TiO_2 (Fig. 5.7b). On exposing the couple to visible light, the VB electrons of the small band gap semiconductor are excited to the CB. The excited electrons then move to the CB of TiO_2, thereby reducing electron–hole recombination. Thus, the couple acts as a visible light active photocatalyst with efficient charge carrier separation. The conditions for efficient use of TiO_2-semiconductor composite are; the semiconductor should be resistant to photo-corrosion and should have band gap corresponding to visible light excitation. Figure 5.7b depicts a schematic of electron injection process observed in composite semiconductors.

Among the composite semiconductor systems, CdS/TiO_2 has been the focal point of extensive research activity and has been investigated in detail [157–159]. However, the fabrication of CdS/TiO_2 composite is a challenging task as oxygen intercalation to chalcogenite (CdS) lattice can alter the band gap of the same and deteriorate the photocatalytic activity. In addition it is photo-corrosive. Therefore, alternative small band gap semiconductors such as CuO, Cu_2O, SnO_2, and Fe_2O_3 have been used to couple with TiO_2. Recently, composite of TiO_2/single-walled carbon nanotubes has been reported to show reduced recombination rate, in turn exhibiting enhanced photocatalytic activity [160].

5.7.3 Dye Sensitization

Sensitization by chemisorbed or physisorbed dyes on semiconductors increases the efficiency of the excitation process, thereby enhancing the photocatalytic activity. Under visible light illumination, the dye electrons undergo excitation and the excited electrons are injected to the conduction band of TiO_2 by virtue of the position of the lowest unoccupied molecular orbital (LUMO) and highest occupied molecular orbital (HOMO) of the dye molecule. The electrons in the CB of TiO_2 can be utilized for H_2 production by the decomposition of organic species as well as to send current through an external circuit (Dye-sensitized solar cells). The process occurs only when the oxidative energy level of the excited dye molecule with respect to the conduction band energy level of TiO_2 is higher. Thus, sensitization expands the wavelength range of excitation for the photocatalyst through the sensitizer, followed by the charge transfer to the semiconductor. Figure 5.8 shows

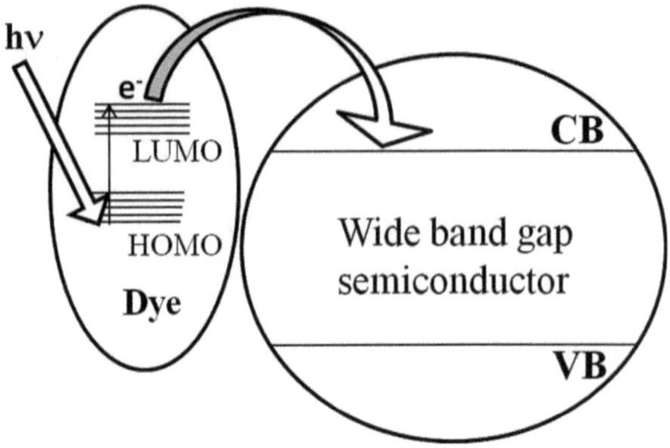

Fig. 5.8 Schematic of Dye sensitization in TiO_2

the schematic of the dye sensitization process. Since electron injection rates differ for various dyes, a variation in photocatalytic rates occurs. This necessitates further studies to understand and compare the dynamics of charge excitation, recombination, and electron injection rates in different dyes.

5.7.4 Cation Doping

Modification of electronic energy band structure by introducing impurity levels in the band gap is another way of making TiO_2 visible light active. This can be easily achieved by metal ion doping. Numerous reports on the modification of the band gap of the materials by cation doping lists a wide range of metals such as Cu, Ni, Au, Cr, Mn, Fe, V [147–150]. Even though it was reported that cation doping decreases the photo threshold energy of TiO_2, several drawbacks such as thermal instability and shorter electron–hole lifetime were observed for the same, thus reducing the overall photocatalytic efficiency [161]. Only a few metal ions like Fe^{3+} and Cu^{2+} were found to inhibit the electron–hole recombination. The concentration of beneficial transition metals dopants needs to be kept low and large concentrations are detrimental. Therefore, the photocatalytic activity is highly sensitive to the cation dopant concentration. In addition, it has been observed that many of the cations do not result in free charge carriers but in bound excitons and have a tendency to generate charge carrier recombination centers, thereby rendering them impractical for photocatalytic applications.

5.7.5 Anion Doping

Another approach to make TiO_2 visible light active is by nonmetal doping. The band gap of the semiconducting material can be narrowed down by introducing donor and acceptor levels below and above the conduction and valence bands, respectively. Doping with anions like C, N, F, B, and S is likely to reduce the recombination centers and shift the absorbance of TiO_2 to visible light region [151–153]. Asahi et al. have studied the substitution effects of N, C, S, F, and P for O in TiO_2,using full potential linearized augmented plane wave method (FLAPW) [162]. They calculated the electronic structures of anatase TiO_2 with the above-mentioned dopants. It was observed that both sulfur and nitrogen create a band gap narrowing in TiO_2. But the size of S is too large to be incorporated into the oxygen sub-lattice of TiO_2. From the results of FLAPW studies,it was found that the mixing of 2p states of nitrogen atom with 2p states of oxygen could shift VB edge upwards leading to narrowing of TiO_2 band gap. Later Chen et al. found that the degree of nitrogen doping is an important factor in enhancing the photocatalytic activity of doped TiO_2 [163]. Since then, N-doping of TiO_2 and its photocatalytic activity has been widely investigated both theoretically and experimentally [163–165]. The synthesis of N-doped TiO_2 can be done by various methods which include both physical and chemical techniques such as vapor deposition, heating with nitrogen precursors and treatment under ammonia gas flow, etc. X-ray photoelectron spectroscopy is the major tool used for the confirmation of nitrogen incorporation into the oxygen sub-lattice. Theoretical work by Valentine et al. revealed that nitrogen can be incorporated into the crystal lattice both interstitially and substitutionally [164]. The model structures for substitutional and interstitial N-doped anatase TiO_2 are depicted in Fig. 5.9.

Reports on the photocatalytic activity of N-doped TiO_2 reveal that both substitutional and interstitial nitrogen doping yield enhanced photocatalytic activity [162, 167–169]. However, most of the researchers believe that substitutional nitrogen doping is the major factor responsible for enhancement in photocatalytic activity. Density functional theory calculations have addressed the variances in the photoactivity measurements, taking both anatase and rutile TiO_2 that have been substitutionally doped with nitrogen [170]. For anatase samples, substitutional N doping results in a decrease in the photon energy necessary to excite the material, whereas for rutile TiO_2 materials, the opposite effect is observed and is attributed to the contraction of the valence band and the stabilization of the N 2p state, thus causing an overall increase in the effective band gap [170, 171] which is in agreement with the work of Diwald et al. [167]. For anatase TiO_2, substitutionally bound nitrogen atoms create localized occupied electronic states (N 2p in character) above the top of the O 2p valence band. The creation of this occupied electronic state has also been shown experimentally [164, 171]. As a result, the mechanism for photoexcitation of N-doped anatase most probably results from direct excitation of

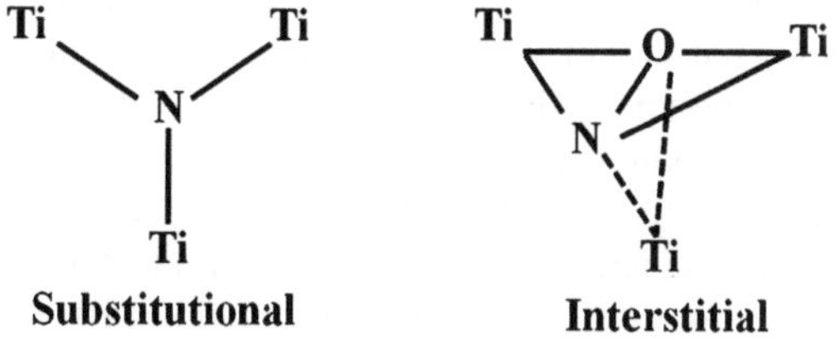

Fig. 5.9 The structure of anatase TiO_2 in which the nitrogen is occupied substitutionally as well as interstitially [166]

electrons from the N 2p state located within the band gap of the TiO_2 into the conduction band. The measured inter-band-gap density of states has been postulated to reduce the rate of electron–hole pair generation, which otherwise readily occurs without the presence of localized N 2p states [164]. For rutile TiO_2 the suppression of the top of the valence band is observed upon N doping [166, 172].

The substitutional and interstitial type impurities in anatase TiO_2 crystal are investigated experimentally (electron paramagnetic resonance spectroscopy and XPS) and theoretically (DFT calculations). Both types of impurities are found to add localized states within the band gap. For substitutional type nitrogen, these states are located 0.14 eV above the valence band, and for interstitial type nitrogen species (referred to as N–O), the localized states are calculated to lie 0.73 eV above the valence band. In addition, these calculations have also found that there is a large decrease in the formation energy of oxygen vacancies. Therefore, oxygen vacancies are most probably induced by N doping of TiO_2 [169].

5.8 Photocatalytic Hydrogen Generation

Presently, the major source of energy for industrial and economic purposes is fossil fuels. Depletion of fossil fuel reserves and pollution caused by combustion of fossil fuels is a major concern. Hydrogen, an effective energy carrier, is considered as a possible alternative to reduce the strong dependence on fossil fuels and environmental pollution. When H_2 becomes the primary energy carrier, the hydrocarbon economy (oil economy) will be replaced by hydrogen economy. Currently 95 % of the H_2 production is from steam reforming of natural gas, which has an indirect dependence on fossil fuels. Hydrogen generation by photo splitting of water, being an environmentally friendly and economically viable process, is being considered as the best alternative. The first report on photocatalytic splitting of water on TiO_2 electrode by Fujishima and Honda created a new paradigm in energy conversion and environmental remediation using solar photons [136]. Currently, various research groups are working on H_2 generation by water splitting using solar energy. Water splitting into H_2 and O_2 is an uphill reaction. It needs the standard Gibb's free energy change (ΔG^0) of 237 kJ/mol or 1.23 V [20], as represented in equation,

$$H_2O \rightarrow \frac{1}{2}O_2 + H_2; \quad \Delta G^0 = +237\,\mathrm{kJ/mol}. \tag{5.3}$$

Therefore, in order to achieve water splitting, the band gap of the semiconducting material should be greater than 1.23 eV. As mentioned previously the reduction and oxidation potential of H_2O should lie within the band gap of the material (Fig. 5.5). Thus, band gap structure of the material is a thermodynamic requirement for photocatalytic hydrogen generation. Life time of the photogenerated electrons, charge separation, and mobility are the other factors which affect the efficiency of photocatalytic hydrogen generation [20]. The processes in the photocatalytic hydrogen generation include absorption of light by semiconductor, generation of charge carriers, separation, migration, and transfer of the charge carriers to water or other molecule. Recombination of charge carriers is a competing process which reduces the hydrogen generation efficiency [9]. Effective charge separation and migration is essential for efficient hydrogen generation. The principle behind photolysis of water using TiO_2 is that, when TiO_2 is in contact with water the photo-excited conduction band electrons can reduce water, to generate hydrogen, if the conduction band level is more negative with respect to the H^+/H_2 reduction potential. Compared to bulk, nano-structured TiO_2 have enhanced photocatalytic properties due to quantum confinement and heightened reactivity associated with their molecular electronic structure and phenomenal increase in surface to volume ratio. When the size of the particles becomes comparable with the de-Broglie wave length of the charge carriers, which lies between 5 and 25 nm for oxide semiconductors like TiO_2, the wave function of charge carriers gets spread over the entire particle and hence they do not need to diffuse anymore to accomplish reactions with species present at the surface, thereby increasing the quantum yield [53]. However, the photocatalytic efficiency of TiO_2 is limited due to the high recombination rate of photo-generated electron–hole pairs. Any other processes which consume the excited electrons should be avoided to increase the efficiency. This can be achieved by surface modification of the semiconducting material. As discussed earlier, noble metal (Pt) loading is the widely adopted method for efficient charge separation and generation of active sites for hydrogen production [138]. The efficient hydrogen generation due to noble metal loading is attributed to the formation of Schottky barriers which enable easy flow of photogenerated electrons from semiconductor to metal [173, 174]. To resolve this issue and enhance the photocatalytic activity, methods like deposition of noble metals and their oxide particles and chemical surface modification have been proposed [175–182]. Noble metals like Pt, Au, and Pd have been reported to be very effective in enhancing the photocatalytic efficiency, when loaded onto TiO_2 [183]. Since the Fermi levels of the noble metals are lower than the TiO_2 conduction band, when loaded onto TiO_2, the photo-excited TiO_2 electrons move to the noble metal Fermi levels leaving the holes in the valence band [184], reducing the electron–hole recombination and thereby enhancing the photocatalytic efficiency. As the number of electrons on the noble metal particles increases, the Fermi level shifts more and more close to the conduction band of the TiO_2 causing more energy levels for electron transfer. The Fermi level of the TiO_2-noble metal composite photocatalyst should be more negative than the H_2 evolution level $\left(E_{H_2/H_2O}\right)$ [185, 186] to enable effective transfer of the accumulated electrons to the protons adsorbed on the noble metal–TiO_2 interfaces and noble metal surfaces. Among the noble metals, Pt is found to be more effective and acts as a catalyst for the hydrogen evolution, by reducing the activation energy for water reduction [187]. Various techniques such as evaporation, irradiation, thermal decomposition, and chemical reduction have been adopted for Pt loading [188–191]. Chemical reduction technique has advantages over other techniques in terms of high yield and narrow particle size distribution, etc. [192].

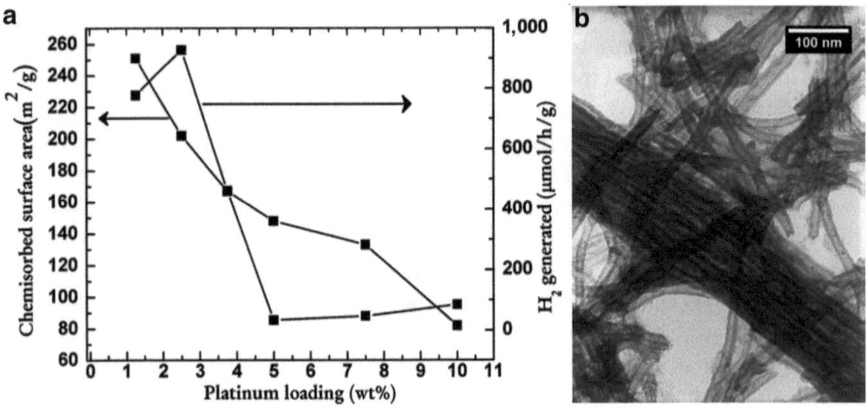

Fig. 5.10 (**a**) The variation of chemisorption surface area and photocatalytic hydrogen generation with different weight percent of Pt loading on TiO_2 and (**b**) the TEM micrograph of the TiO_2 nanotube powders fabricated by electrochemical anodization. Reprinted with permission from [194], Antony et al., *International Journal of Hydrogen Energy* **37** (2012)

Our group has reported the fabrication of TiO_2 nanotube powders by rapid breakdown anodization technique [193] and modified with different weight percent of Pt for making photocatalyst for hydrogen generation [194]. The nanotubes were having anatase phase in the pristine stage and were having a surface area of 154 m^2/g. The photocatalytic hydrogen generation measurement studies in ethanol water mixture (ethanol used as hole scavenger) revealed that TiO_2 with 2 mol% of Pt loading produced 934 μmol/h/g of H_2 under near UV irradiation (380 nm wavelength) and 30 mmol/h/g under UV light irradiation (254 nm). The variation of H_2 generation with change in Pt loading and the corresponding change in the chemisorption surface area is shown as Fig. 5.10a. Figure 5.10b represents the TEM micrograph of the TiO_2 nanotube powders fabricated by electrochemical anodization. The turnover frequency of the Pt-loaded TiO_2 was also observed to be higher for the low weight percent Pt-loaded TiO_2 both under near UV and UV irradiation. But the major drawback with Pt co-catalyst is its high cost.

The challenges in the semiconductor-based photocatalytic hydrogen generation are, synthesis of solar light-absorbing one-dimensional high surface area, long (2–5 μ), thin-walled (~10 nm) tubular TiO_2-based nanostructures [195], replacement of Pt by cost-effective equally competent metal/oxide co-catalyst for effective prevention of recombination, and transfer of electrons to protons. Visible light activity can be achieved by suitable doping of TiO_2. Considering the superior electron mobility and high surface area, Graphene [196, 197] can be utilized as an alternative to the metal co-catalyst, especially Pt, to make the photo catalyst cost-effective and improve photocatalytic hydrogen production activity. For this high quality graphene oxide/graphene are obtained by further exfoliation using sonication, thermal, and microwave-assisted reactions [198]. Cu_2O is found to be another interesting co-catalyst which is reported to enhance the photocatalytic hydrogen generation when loaded onto TiO_2 [199]. Since the potential of

$Cu(OH)_2/Cu$ couple ($Cu(OH)_2 + 2e^- \rightarrow Cu + 2OH^-$, $E^0 = -0.222$ V vs. SHE) is more positive than the CB (−0.260 V) of TiO_2 and higher than the reduction potential of H^+/H_2($E^0 = 0.000$ V, vs. SHE), the excited electrons from CB of TNTs easily transfer to $Cu(OH)_2$ to reduce H^+, hindering the recombination of photo-induced electron/hole pairs. Recently compounds such as C_3N_4, MoS_2 and Graphene have been found to be an efficient co-catalyst for photocatalytic hydrogen generation [200–203].

5.9 Dye-Sensitized Solar Cell

Dye-sensitized solar cells (DSSC) belong to the class of third-generation solar cells which include liquid-based solar cells, Forster resonance energy transfer-based solar cells, and organic solar cells. Since the original report on the fabrication of DSSCs by O'regan and Graetzel [10], DSSCs continued to be an exciting field of research because of its regenerative low cost, ease of fabrication, and possibility of large scale production.

5.9.1 Theory of DSSC

In DSSCs, the electrons in the dyes are excited by light absorption and injected to the conduction band of the semiconducting material (generally TiO_2). The injected electrons then diffuse through the semiconducting matrix and reach the conducting electrode, subsequently to the counter electrode through the external circuit. At the counter electrode (Pt-coated conducting glass), the electrons react with the tri-iodide ions of the electrolyte to generate iodide ions ($I_3^- + 2e^- \leftrightarrow 3I^-$). The iodide ions in contact with the dye supply the electrons to the HOMO of the dye molecules regenerating triodide ions ($3I^- \leftrightarrow I_3^- + 2e^-$). The $I_3^- + 2e^- \leftrightarrow 3I^-$ reaction is catalyzed by the Pt nanoparticles of the counter electrode. Titanium dioxide acts as the photo anode and the nanostructure increases the available surface area for the dye attachment. The light harvesting component of the DSSC is the dye, generally a ruthenium bi-pyridyl dye, adsorbed on TiO_2 [204, 205]. The enhanced visible light absorption and power conversion efficiencies have made DSSCs a strong competitor to solar cell based on other approaches. The schematic of the working principle of the DSSC is depicted in Fig. 5.11.

The efficiency can be calculated from the *I–V* characteristics of the DSSC which is discussed in Chap. 4. The photocurrent produced in a DSSC is directly influenced by the properties of the sensitizer. The energy gap between the HOMO and LUMO of the sensitizer (which corresponds to the band gap, E_g, for inorganic semiconductor materials) determines the photoresponse range of the DSSC. Absorption over a wide range of wavelengths, extending into the near-IR region due to a small HOMO–LUMO energy gap, helps in harvesting a large fraction of the solar

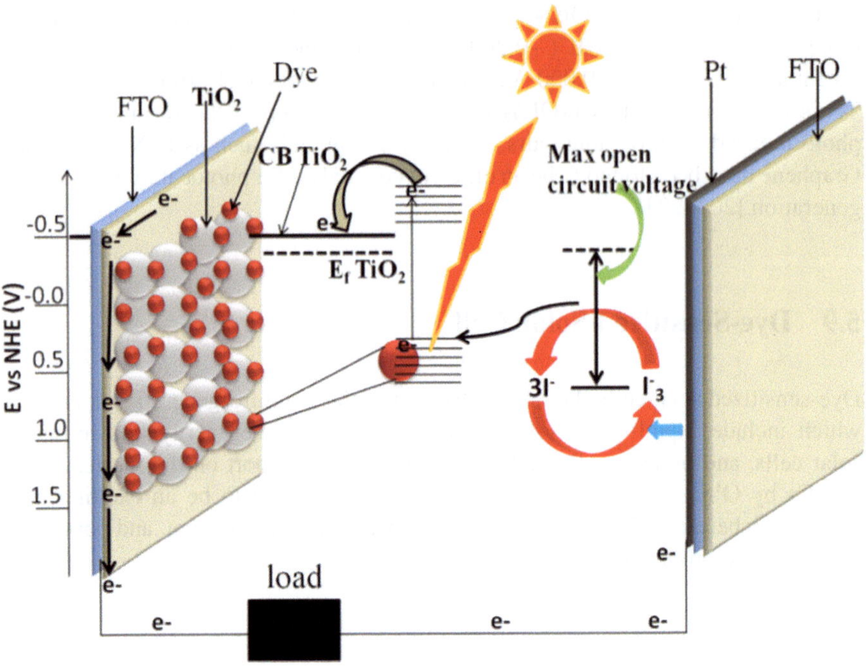

Fig. 5.11 Schematic representation of working principle of Dye-sensitized solar cell

spectrum. This in turn produces large photocurrent which enhances the device performance. In addition, the energy levels of the HOMO and LUMO must match the I_3^-/I^- redox potential and the CB of the TiO_2 electrode (Figs. 5.6 and 5.11). For efficient electron injection, the LUMO must be sufficiently more negative (higher energy) than the TiO_2 CB, which is the driving force for electron injection. Thus, the molecular structure of the sensitizer must be strategically designed so that its properties are optimal for efficient DSSC performance. Traditionally used sensitizing dyes are made from ruthenium-based complexes (N3, N719, C106, and CYB11), which possess broad absorption spectra ($\Delta\lambda = 350$ nm) but low molar extinction coefficient [204, 205]. These are generally packed with a density of 0.5–1 dye molecules per square nanometer. Due to the low molar extinction coefficient of ruthenium-based complexes, great strides have been made during the past decade, to understand and design new dyes. The HOMO of the dye molecule must be sufficiently more positive (lower energy) than the redox potential of $I_3^-/3I^-$ to accept electrons effectively [206, 207]. The open circuit potential of the DSSC corresponds to the difference between the redox potential of the mediator and the Fermi level of TiO_2. The iodide/triodide system has been particularly used in DSSC because of the slow recombination kinetics between electrons in TiO_2 with the oxidized dye and tri iodide in the electrolyte, thereby increasing the electron life time between 1 ms to 1 s [208, 209]. The small size of the I^-/I_3^- redox component allows fast diffusion and provides two electrons per molecule to the counter electrode. Unfortunately the redox system is corrosive and dissolves sealants and contact materials in the long run. Alternative redox system such as Co^{2+}/Co^{3+}- and Fc/Fc^{+}-based electrolytes have found to give good performance [210, 211].

The efficiency η of the DSSC fabricated were calculated, from the *I–V* characteristics, using the equation

$$\eta = \frac{J_{sc} V_{oc} \mathrm{FF}}{P_{int}}, \tag{5.4}$$

where J_{sc} is the short circuit current density, V_{oc} the open circuit potential, FF the fill factor, and P_{int} is the incident light intensity in mW/cm^2. The fill factor is calculated by the equation

$$\mathrm{FF} = \frac{p_{max}}{I_{sc} V_{oc}}, \tag{5.5}$$

where I_{sc} is the short circuit current and P_{max} the maximum power obtained from the *I–V* plot. The incident photon to current conversion efficiency (IPCE) was calculated using the equation

$$\mathrm{IPCE} = \frac{1,239 J_{sc}}{P_{int} \lambda}. \tag{5.6}$$

where λ is the wavelength of the light energy used.

Different forms of TiO_2 nano-architectures have been used as electrodes in third-generation solar cells (dye-sensitized solar cells, DSSCs), which includes quantum well, nanorods, nanowires, nanotubes, and nanoparticles [212–216].

Normally the TiO_2-based photo anodes are prepared by making colloidal suspensions of TiO_2 nanoparticles and doctor blading the same on the conducting glasses. But the efficiencies were limited due to the recombination of electrons due to defects, surface states, and grain boundaries [217]. One among the various steps adopted was to modify the morphology of the titania nanostructures to enhance a better electron transport and minimize the charge transfer resistance [213].

An alternative to nanoparticles can be one-dimensional nanorods of TiO_2 which can give better electron transport properties favorable for DSSCs. Adachi et al. in 2004 reported an efficient highly crystalline anatase TiO_2 nanorod-based photoanode which gave an efficiency of 7.29 % [213]. One-dimensional nanorods in randomly oriented and vertically aligned form have been synthesized for using as photoanode in DSSCs and found to show better efficiency than the nanoparticulate TiO_2 materials.

Vertically aligned nanotube arrays offer distinct advantages such as high surface area for dye adsorption, thereby enhancing light harvesting, readily accessible space for intercalation of redox electrolyte, and easy electron transfer from the dye to the electrode and electrode to the electrical contact due to their vectorial nature [218]. The nanotube arrays are normally synthesized by electrochemical anodization technique which was described earlier in section. However, the DSSCs fabricated using vertically aligned TNTA on Ti foils, as semiconducting electrodes [219, 220], are limited by the need for illumination from the counter electrode side which reduces light intensity. In order to resolve this issue, several research groups synthesized transparent TNTA films by anodizing Ti thin films [124, 221, 222], especially those deposited on FTO-coated conducting glasses. In this case, the Ti films deposited on ITO/FTO-coated glasses are anodized to obtain TNTA thin films. The preparation method is challenging as the deposited Ti thin films are anodized at high positive potentials, which can lead to detachment of Ti film and degradation of the ITO/FTO coatings, thereby reducing the overall conductivity. Several reports discuss these issues and their effect on the photovoltaic performance [125, 223–225].

All these materials in the pristine state show a limited performance due to fast recombination of electron hole pairs. In order to avoid the electron hole recombination and increase the conductivity, different kinds of modification have been adopted by using composite of TiO_2 with other materials such as ZnO, SnO_2, Graphene, N-doped Graphene, CNT, etc. All these materials have shown to increase the DSSC performance by either reducing the recombination rate or improving the electron transfer to the back contact [202, 203, 226–228].

It should be pointed out at this juncture that there are other factors such as the efficient performance of the counter electrodes, electrolyte, and dye that plays a major role in enhancing the efficiency of the dye-sensitized solar cell, which is not mentioned in this section. Researchers are still going on to improve the performance of above-mentioned factors to make efficient DSSCs.

References

1. A.N. Banerjee, The design, fabrication, and photocatalytic utility of nanostructured semiconductors: focus on TiO_2-based nanostructures. Nanotech. Sci. App. **4**, 35–65 (2011)
2. A. Weir, P. Westerhoff, L. Fabricius, K. Hristovski, N. von Goetz, Titanium dioxide nanoparticles in food and personal care products. Environ. Sci. Technol. **46**(4), 2242–2250 (2012)
3. W. Clark, P. Broadhead, Optical absorption and photochromism in iron-doped rutile. J. Phys. C Solid State Phys. **3**(5), 1047 (1970)
4. P. Layman, Outlook brightens for titanium dioxide following recent business pickup. Chem. Eng. News Arch. **74**(4), 13–14 (1996)
5. A.W. Harris, B.P. Ludden, D.C.W. Blaikley, The production of titania thin film waveguides by DC reactive magnetron sputtering. Vacuum **43**(1–2), 143–147 (1992)
6. S.N. Frank, A.J. Bard, Heterogeneous photocatalytic oxidation of cyanide ion in aqueous solutions at titanium dioxide powder. J. Am. Chem. Soc. **99**(1), 303–304 (1977)
7. Y. Ohko, K. Hashimoto, A. Fujishima, Kinetics of photocatalytic reactions under extremely low-intensity UV illumination on titanium dioxide thin films. J. Phys. Chem. A **101**(43), 8057–8062 (1997)
8. T. Sakata, T. Kawai, Heterogeneous photocatalytic production of hydrogen and methane from ethanol and water. Chem. Phys. Lett. **80**(2), 341–344 (1981)

9. A.L. Linsebigler, G. Lu, J.T. Yates, Photocatalysis on TiO2 surfaces: principles, mechanisms, and selected results. Chem. Rev. **95**(3), 735–758 (1995)
10. B. O'Regan, M. Gratzel, A low-cost, high-efficiency solar cell based on dye-sensitized colloidal TiO_2 films. Nature **353**(6346), 737–740 (1991)
11. P. Bonhote, J.E. Moser, N. Vlachopoulos, L. Walder, S.M. Zakeeruddin, R. Humphry-Baker, P. Pechy, M. Gratzel, Photoinduced electron transfer and redox-type photochromism of a TiO_2-anchored molecular diad. Chem. Commun. **10**, 1163–1164 (1996)
12. R. Wang, K. Hashimoto, A. Fujishima, M. Chikuni, E. Kojima, A. Kitamura, M. Shimohigoshi, T. Watanabe, Light-induced amphiphilic surfaces. Nature **388**(6641), 431–432 (1997)
13. I.A. Al-Homoudi, J.S. Thakur, R. Naik, G.W. Auner, G. Newaz, Anatase TiO2 films based CO gas sensor: film thickness, substrate and temperature effects. Appl. Surf. Sci. **253**(21), 8607–8614 (2007)
14. G.K. Mor, M.A. Carvalho, O.K. Varghese, M.V. Pishko, C.A. Grimes, A room-temperature TiO_2-nanotube hydrogen sensor able to self-clean photoactively from environmental contamination. J. Mater. Res. **19**(02), 628–634 (2004)
15. A.R. Bally, K. Prasad, R. Sanjinés, P.E. Schmid, F. Lévy, J. Benoit, C. Barthou, P. Benalloul, TiO2:Ce/CeO2 high performance insulators for thin film electroluminescent devices. MRS Online Proc. Libr. **424**, 471 (1996). doi:10.1557/PROC-424-471
16. M. Gratzel, K. Kalyanasundaram, Curr. Sci. **66**, 706–714 (1994)
17. H. Nishikiori, W. Qian, M.A. El-Sayed, N. Tanaka, T. Fujii, Change in titania structure from amorphousness to crystalline increasing photoinduced electron-transfer rate in dye-titania system. J. Phys. Chem. C **111**(26), 9008–9011 (2007)
18. J.N. Kondo, K. Domen, Crystallization of mesoporous metal oxides. Chem. Mater. **20**(3), 835–847 (2007)
19. U. Diebold, The surface science of titanium dioxide. Surf. Sci. Rep. **48**(5–8), 53–229 (2003)
20. X. Chen, S.S. Mao, Titanium dioxide nanomaterials: synthesis, properties, modifications, and applications. Chem. Rev. **107**(7), 2891–2959 (2007)
21. A. Fujishima, X. Zhang, D.A. Tryk, TiO2 photocatalysis and related surface phenomena. Surf. Sci. Rep. **63**(12), 515–582 (2008)
22. G.R. Patzke, F. Krumeich, R. Nesper, Oxidic nanotubes and nanorods—anisotropic modules for a future nanotechnology. Angew. Chem. Int. Ed. **41**(14), 2446–2461 (2002)
23. H. Tang, K. Prasad, R. Sanjinés, F. Lévy, TiO_2 anatase thin films as gas sensors. Sens. Actuators B **26**(1–3), 71–75 (1995)
24. L.E. Brus, Electron–electron and electron-hole interactions in small semiconductor crystallites: the size dependence of the lowest excited electronic state. J. Chem. Phys. **80**(9), 4403–4409 (1984)
25. A.I.L. Efros, A.L. Efros, Sov. Phys. Semicond. **16**, 772–775 (1982)
26. B. Enright, D. Fitzmaurice, Spectroscopic determination of electron and hole effective masses in a nanocrystalline semiconductor film. J. Phys. Chem. **100**(3), 1027–1035 (1996)
27. A.J. Nozik, R. Memming, Physical chemistry of semiconductor – liquid interfaces. J. Phys. Chem. **100**(31), 13061–13078 (1996)
28. M.A. Grela, M.A. Brusa, A.J. Colussi, Harnessing excess photon energy in photoinduced surface electron transfer between salicylate and illuminated titanium dioxide nanoparticles. J. Phys. Chem. B **101**(51), 10986–10989 (1997)
29. W. Zhou, H. Liu, R.I. Boughton, G. Du, J. Lin, J. Wang, D. Liu, One-dimensional single-crystalline Ti-O based nanostructures: properties, synthesis, modifications and applications. J. Mater. Chem. **20**(29), 5993–6008 (2010)
30. C.-C. Tsai, H. Teng, Structural features of nanotubes synthesized from NaOH treatment on TiO_2 with different post-treatments. Chem. Mater. **18**(2), 367–373 (2005)
31. J. Wang, D.N. Tafen, J.P. Lewis, Z. Hong, A. Manivannan, M. Zhi, M. Li, N. Wu, Origin of photocatalytic activity of nitrogen-doped TiO_2 nanobelts. J. Am. Chem. Soc. **131**(34), 12290–12297 (2009)

32. J. Joo, S.G. Kwon, T. Yu, M. Cho, J. Lee, J. Yoon, T. Hyeon, Large-scale synthesis of TiO_2 nanorods via nonhydrolytic sol–gel ester elimination reaction and their application to photocatalytic inactivation of *E. coli*. J. Phys. Chem. B **109**(32), 15297–15302 (2005)
33. Y. Lin, G.S. Wu, X.Y. Yuan, T. Xie, L.D. Zhang, Fabrication and optical properties of TiO_2 nanowire arrays made by sol–gel electrophoresis deposition into anodic alumina membranes. J. Phys. Condens. Matter **15**(17), 2917 (2003)
34. C. Burda, X. Chen, R. Narayanan, M.A. El-Sayed, Chemistry and properties of nanocrystals of different shapes. Chem. Rev. **105**(4), 1025–1102 (2005)
35. C.-C. Tsai, H. Teng, Regulation of the physical characteristics of titania nanotube aggregates synthesized from hydrothermal treatment. Chem. Mater. **16**(22), 4352–4358 (2004)
36. A. Riss, M.J. Elser, J. Bernardi, O. Diwald, Stability and photoelectronic properties of layered titanate nanostructures. J. Am. Chem. Soc. **131**(17), 6198–6206 (2009)
37. Y. Xia, P. Yang, Y. Sun, Y. Wu, B. Mayers, B. Gates, Y. Yin, F. Kim, H. Yan, One-dimensional nanostructures: synthesis, characterization, and applications. Adv. Mater. **15**(5), 353–389 (2003)
38. M. Adachi, Y. Murata, M. Harada, S. Yoshikawa, Formation of titania nanotubes with high photo-catalytic activity. Chem. Lett. **29**(8), 942–943 (2000)
39. X. Quan, S. Yang, X. Ruan, H. Zhao, Preparation of titania nanotubes and their environmental applications as electrode. Environ. Sci. Technol. **39**(10), 3770–3775 (2005)
40. S. Uchida, R. Chiba, M. Tomiha, N. Masaki, M. Shirai, Electrochemistry **70**, 418–420 (2002)
41. M. Adachi, I. Okada, S. Ngamsinlapasathian, Y. Murata, S. Yoshikawa, Electrochemistry **70**, 449–452 (2002)
42. R.R. Hasiguti, The structure of defects in solids. Ann. Rev. Mater. Sci. **2**(1), 69–92 (1972)
43. D.C. Cronemeyer, Electrical and optical properties of rutile single crystals. Phys. Rev. **87**(5), 876–886 (1952)
44. R.W.G. Wycoff, *The Structure of Crystals* (The Chemical Catalog Company Inc., New York, 1931)
45. V.E. Henrich, The surfaces of metal oxides. Rep. Prog. Phys. **48**(11), 1481 (1985)
46. L. Thiên-Nga, A.T. Paxton, Electronic structure of transition metals adsorbed on the stoichiometric (110) rutile surface. Phys. Rev. B **58**(19), 13233–13241 (1998)
47. M. Landmann, E. Rauls, W.G. Schmidt, The electronic structure and optical response of rutile, anatase and brookite TiO_2. J. Phys. Condens. Matter **24**(19), 195503 (2012)
48. A. Navrotsky, J.C. Jamieson, O.J. Kleppa, Enthalpy of transformation of a high-pressure polymorph of titanium dioxide to the rutile modification. Science **158**(3799), 388–389 (1967)
49. J.C. Jamieson, B. Olinger, Pressure-temperature studies of anatase, brookite, rutile, and TiO_2(II): a discussion. Am. Min. **54**, 1477 (1969)
50. A. Daβler, A. Feltz, J. Jung, W. Ludwig, E. Kaisersberger, Characterization of rutile and anatase powders by thermal analysis. J. Therm. Anal. **33**(3), 803–809 (1988)
51. D.H. Hanaor, C. Sorrell, Review of the anatase to rutile phase transformation. J. Mater. Sci. **46**(4), 855–874 (2011)
52. M. Horn, C.F. Schwebdtfeger, E.P. Meagher, Refinement of the structure of anatase at several temperatures. Z. Krist. **136**, 273 (1972)
53. O. Carp, C.L. Huisman, A. Reller, Photoinduced reactivity of titanium dioxide. Prog. Solid State Chem. **32**(1–2), 33–177 (2004)
54. R. Asahi, Y. Taga, W. Mannstadt, A.J. Freeman, Electronic and optical properties of anatase TiO_2. Phys. Rev. B **61**(11), 7459–7465 (2000)
55. Z.Y. Wu, G. Ouvrard, P. Gressier, C.R. Natoli, Ti and O K edges for titanium oxides by multiple scattering calculations: comparison to XAS and EELS spectra. Phys. Rev. B **55**(16), 10382–10391 (1997)
56. R. Brydson, H. Sauer, W. Engel, J.M. Thomass, E. Zeitler, N. Kosugi, H. Kuroda, Electron energy loss and X-ray absorption spectroscopy of rutile and anatase: a test of structural sensitivity. J. Phys. Condens. Matter **1**(4), 797 (1989)

57. R. Brydson, B.G. Williams, W. Engel, H. Sauer, E. Zeitler, J.M. Thomas, Electron energy-loss spectroscopy (EELS) and the electronic structure of titanium dioxide. Solid State Commun. **64**(4), 609–612 (1987)
58. U. Balachandran, N.G. Eror, Electrical conductivity in non-stoichiometric titanium dioxide at elevated temperatures. J. Mater. Sci. **23**(8), 2676–2682 (1988)
59. T. Sekiya, K. Ichimura, M. Igarashi, S. Kurita, Absorption spectra of anatase TiO2 single crystals heat-treated under oxygen atmosphere. J. Phys. Chem. Solid **61**(8), 1237–1242 (2000)
60. R.R. Hasiguti, E. Yagi, Electrical conductivity below 3 K of slightly reduced oxygen-deficient rutile TiO_2. Phys. Rev. B **49**(11), 7251–7256 (1994)
61. T. Kasuga, M. Hiramatsu, A. Hoson, T. Sekino, K. Niihara, Formation of titanium oxide nanotube. Langmuir **14**(12), 3160–3163 (1998)
62. Y. Zhu, H. Li, Y. Koltypin, Y.R. Hacohen, A. Gedanken, Sonochemical synthesis of titania whiskers and nanotubes. Chem. Commun. (24), 2616–2617 (2001)
63. T. Kasuga, M. Hiramatsu, A. Hoson, T. Sekino, K. Niihara, Titania nanotubes prepared by chemical processing. Adv. Mater. **11**(15), 1307–1311 (1999)
64. J.H. Jung, H. Kobayashi, K.J.C. van Bommel, S. Shinkai, T. Shimizu, Creation of novel helical ribbon and double-layered nanotube TiO_2 structures using an organogel template. Chem. Mater. **14**(4), 1445–1447 (2002)
65. Q. Zhang, L. Gao, Preparation of oxide nanocrystals with tunable morphologies by the moderate hydrothermal method: insights from rutile TiO2. Langmuir **19**(3), 967–971 (2002)
66. X. Feng, K. Shankar, O.K. Varghese, M. Paulose, T.J. Latempa, C.A. Grimes, Vertically aligned single crystal TiO_2 nanowire arrays grown directly on transparent conducting oxide coated glass: synthesis details and applications. Nano Lett. **8**(11), 3781–3786 (2008)
67. X. Feng, J. Zhai, L. Jiang, The fabrication and switchable superhydrophobicity of TiO_2 nanorod films. Angew. Chem. Int. Ed. **44**(32), 5115–5118 (2005)
68. S.K. Pradhan, P.J. Reucroft, F. Yang, A. Dozier, Growth of TiO_2 nanorods by metalorganic chemical vapor deposition. J. Cryst. Growth **256**(1–2), 83–88 (2003)
69. J.-J. Wu, C.-C. Yu, Aligned TiO_2 nanorods and nanowalls. J. Phys. Chem. B **108**(11), 3377–3379 (2004)
70. J.-M. Wu, H.C. Shih, W.-T. Wu, Y.-K. Tseng, I.C. Chen, Thermal evaporation growth and the luminescence property of TiO_2 nanowires. J. Cryst. Growth **281**(2–4), 384–390 (2005)
71. Y. Lei, L.D. Zhang, J.C. Fan, Fabrication, characterization and Raman study of TiO_2 nanowire arrays prepared by anodic oxidative hydrolysis of $TiCl_3$. Chem. Phys. Lett. **338** (4–6), 231–236 (2001)
72. S. Liu, K. Huang, Straightforward fabrication of highly ordered TiO_2 nanowire arrays in AAM on aluminum substrate. Solar Energy Mater. Solar Cell **85**(1), 125–131 (2005)
73. D. Gong, C.A. Grimes, O.K. Varghese, W. Hu, R.S. Singh, Z. Chen, E.C. Dickey, Titanium oxide nanotube arrays prepared by anodic oxidation. J. Mater. Res. **16**(12), 3331–3334 (2001)
74. Y. Bessekhouad, D. Robert, J.V. Weber, Synthesis of photocatalytic TiO_2 nanoparticles: optimization of the preparation conditions. J. Photochem. Photobiol. A Chem. **157**(1), 47–53 (2003)
75. Y. Chen, J.C. Crittenden, S. Hackney, L. Sutter, D.W. Hand, Preparation of a novel TiO_2-based $p-n$ junction nanotube photocatalyst. Environ. Sci. Technol. **39**(5), 1201–1208 (2005)
76. B. Koo, J. Park, Y. Kim, S.-H. Choi, Y.-E. Sung, T. Hyeon, Simultaneous phase- and size-controlled synthesis of TiO_2 nanorods via non-hydrolytic sol–gel reaction of syringe pump delivered precursors. J. Phys. Chem. B **110**(48), 24318–24323 (2006)
77. A. Chemseddine, T. Moritz, Nanostructuring titania: control over nanocrystal structure, size, shape, and organization. Eur. J. Inorg. Chem. **1999**(2), 235–245 (1999)
78. B.B. Lakshmi, P.K. Dorhout, C.R. Martin, Sol–gel template synthesis of semiconductor nanostructures. Chem. Mater. **9**(3), 857–862 (1997)

79. S.J. Limmer, S. Seraji, M.J. Forbess, Y. Wu, T.P. Chou, C. Nguyen, G.Z. Cao, Electrophoretic growth of lead zirconate titanate nanorods. Adv. Mater. **13**(16), 1269–1272 (2001)
80. K. Matsui, T. Kyotani, A. Tomita, Hydrothermal synthesis of single-crystal $Ni(OH)_2$ nanorods in a carbon-coated anodic alumina film. Adv. Mater. **14**(17), 1216–1219 (2002)
81. W. Wang, B. Gu, L. Liang, W.A. Hamilton, D.J. Wesolowski, Synthesis of rutile (α-TiO2) nanocrystals with controlled size and shape by low-temperature hydrolysis: effects of solvent composition. J. Phys. Chem. B **108**(39), 14789–14792 (2004)
82. Y. Li, Y. Fan, Y. Chen, A novel method for preparation of nanocrystalline rutile TiO_2 powders by liquid hydrolysis of $TiCl_4$. J. Mater. Chem. **12**(5), 1387–1390 (2002)
83. C.R. Martin, Nanomaterials: a membrane-based synthetic approach. Science **266**(5193), 1961–1966 (1994)
84. Z. Miao, D. Xu, J. Ouyang, G. Guo, X. Zhao, Y. Tang, Electrochemically induced sol–gel preparation of single-crystalline TiO_2 nanowires. Nano Lett. **2**(7), 717–720 (2002)
85. Z. Liu, Z. Hu, J. Liang, S. Li, Y. Yang, S. Peng, Y. Qian, Size-controlled synthesis and growth mechanism of monodisperse tellurium nanorods by a surfactant-assisted method. Langmuir **20**(1), 214–218 (2003)
86. X. Peng, L. Manna, W. Yang, J. Wickham, E. Scher, A. Kadavanich, A.P. Alivisatos, Shape control of CdSe nanocrystals. Nature **404**(6773), 59–61 (2000)
87. Y. Sun, Y. Xia, Shape-controlled synthesis of gold and silver nanoparticles. Science **298**(5601), 2176–2179 (2002)
88. L. Manna, E.C. Scher, A.P. Alivisatos, Synthesis of soluble and processable rod-, arrow-, teardrop-, and tetrapod-shaped CdSe nanocrystals. J. Am. Chem. Soc. **122**(51), 12700–12706 (2000)
89. L. Manna, E.C. Scher, L.-S. Li, A.P. Alivisatos, Epitaxial growth and photochemical annealing of graded CdS/ZnS shells on colloidal CdSe nanorods. J. Am. Chem. Soc. **124**(24), 7136–7145 (2002)
90. P.D. Cozzoli, A. Kornowski, H. Weller, Low-temperature synthesis of soluble and processable organic-capped anatase TiO_2 nanorods. J. Am. Chem. Soc. **125**(47), 14539–14548 (2003)
91. W. Wang, O.K. Varghese, M. Paulose, C.A. Grimes, Q. Wang, E.C. Dickey, A study on the growth and structure of titania nanotubes. J. Mater. Res. **19**(02), 417–422 (2004)
92. B.D. Yao, Y.F. Chan, X.Y. Zhang, W.F. Zhang, Z.Y. Yang, N. Wang, Formation mechanism of TiO_2 nanotubes. Appl. Phys. Lett. **82**(2), 281–283 (2003)
93. Z.Z. Ye, J.Y. Huang, W.Z. Xu, J. Zhou, Z.L. Wang, Catalyst-free MOCVD growth of aligned ZnO nanotip arrays on silicon substrate with controlled tip shape. Solid State Commun. **141**(8), 464–466 (2007)
94. G. Melcarne, L. De Marco, E. Carlino, F. Martina, M. Manca, R. Cingolani, G. Gigli, G. Ciccarella, Surfactant-free synthesis of pure anatase TiO_2 nanorods suitable for dye-sensitized solar cells. J. Mater. Chem. **20**(34), 7248–7254 (2010)
95. K. Das, S.K. Panda, S. Chaudhuri, Solvent-controlled synthesis of TiO_2 1D nanostructures: growth mechanism and characterization. J. Cryst. Growth **310**(16), 3792–3799 (2008)
96. P.D. Christy, N. Melikechi, N.S. Nirmala Jothi, A.R. Baby Suganthi, P. Sagayaraj, Synthesis of TiO_2 nanorods by oriented attachment using EDTA modifier: a novel approach towards 1D nanostructure development. J. Nanopart. Res. **12**(8), 2875–2882 (2010)
97. M.N. Tahir, P. Theato, P. Oberle, G. Melnyk, S. Faiss, U. Kolb, A. Janshoff, M. Stepputat, W. Tremel, Facile synthesis and characterization of functionalized, monocrystalline rutile TiO_2 nanorods. Langmuir **22**(12), 5209–5212 (2006)
98. T.A. Kandiel, A. Feldhoff, L. Robben, R. Dillert, D.W. Bahnemann, Tailored titanium dioxide nanomaterials: anatase nanoparticles and brookite nanorods as highly active photocatalysts. Chem. Mater. **22**(6), 2050–2060 (2010)
99. Z.-W. Fu, J. Ma, Q.-Z. Qin, Nanostructured $LiCoO_2$ and $LiMn_2O_4$ fibers fabricated by a high frequency electrospinning. Solid State Ion. **176**(17–18), 1635–1640 (2005)

100. D. Li, Y. Xia, Fabrication of titania nanofibers by electrospinning. Nano Lett. **3**(4), 555–560 (2003)
101. D. Hongqin, G. Jian, K. Hakyong, L. Doukrae, A novel method for preparing ultra-fine alumina-borate oxide fibres via an electrospinning technique. Nanotechnology **13**(5), 674 (2002)
102. P. Suresh Kumar, S. Jayaraman, S. Subramanian, J.B. Veluru, G. Singh, S.I. Allakhverdiev, S. Ramakrishna, Hierarchical electrospun nanofibers for energy harvesting, production and environmental remediation. Energy Environ. Sci. **7**(10), 3192–3222 (2014)
103. H. Masuda, K. Fukuda, Ordered metal nanohole arrays made by a two-step replication of honeycomb structures of anodic alumina. Science **268**(5216), 1466–1468 (1995)
104. G.E. Possin, A method for forming very small diameter wires. Rev. Sci. Instrum. **41**(5), 772–774 (1970)
105. D. AlMawlawi, N. Coombs, M. Moskovits, Magnetic properties of Fe deposited into anodic aluminum oxide pores as a function of particle size. J. Appl. Phys. **70**(8), 4421–4425 (1991)
106. S. Kawai, R. Ueda, Magnetic properties of anodic oxide coatings on aluminum containing electrodeposited Co and Co-Ni. J. Electrochem. Soc. **122**(1), 32–36 (1975)
107. C.E. McGarvey, D.A. Holden, Chemistry in adsorbed monolayers. 2. Thermal and photochemical grafting reactions at the polymer-filler interface. Langmuir **6**(6), 1123–1132 (1990)
108. V. Zwilling, E. Darque-Ceretti, A. Boutry-Forveille, D. David, M.Y. Perrin, M. Aucouturier, Structure and physicochemistry of anodic oxide films on titanium and TA6V alloy. Surf. Interface Anal. **27**(7), 629–637 (1999)
109. C.A. Grimes, Synthesis and application of highly ordered arrays of TiO_2 nanotubes. J. Mater. Chem. **17**(15), 1451–1457 (2007)
110. G.K. Mor, O.K. Varghese, M. Paulose, K. Shankar, C.A. Grimes, A review on highly ordered, vertically oriented TiO_2 nanotube arrays: fabrication, material properties, and solar energy applications. Solar Energy Mater. Solar Cell **90**(14), 2011–2075 (2006)
111. J.W. Schultze, M.M. Lohrengel, Stability, reactivity and breakdown of passive films. Problems of recent and future research. Electrochim. Acta **45**(15–16), 2499–2513 (2000)
112. M. Pourbaix, *Atlas d'Equilibres Electrochimiques* (GauthierVillars, Paris, 1963), p. 644
113. V.E. Carter, *Metallic Coatings for Corrosion Control* (Butterworth, Hants, 1977), p. 27
114. A. Caprani, I. Epelboin, P. Morel, Valence de dissolution du titane en milieu sulfurique fluoré. J. Electroanal. Chem. Interfacial Electrochem. **43**(1), A2–A9 (1973)
115. M.J. Mandry, G. Rosenblatt, The effect of fluoride ion on the anodic behavior of titanium in sulfuric acid. J. Electrochem. Soc. **119**(1), 29–33 (1972)
116. J.J. Kelly, The influence of fluoride ions on the passive dissolution of titanium. Electrochim. Acta **24**(12), 1273–1282 (1979)
117. J.M. Macak, H. Tsuchiya, A. Ghicov, K. Yasuda, R. Hahn, S. Bauer, P. Schmuki, TiO_2 nanotubes: self-organized electrochemical formation, properties and applications. Curr. Opin. Solid State Mater. Sci. **11**(1–2), 3–18 (2007)
118. H.E. Prakasam, K. Shankar, M. Paulose, O.K. Varghese, C.A. Grimes, A new benchmark for TiO_2 nanotube array growth by anodization. J. Phys. Chem. C **111**(20), 7235–7241 (2007)
119. R. Narayanan, T.-Y. Kwon, K.-H. Kim, Anodic TiO_2 from stirred Na_2SO_4/NaF electrolytes: effect of applied voltage and stirring. Mater. Lett. **63**(23), 2003–2006 (2009)
120. G.K. Mor, O.K. Varghese, M. Paulose, K.G. Ong, C.A. Grimes, Fabrication of hydrogen sensors with transparent titanium oxide nanotube-array thin films as sensing elements. Thin Solid Films **496**(1), 42–48 (2006)
121. Q. Cai, M. Paulose, O.K. Varghese, C.A. Grimes, The effect of electrolyte composition on the fabrication of self-organized titanium oxide nanotube arrays by anodic oxidation. J. Mater. Res. **20**(01), 230–236 (2005)
122. M. Paulose, K. Shankar, S. Yoriya, H.E. Prakasam, O.K. Varghese, G.K. Mor, T.A. Latempa, A. Fitzgerald, C.A. Grimes, Anodic growth of highly ordered TiO_2 nanotube arrays to 134 μm in length. J. Phys. Chem. B **110**(33), 16179–16184 (2006)

123. G.K. Mor, H.E. Prakasam, O.K. Varghese, K. Shankar, C.A. Grimes, Vertically oriented Ti – Fe – O nanotube array films: toward a useful material architecture for solar spectrum water photoelectrolysis. Nano Lett. **7**(8), 2356–2364 (2007)
124. J.M. Macak, H. Tsuchiya, S. Berger, S. Bauer, S. Fujimoto, P. Schmuki, On wafer TiO_2 nanotube-layer formation by anodization of Ti-films on Si. Chem. Phys. Lett. **428**(4–6), 421–425 (2006)
125. O.K. Varghese, M. Paulose, C.A. Grimes, Long vertically aligned titania nanotubes on transparent conducting oxide for highly efficient solar cells. Nat. Nanotechnol. **4**(9), 592–597 (2009)
126. R. Hahn, J.M. Macak, P. Schmuki, Rapid anodic growth of TiO_2 and WO_3 nanotubes in fluoride free electrolytes. Electrochem. Commun. **9**(5), 947–952 (2007)
127. N.F. Fahim, T. Sekino, A novel method for synthesis of titania nanotube powders using rapid breakdown anodization. Chem. Mater. **21**(9), 1967–1979 (2009)
128. J.-M. Herrmann, Heterogeneous photocatalysis: fundamentals and applications to the removal of various types of aqueous pollutants. Catal. Today **53**(1), 115–129 (1999)
129. Y. Nosaka, M.A. Fox, Kinetics for electron transfer from laser-pulse irradiated colloidal semiconductors to adsorbed methylviologen: dependence of the quantum yield on incident pulse width. J. Phys. Chem. **92**(7), 1893–1897 (1988)
130. M. Ni, M.K.H. Leung, D.Y.C. Leung, K. Sumathy, A review and recent developments in photocatalytic water-splitting using for hydrogen production. Renew. Sustain. Energy Rev. **11**(3), 401–425 (2007)
131. R. Munter, Proc. Estonian Acad. Sci. Chem. **50**, 59–80 (2001)
132. R.A. Greenwald (ed.), *CRC Handbook of Methods for Oxygen Radical Research* (CRC Press, Boca Raton, 1985)
133. J.A. Dean, *Lange's Handbook of Chemistry*, 11th edn. (McGraw-Hill, New York, 1973)
134. M. Gratzel, *Heterogeneous Electron Transfer* (CRC, Boca Raton, FL, 1989)
135. M.R. Hoffmann, S.T. Martin, W. Choi, D.W. Bahnemann, Environmental applications of semiconductor photocatalysis. Chem. Rev. **95**(1), 69–96 (1995)
136. A. Fujishima, K. Honda, Electrochemical photolysis of water at a semiconductor electrode. Nature **238**(5358), 37–38 (1972)
137. G. Wu, T. Chen, G. Zhou, X. Zong, C. Li, H2 production with low CO selectivity from photocatalytic reforming of glucose on metal/TiO_2 catalysts. Sci. China Ser. B Chem. **51**(2), 97–100 (2008)
138. S. Sato, J.M. White, Photodecomposition of water over Pt/TiO_2 catalysts. Chem. Phys. Lett. **72**(1), 83–86 (1980)
139. V. Idakiev, Z.Y. Yuan, T. Tabakova, B.L. Su, Titanium oxide nanotubes as supports of nano-sized gold catalysts for low temperature water-gas shift reaction. Appl. Catal. Gen. **281**(1–2), 149–155 (2005)
140. N. Vlachopoulos, P. Liska, J. Augustynski, M. Graetzel, Very efficient visible light energy harvesting and conversion by spectral sensitization of high surface area polycrystalline titanium dioxide films. J. Am. Chem. Soc. **110**(4), 1216–1220 (1988)
141. P.V. Kamat, M.A. Fox, Photosensitization of TiO_2 colloids by Erythrosin B in acetonitrile. Chem. Phys. Lett. **102**(4), 379–384 (1983)
142. J. Desilvestro, M. Graetzel, L. Kavan, J. Moser, J. Augustynski, Highly efficient sensitization of titanium dioxide. J. Am. Chem. Soc. **107**(10), 2988–2990 (1985)
143. T. Peng, K. Dai, H. Yi, D. Ke, P. Cai, L. Zan, Photosensitization of different ruthenium (II) complex dyes on TiO_2 for photocatalytic H_2 evolution under visible-light. Chem. Phys. Lett. **460**(1–3), 216–219 (2008)
144. J.S. Jang, W. Li, S.H. Oh, J.S. Lee, Fabrication of CdS/TiO_2 nano-bulk composite photocatalysts for hydrogen production from aqueous H2S solution under visible light. Chem. Phys. Lett. **425**(4–6), 278–282 (2006)
145. R. Sasikala, A. Shirole, V. Sudarsan, T. Sakuntala, C. Sudakar, R. Naik, S.R. Bharadwaj, Highly dispersed phase of SnO_2 on TiO_2 nanoparticles synthesized by polyol-mediated route: photocatalytic activity for hydrogen generation. Int. J. Hydrogen Energy **34**(9), 3621–3630 (2009)

146. T. Hirai, K. Suzuki, I. Komasawa, Preparation and photocatalytic properties of composite CdS nanoparticles–titanium dioxide particles. J. Colloid Interface Sci. **244**(2), 262–265 (2001)
147. H. Kato, A. Kudo, Visible-light-response and photocatalytic activities of TiO_2 and $SrTiO_3$ photocatalysts codoped with antimony and chromium. J. Phys. Chem. B **106**(19), 5029–5034 (2002)
148. J.W. Liu, G. Chen, Z.H. Li, Z.G. Zhang, Electronic structure and visible light photocatalysis water splitting property of chromium-doped $SrTiO_3$. J. Solid State Chem. **179**(12), 3704–3708 (2006)
149. D.W. Hwang, H.G. Kim, J.S. Lee, J. Kim, W. Li, S.H. Oh, Photocatalytic hydrogen production from water over M-doped $La_2Ti_2O_7$ (M = Cr, Fe) under visible light irradiation ($\lambda > 420$ nm). J. Phys. Chem. B **109**(6), 2093–2102 (2004)
150. E. Borgarello, J. Kiwi, M. Graetzel, E. Pelizzetti, M. Visca, Visible light induced water cleavage in colloidal solutions of chromium-doped titanium dioxide particles. J. Am. Chem. Soc. **104**(11), 2996–3002 (1982)
151. T. Lindgren, J.M. Mwabora, E. Avendaño, J. Jonsson, A. Hoel, C.-G. Granqvist, S.-E. Lindquist, Photoelectrochemical and optical properties of nitrogen doped titanium dioxide films prepared by reactive DC magnetron sputtering. J. Phys. Chem. B **107**(24), 5709–5716 (2003)
152. X. Chen, C. Burda, The electronic origin of the visible-light absorption properties of C-, N- and S-doped TiO_2 nanomaterials. J. Am. Chem. Soc. **130**(15), 5018–5019 (2008)
153. T. Ohno, T. Mitsui, M. Matsumura, Photocatalytic activity of S-doped TiO_2 photocatalyst under visible light. Chem. Lett. **32**(4), 364–365 (2003)
154. C. Le Paven-Thivet, A. Ishikawa, A. Ziani, L. Le Gendre, M. Yoshida, J. Kubota, F. Tessier, K. Domen, Photoelectrochemical properties of crystalline perovskite lanthanum titanium oxynitride films under visible light. J. Phys. Chem. C **113**(15), 6156–6162 (2009)
155. D.B. Ingram, S. Linic, Water splitting on composite plasmonic-metal/semiconductor photoelectrodes: evidence for selective plasmon-induced formation of charge carriers near the semiconductor surface. J. Am. Chem. Soc. **133**(14), 5202–5205 (2011)
156. L. Du, A. Furube, K. Yamamoto, K. Hara, R. Katoh, M. Tachiya, Plasmon-induced charge separation and recombination dynamics in gold – TiO2 nanoparticle systems: dependence on TiO_2 particle size. J. Phys. Chem. C **113**(16), 6454–6462 (2009)
157. P.A. Sant, P.V. Kamat, Interparticle electron transfer between size-quantized CdS and TiO_2 semiconductor nanoclusters. Phys. Chem. Chem. Phys. **4**(2), 198–203 (2002)
158. D.R. Baker, P.V. Kamat, Photosensitization of TiO_2 nanostructures with CdS quantum dots: particulate versus tubular support architectures. Adv. Funct. Mater. **19**(5), 805–811 (2009)
159. J.E. Evans, K.W. Springer, J.Z. Zhang, Femtosecond studies of interparticle electron transfer in a coupled CdS–TiO_2 colloidal system. J. Chem. Phys. **101**(7), 6222–6225 (1994)
160. Y. Yao, G. Li, S. Ciston, R.M. Lueptow, K.A. Gray, Photoreactive TiO_2/carbon nanotube composites: synthesis and reactivity. Environ. Sci. Technol. **42**(13), 4952–4957 (2008)
161. W. Choi, A. Termin, M.R. Hoffmann, The role of metal ion dopants in quantum-sized TiO_2: correlation between photoreactivity and charge carrier recombination dynamics. J. Phys. Chem. **98**(51), 13669–13679 (1994)
162. R. Asahi, T. Morikawa, T. Ohwaki, K. Aoki, Y. Taga, Visible-light photocatalysis in nitrogen-doped titanium oxides. Science **293**(5528), 269–271 (2001)
163. X. Chen, Y.B. Lou, A.C.S. Samia, C. Burda, J.L. Gole, Formation of oxynitride as the photocatalytic enhancing site in nitrogen-doped titania nanocatalysts: comparison to a commercial nanopowder. Adv. Funct. Mater. **15**(1), 41–49 (2005)
164. R. Nakamura, T. Tanaka, Y. Nakato, Mechanism for visible light responses in anodic photocurrents at N-doped TiO_2 film electrodes. J. Phys. Chem. B **108**(30), 10617–10620 (2004)
165. C. Di Valentin, E. Finazzi, G. Pacchioni, A. Selloni, S. Livraghi, M.C. Paganini, E. Giamello, N-doped TiO_2: theory and experiment. Chem. Phys. **339**(1–3), 44–56 (2007)

166. C.W. Dunnill, I.P. Parkin, Nitrogen-doped TiO_2 thin films: photocatalytic applications for healthcare environments. Dalton Trans. **40**(8), 1635–1640 (2011)
167. O. Diwald, T.L. Thompson, E.G. Goralski, S.D. Walck, J.T. Yates, The effect of nitrogen ion implantation on the photoactivity of TiO_2 rutile single crystals. J. Phys. Chem. B **108**(1), 52–57 (2003)
168. X. Chen, C. Burda, Photoelectron spectroscopic investigation of nitrogen-doped titania nanoparticles. J. Phys. Chem. B **108**(40), 15446–15449 (2004)
169. C. Di Valentin, G. Pacchioni, A. Selloni, S. Livraghi, E. Giamello, Characterization of paramagnetic species in N-doped TiO_2 powders by EPR spectroscopy and DFT calculations. J. Phys. Chem. B **109**(23), 11414–11419 (2005)
170. C. Di Valentin, G. Pacchioni, A. Selloni, Origin of the different photoactivity of doped anatase and rutile TiO_2. Phys. Rev. B **70**(8), 085116 (2004)
171. M. Miyauchi, A. Ikezawa, H. Tobimatsu, H. Irie, K. Hashimoto, Zeta potential and photocatalytic activity of nitrogen doped TiO_2 thin films. Phys. Chem. Chem. Phys. **6**(4), 865–870 (2004)
172. T.L. Thompson, J.T. Yates, Surface science studies of the photoactivation of TiO_2 new photochemical processes. Chem. Rev. **106**(10), 4428–4453 (2006)
173. A.V. Korzhak, N.I. Ermokhina, A.L. Stroyuk, V.K. Bukhtiyarov, A.E. Raevskaya, V.I. Litvin, S.Y. Kuchmiy, V.G. Ilyin, P.A. Manorik, Photocatalytic hydrogen evolution over mesoporous TiO_2/metal nanocomposites. J. Photochem. Photobiol. A Chem. **198**(2–3), 126–134 (2008)
174. K. Gurunathan, Photocatalytic hydrogen production using transition metal ions-doped γ-Bi_2O_3 semiconductor particles. Int. J. Hydrogen Energy **29**(9), 933–940 (2004)
175. G.R. Bamwenda, S. Tsubota, T. Nakamura, M. Haruta, Photoassisted hydrogen production from a water-ethanol solution: a comparison of activities of Au/TiO_2 and Pt/TiO_2. J. Photochem. Photobiol. A Chem. **89**(2), 177–189 (1995)
176. T. Sreethawong, S. Yoshikawa, Enhanced photocatalytic hydrogen evolution over Pt supported on mesoporous prepared by single-step sol–gel process with surfactant template. Int. J. Hydrogen Energy **31**(6), 786–796 (2006)
177. K. Wilke, H.D. Breuer, The influence of transition metal doping on the physical and photocatalytic properties of titania. J. Photochem. Photobiol. A Chem. **121**(1), 49–53 (1999)
178. S.U.M. Khan, M. Al-Shahry, W.B. Ingler, Efficient photochemical water splitting by a chemically modified n-TiO_2. Science **297**(5590), 2243–2245 (2002)
179. H. Park, W. Choi, Effects of TiO2 surface fluorination on photocatalytic reactions and photoelectrochemical behaviors. J. Phys. Chem. B **108**(13), 4086–4093 (2004)
180. W. Choi, Pure and modified TiO_2 photocatalysts and their environmental applications. Catal. Surv. Asia **10**(1), 16–28 (2006)
181. Y. Li, G. Lu, S. Li, Photocatalytic transformation of rhodamine B and its effect on hydrogen evolution over Pt/TiO_2 in the presence of electron donors. J. Photochem. Photobiol. A Chem. **152**(1–3), 219–228 (2002)
182. T. Tachikawa, S. Tojo, M. Fujitsuka, T. Majima, Influence of metal ions on the charge recombination processes during TiO_2 photocatalytic one-electron oxidation reactions. J. Phys. Chem. B **108**(30), 11054–11061 (2004)
183. M.K. Seery, R. George, P. Floris, S.C. Pillai, Silver doped titanium dioxide nanomaterials for enhanced visible light photocatalysis. J. Photochem. Photobiol. A Chem. **189**(2–3), 258–263 (2007)
184. R. Priya, K.V. Baiju, S. Shukla, S. Biju, M.L.P. Reddy, K.R. Patil, K.G.K. Warrier, Enhanced solar-radiation induced photocatalytic activity of surface-modified nanocrystalline anatase-titania. Catal. Lett. **128**(1–2), 137–143 (2009)
185. E. Kowalska, O.O.P. Mahaney, R. Abe, B. Ohtani, Visible-light-induced photocatalysis through surface plasmon excitation of gold on titania surfaces. Phys. Chem. Chem. Phys. **12**(10), 2344–2355 (2010)

186. V. Subramanian, E.E. Wolf, P.V. Kamat, Green emission to probe photoinduced charging events in ZnO – Au nanoparticles. Charge distribution and fermi-level equilibration. J. Phys. Chem. B **107**(30), 7479–7485 (2003)
187. R. Baba, S. Nakabayashi, A. Fujishima, K. Honda, Investigation of the mechanism of hydrogen evolution during photocatalytic water decomposition on metal-loaded semiconductor powders. J. Phys. Chem. **89**(10), 1902–1905 (1985)
188. M. Jakob, H. Levanon, P.V. Kamat, Charge distribution between UV-irradiated TiO2 and gold nanoparticles: determination of shift in the fermi level. Nano Lett. **3**(3), 353–358 (2003)
189. C.G. Granqvist, R.A. Buhrman, Log-normal size distributions of ultrafine metal particles. Solid State Commun. **18**(1), 123–126 (1976)
190. J. Charvolin, C. Froidevaux, C. Taupin, J.M. Winter, NMR in minute metallic particles. Solid State Commun. **4**(7), 357–359 (1966)
191. J.E. Benson, M. Boudart, Hydrogen-oxygen titration method for the measurement of supported platinum surface areas. J. Catal. **4**(6), 704–710 (1965)
192. P.R. Van Rheenen, M.J. McKelvy, W.S. Glaunsinger, Synthesis and characterization of small platinum particles formed by the chemical reduction of chloroplatinic acid. J. Solid State Chem. **67**(1), 151–169 (1987)
193. R.P. Antony, T. Mathews, A. Dasgupta, S. Dash, A.K. Tyagi, B. Raj, Rapid breakdown anodization technique for the synthesis of high aspect ratio and high surface area anatase TiO_2 nanotube powders. J. Solid State Chem. **184**(3), 624–632 (2011)
194. R.P. Antony, T. Mathews, C. Ramesh, N. Murugesan, A. Dasgupta, S. Dhara, S. Dash, A.K. Tyagi, Efficient photocatalytic hydrogen generation by Pt modified TiO_2 nanotubes fabricated by rapid breakdown anodization. Int. J. Hydrogen Energy **37**(10), 8268–8276 (2012)
195. H.-H. Ou, S.-L. Lo, Review of titania nanotubes synthesized via the hydrothermal treatment: fabrication, modification, and application. Sep. Purif. Technol. **58**(1), 179–191 (2007)
196. K.S. Novoselov, A.K. Geim, S.V. Morozov, D. Jiang, M.I. Katsnelson, I.V. Grigorieva, S.V. Dubonos, A.A. Firsov, Two-dimensional gas of massless Dirac fermions in graphene. Nature **438**(7065), 197–200 (2005)
197. Q. Xiang, J. Yu, Graphene-based photocatalysts for hydrogen generation. J. Phys. Chem. Lett. **4**(5), 753–759 (2013)
198. Y. Zhu, S. Murali, W. Cai, X. Li, J.W. Suk, J.R. Potts, R.S. Ruoff, Graphene and graphene oxide: synthesis, properties, and applications. Adv. Mater. **22**(35), 3906–3924 (2010)
199. S. Xu, D.D. Sun, Significant improvement of photocatalytic hydrogen generation rate over TiO_2 with deposited CuO. Int. J. Hydrogen Energy **34**(15), 6096–6104 (2009)
200. Y. Zang, L. Li, Y. Xu, Y. Zuo, G. Li, Hybridization of brookite TiO_2 with g-C_3N_4- visible-light-driven photocatalyst for As^{3+} oxidation, MO degradation and water splitting for hydrogen evolution. J. Mater. Chem. A **2**, 15774–15780 (2014)
201. Q. Xiang, J. Yu, M. Jaroniec, Synergetic effect of MoS_2 and graphene as cocatalysts for enhanced photocatalytic H_2 production activity of TiO_2 nanoparticles. J. Am. Chem. Soc. **134** (15), 6575–6578 (2012)
202. K. Woan, G. Pyrgiotakis, W. Sigmund, Photocatalytic carbon-nanotube–TiO_2 composites. Adv. Mater. **21**(21), 2233–2239 (2009)
203. Y. Liang, H. Wang, H. Sanchez Casalongue, Z. Chen, H. Dai, TiO_2 nanocrystals grown on graphene as advanced photocatalytic hybrid materials. Nano Res. **3**(10), 701–705 (2010)
204. N. Robertson, Optimizing dyes for dye-sensitized solar cells. Angew. Chem. Int. Ed. **45**(15), 2338–2345 (2006)
205. A. Mishra, M.K.R. Fischer, P. Bäuerle, Metal-free organic dyes for dye-sensitized solar cells: from structure: property relationships to design rules. Angew. Chem. Int. Ed. **48**(14), 2474–2499 (2009)
206. G. Oskam, B.V. Bergeron, G.J. Meyer, P.C. Searson, Pseudohalogens for dye-sensitized TiO2 photoelectrochemical cells. J. Phys. Chem. B **105**(29), 6867–6873 (2001)

207. H. Nusbaumer, J.-E. Moser, S.M. Zakeeruddin, M.K. Nazeeruddin, M. Grätzel, CoII(dbbip)$_2$$^{2+}$ complex rivals tri-iodide/iodide redox mediator in dye-sensitized photovoltaic cells. J. Phys. Chem. B **105**(43), 10461–10464 (2001)
208. S.Y. Huang, G. Schlichthörl, A.J. Nozik, M. Grätzel, A.J. Frank, Charge recombination in dye-sensitized nanocrystalline TiO_2 solar cells. J. Phys. Chem. B **101**(14), 2576–2582 (1997)
209. G. Boschloo, A. Hagfeldt, Characteristics of the iodide/triiodide redox mediator in dye-sensitized solar cells. Acc. Chem. Res. **42**(11), 1819–1826 (2009)
210. Z. Zhang, P. Chen, T.N. Murakami, S.M. Zakeeruddin, M. Grätzel, The 2,2,6,6-tetramethyl-1-piperidinyloxy radical: an efficient, iodine- free redox mediator for dye-sensitized solar cells. Adv. Funct. Mater. **18**(2), 341–346 (2008)
211. M. Wang, N. Chamberland, L. Breau, J.-E. Moser, R. Humphry-Baker, B. Marsan, S.M. Zakeeruddin, M. Grätzel, An organic redox electrolyte to rival triiodide/iodide in dye-sensitized solar cells. Nat. Chem. **2**(5), 385–389 (2010)
212. Y.-L. Lee, Y.-S. Lo, Highly efficient quantum-dot-sensitized solar cell based on co-sensitization of CdS/CdSe. Adv. Funct. Mater. **19**(4), 604–609 (2009)
213. M. Adachi, Y. Murata, J. Takao, J. Jiu, M. Sakamoto, F. Wang, Highly efficient dye-sensitized solar cells with a titania thin-film electrode composed of a network structure of single-crystal-like TiO_2 nanowires made by the "oriented attachment" mechanism. J. Am. Chem. Soc. **126**(45), 14943–14949 (2004)
214. B. Tan, Y. Wu, Dye-sensitized solar cells based on anatase TiO_2 nanoparticle/nanowire composites. J. Phys. Chem. B **110**(32), 15932–15938 (2006)
215. G.K. Mor, K. Shankar, M. Paulose, O.K. Varghese, C.A. Grimes, Use of highly-ordered TiO_2 nanotube arrays in dye-sensitized solar cells. Nano Lett. **6**(2), 215–218 (2005)
216. S. Muduli, W. Lee, V. Dhas, S. Mujawar, M. Dubey, K. Vijayamohanan, S.-H. Han, S. Ogale, Enhanced conversion efficiency in dye-sensitized solar cells based on hydrothermally synthesized $TiO_2 - $MWCNT nanocomposites. ACS Appl. Mater. Interfaces **1**(9), 2030–2035 (2009)
217. Y.-W. Jun, M.F. Casula, J.-H. Sim, S.Y. Kim, J. Cheon, A.P. Alivisatos, Surfactant-assisted elimination of a high energy facet as a means of controlling the shapes of TiO_2 nanocrystals. J. Am. Chem. Soc. **125**(51), 15981–15985 (2003)
218. P. Maggie, S. Karthik, K.V. Oomman, K.M. Gopal, A.G. Craig, Application of highly-ordered TiO_2 nanotube-arrays in heterojunction dye-sensitized solar cells. J. Phys. D Appl. Phys. **39**(12), 2498 (2006)
219. S. Karthik, K.M. Gopal, E.P. Haripriya, Y. Sorachon, P. Maggie, K.V. Oomman, A.G. Craig, Highly-ordered TiO_2 nanotube arrays up to 220 μm in length: use in water photoelectrolysis and dye-sensitized solar cells. Nanotechnology **18**(6), 065707 (2007)
220. K. Shankar, J. Bandara, M. Paulose, H. Wietasch, O.K. Varghese, G.K. Mor, T.J. LaTempa, M. Thelakkat, C.A. Grimes, Highly efficient solar cells using TiO_2 nanotube arrays sensitized with a donor-antenna dye. Nano Lett. **8**(6), 1654–1659 (2008)
221. G.K. Mor, O.K. Varghese, M. Paulose, C.A. Grimes, Transparent highly ordered TiO_2 nanotube arrays via anodization of titanium thin films. Adv. Funct. Mater. **15**(8), 1291–1296 (2005)
222. H. Zheng, A.Z. Sadek, M. Breedon, D. Yao, K. Latham, J.D. Plessis, K. Kalantar-Zadeh, Fast formation of thick and transparent titania nanotubular films from sputtered Ti. Electrochem. Commun. **11**(6), 1308–1311 (2009)
223. J.Y. Kim, J.H. Noh, K. Zhu, A.F. Halverson, N.R. Neale, S. Park, K.S. Hong, A.J. Frank, General strategy for fabricating transparent TiO_2 nanotube arrays for dye-sensitized photoelectrodes: illumination geometry and transport properties. ACS Nano **5**(4), 2647–2656 (2011)
224. A.Z. Sadek, H. Zheng, K. Latham, W. Wlodarski, K. Kalantar-zadeh, Anodization of Ti thin film deposited on ITO. Langmuir **25**(1), 509–514 (2008)
225. J. Weickert, C. Palumbiny, M. Nedelcu, T. Bein, L. Schmidt-Mende, Controlled growth of TiO_2 nanotubes on conducting glass. Chem. Mater. **23**(2), 155–162 (2010)

226. K. Vinodgopal, I. Bedja, P.V. Kamat, Nanostructured semiconductor films for photocatalysis. Photoelectrochemical behavior of SnO_2/TiO_2 composite systems and its role in photocatalytic degradation of a textile Azo dye. Chem. Mater. **8**(8), 2180–2187 (1996)
227. H.Y. Yang, S.F. Yu, S.P. Lau, X. Zhang, D.D. Sun, G. Jun, Direct growth of ZnO nanocrystals onto the surface of porous TiO_2 nanotube arrays for highly efficient and recyclable photocatalysts. Small **5**(20), 2260–2264 (2009)
228. I.V. Kostedt, A.A. Ismail, D.W. Mazyck, Impact of heat treatment and composition of ZnO–TiO_2 nanoparticles for photocatalytic oxidation of an Azo dye. Ind. Eng. Chem. Res. **47**(5), 1483–1487 (2008)

Chapter 6
State-of-the-Art Thin Film Electrolytes for Solid Oxide Fuel Cells

Manjula I. Nandasiri and Suntharampillai Thevuthasan

Abstract State-of-the-art solid oxide fuel cells (SOFC) are among the main candidates for clean energy technology due to their high efficiency, fuel flexibility, low air pollution, and minimal greenhouse gas emission. However, high operational temperature of SOFC is a greater challenge in commercialization of these devices for low cost and portable applications. High temperature operation of SOFC degrades its performance with aging, limits the selection of materials for fuel cell components, and increases the fabrication cost. Thus, there have been enormous efforts to improve the properties of existing materials and develop new materials for SOFC components in order to lower the operating temperature of SOFC. Recent advances in thin film technology have also been utilized to develop new materials with improved properties for SOFC. One of the key components in SOFC is the electrolyte and several research groups are working on developing new electrolyte materials. In this chapter, we will discuss the recent advances in thin film SOFC electrolytes. This extensive discussion includes the evolution of doped ceria, doped zirconia, and multilayer hetero-structured thin film electrolytes. The newly developed nanoscale thin films and multilayer hetero-structures with improved oxygen ionic conductivity will have significant impact on SOFC devices.

6.1 Introduction

6.1.1 Background of Solid Oxide Fuel Cells (SOFC)

The latest U.S. energy reviews revealed that more than 80 % of the energy consumed in the world is still produced by non-renewable energy sources such as petroleum, natural gas, and coal [1]. Moreover, it is projected that world energy consumption will grow by 56 % from 2010 to 2040. Therefore, from a long-term energy perspective, clean energy technologies are needed to utilize primary energy

M.I. Nandasiri (✉) • S. Thevuthasan
Environmental Molecular Sciences Laboratory (EMSL), Pacific Northwest National Laboratory (PNNL), Richland, WA 99354, USA
e-mail: manjula@pnnl.gov

S. Babu Krishna Moorthy (ed.), *Thin Film Structures in Energy Applications*,
DOI 10.1007/978-3-319-14774-1_6

sources in a sustainable and an efficient manner. Clean energy technologies include a variety of devices, services, and processes that harness clean energy sources and renewable materials; reduce use of fossil fuels; and minimize greenhouse gas emissions and wastes. State-of-the-art solid oxide fuel cell (SOFC) is considered as a critical clean energy technology due to their high efficiency, multi-fuel capability, long life time, and reduced greenhouse gas emissions [2–6]. SOFC is one of the main fuel cell types, which can convert chemical energy into electricity using multiple fuels including hydrogen. They have been used in a wide range of power generating applications from portable devices to distributed power plants [6].

In SOFC, a dense solid state electrolyte is sandwiched between the porous cathode and anode. The cathode and anode are also called the air and fuel electrode, respectively. A typical SOFC consists of yttria-stabilized zirconia (YSZ) electrolyte, Ni-YSZ anode, and Sr-doped $LaMnO_3$ cathode. When the cathode is exposed to air, oxygen molecules migrate through the porous electrode and acquire electrons from the cathode to form oxygen ions in the cathode, which can be transported to the cathode/electrolyte interface. The dense electrolyte does not allow the migration of oxygen gas through it. However, the oxygen ions diffuse through the electrolyte and react with the fuel (hydrogen) at the electrolyte/anode interface producing electrons, water, and heat. The electrons released at the electrolyte/anode interface travel from anode to cathode through an external circuit to produce electricity. The schematic diagram shown in Fig. 6.1 summarizes the operation of SOFC indicating

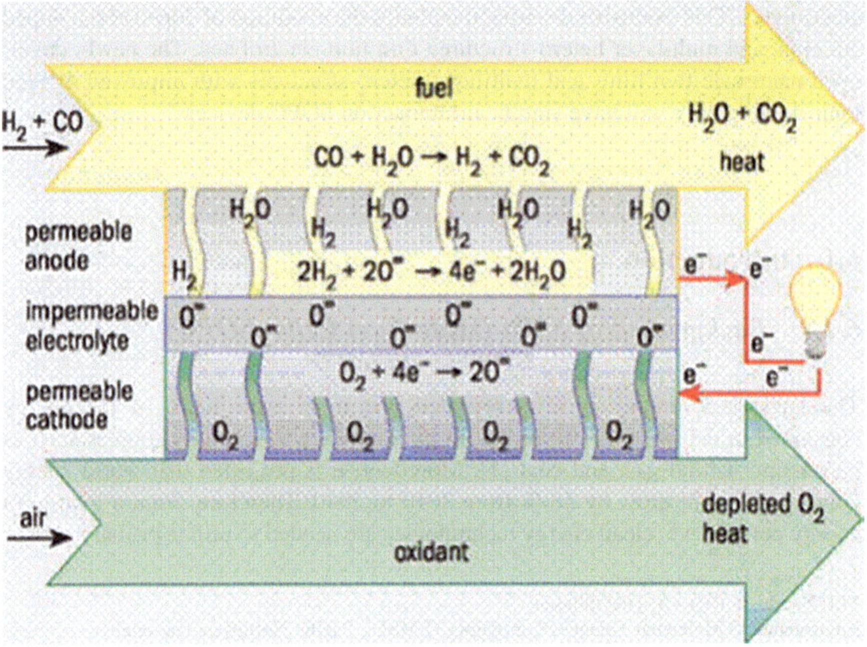

Fig. 6.1 The schematic diagram of the solid oxide fuel cell illustrating the operating principle and electrochemical reactions. Reprinted with permission from [7], Subhash C. Singhal, *Electrochemical Society Interface* **41** (2007). © 2007, Electrochemical society

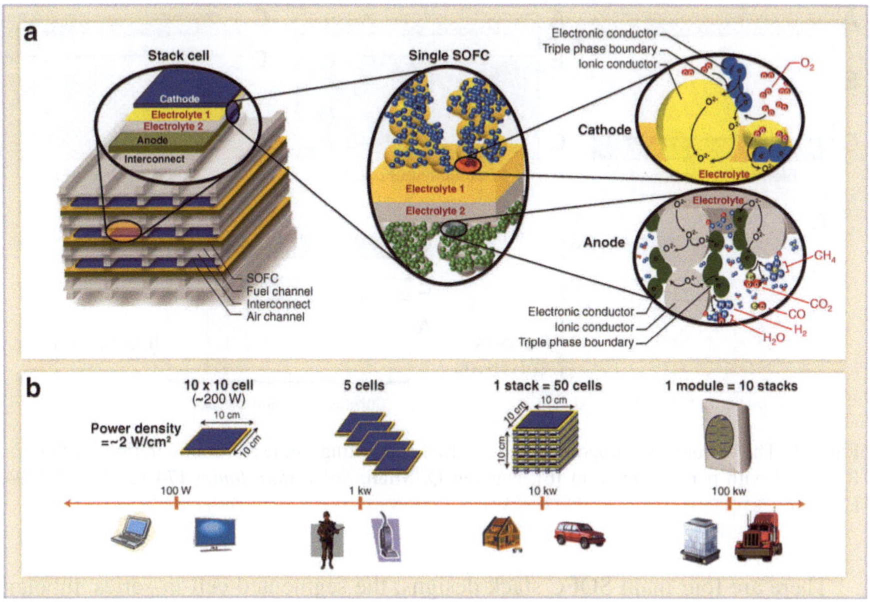

Fig. 6.2 The schematic diagrams of (**a**) low temperature SOFC stack and (**b**) estimated power density of SOFC from a single cell to a module with their respective applications. Reprinted with permission from [8], Eric D. Wachsman et al., *Science* **334** (2011). © 2011, Association for the advancement of Science

the reactions occurring at the cathode/electrolyte and electrolyte/anode interfaces [7].

For a successful and long-term operation of SOFCs, the materials used for different SOFC components should have suitable electrical properties, high chemical and structural stability, minimal inter-diffusion and reactivity between each components, and comparable thermal expansion coefficients [3]. Especially, the electrolyte material should have high ionic conductivity to transport oxygen ions and negligible electronic conductivity to avoid the short circuit of SOFC. In contrast to electrolytes, the cathode and anode materials should exhibit high electronic conductivity. The voltage output of a single cell is only 0.7 V. Thus, they have been arranged as a stack of multiple cells in series as shown in Fig. 6.2 [8], which are connected by interconnects to get a high voltage output [4, 9]. Figure 6.2 also illustrates the power densities of different sizes of cell stacks and their respective applications [8]. The selection of interconnect materials mainly depends on the operating temperature and stack design of SOFC. The interconnect materials should also have high electronic conductivity. Thus, conductive ceramic materials and metallic alloys have been used as the interconnects in high temperature and intermediate temperature SOFC, respectively. Doped lanthanum and yttrium chromite perovskites are the most commonly used ceramic materials for interconnects at high temperatures.

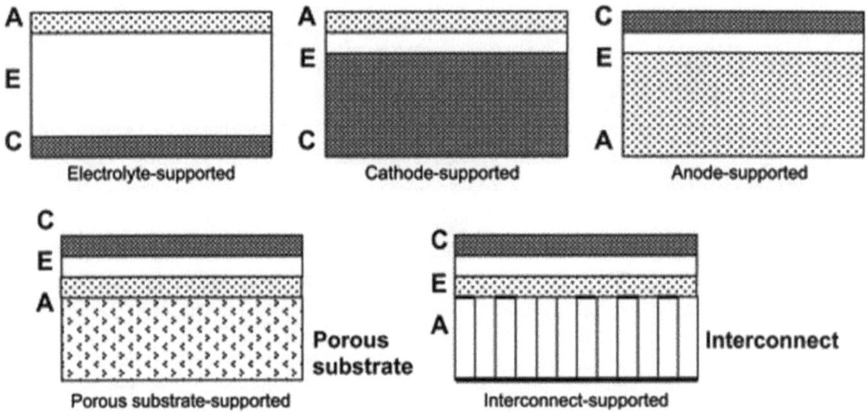

Fig. 6.3 The various self-supporting and external-supporting single cell configurations of SOFCs. Reprinted with permission from [6], Nguyen Q. Minh, *Solid State Ionics*,**174** (2004). © 2004, Elsevier

There are four main SOFC stack designs: the segmented-cell-in-series, tubular, monolithic, and planar [2, 6]. Among these main stack designs, tubular and planar are the two most commonly used designs [5]. In the tubular design, the cell is configured as a tube and the stack is composed of a bundle of single cell tubes. This is the most advanced SOFC stack design; however, low power densities and high fabrication cost are major drawbacks of tubular SOFCs [2]. In the planar SOFCs, the single cell is configured as thin flat plates in rectangular or circular shape, which are electrically connected in series. This is the most widely used and the cheapest stack design, which offers high power densities [2, 5]. However, sealing materials are needed in planar design in contrast to tabular design [10]. In SOFCs, a single cell can be manufactured in various configurations based on the specific stack design, which fall into two main categories: self-supporting and external-supporting. In the self-supporting configuration, the electrolyte, or cathode or anode acts as the supporting structure of the cell as shown in Fig. 6.3 [6]. However, the anode-supported cell is the most widely used design in SOFC. Figure 6.4 shows a cross-sectional SEM micrograph of the Ni-GDC anode-supported SOFC with a 3 μm thick GDC electrolyte and $La_{0.8}Sr_{0.2}Co_{0.8}Fe_{0.2}O_3$ (LSCF) cathode [11]. In the external-supporting configuration, thin layers of the electrolyte, cathode, and anode are supported on the interconnect or a porous substrate.

Despite all the advantages, high operational temperature of SOFC is a greater challenge in selection of materials for fuel cell components [12]. It is also related to the long-term stability and cost of the SOFC. The stringent requirements on materials for high temperature SOFC [2] led to enormous attempts in decreasing the operational temperature. However at lower temperatures, there is a significant increase in the cell resistance which degrades the SOFC performance. In order to lower the operational temperature of SOFC, we need to enhance the oxygen ionic conductivity of existing electrolytes at low temperatures, which allows the use of

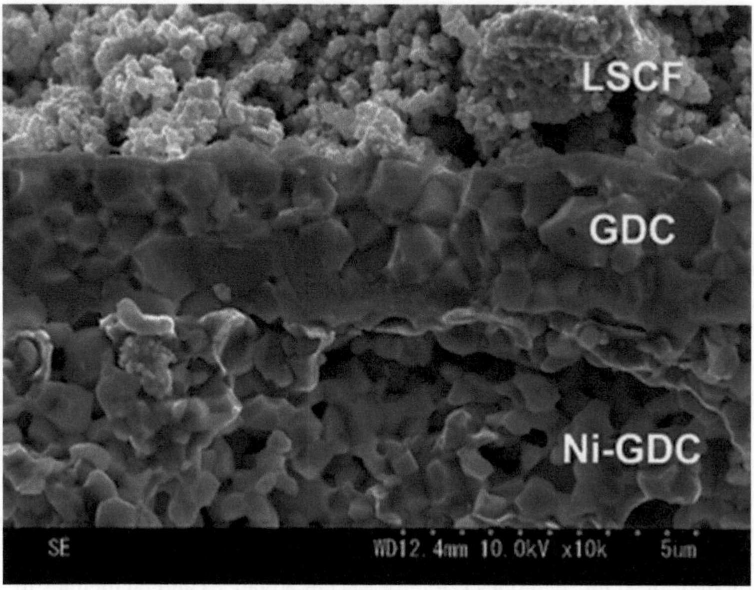

Fig. 6.4 The cross-sectional view of the microstructure of an anode supported single SOFC with thin $Gd_{0.1}Ce_{0.9}O_{1.95}$ (GDC) electrolyte film of 3 μm in thickness, Ni-GDC anode, and nanostructured $La_{0.8}Sr_{0.2}Co_{0.8}Fe_{0.2}O_3$ (LSCF) cathode. Reprinted with permission from [11], Changsheng Ding et al., *Energy and Environmental Science* **3** (2010). © 2010, Royal Society of Chemistry [11]

inexpensive materials for cell components [10]. Furthermore, it will increase the efficiency and lifetime of SOFC [9]. The high temperature operation also increases the polarization resistance losses of the electrodes. Thus, it is really important to optimize the existing materials or develop new materials for SOFC components. The development of new materials for low temperature SOFC may also introduce additional challenges such as compatibility issues between different components. Hence, the continuous engineering and optimization of SOFC materials is required to address these challenges.

6.1.2 *Role of Thin Films in State-of-the-Art SOFC*

There has been a considerable interest in thin film fabrication and characterization of SOFC materials to study and understand their fundamental properties. The decrease in thickness of different SOFC components results in improved performance at low temperatures, which is especially utilized in micro-SOFCs [8, 13–15]. The epitaxial thin films [16] and multilayer hetero-structured thin films [17, 18] are also of major interests in SOFC research community. Thus, the advanced thin film technologies have been employed to improve the properties of materials at lower temperatures for SOFC components including anode, cathode, electrolyte, and interconnects. These thin film SOFC components have been fabricated by various deposition techniques including physical vapor deposition, chemical vapor deposition, spray deposition, spin coating, and electrophoretic deposition. A comprehensive review of these thin film deposition techniques can be found elsewhere [13].

In the SOFC cathode, oxygen is reduced to oxygen ions using the electrons externally transported from anode and these oxygen ions migrate through the electrolyte. Thus, cathode materials should have high electrical conductivity, chemical and thermal stability, and compatibility with the electrolyte. Sr-doped $LaMnO_3$ (LSM) is the most common SOFC cathode material due to high electronic conductivity, high catalytic activity for the O_2 reduction, good thermal and structural stability, and chemical compatibility with YSZ electrolyte [19]. However, LSM is not a suitable cathode material for intermediate temperature SOFC due to the increase in polarization resistance for the O_2 reduction reaction [20]. Thus, YSZ and GDC have also been added to LSM cathode to reduce the polarization resistance at lower temperatures [21]. The materials in the $LaCoO_3$ and $LaFeO_3$ families such as Sr-doped $LaCoO_3$ (LSCO) [22–24], $LaFeO_3$ (LSF) [25–27], and $LaFe_{0.8}Co_{0.2}O_3$ (LSCF) [20, 28] have also used as alternative cathode materials. Ba- and Sr-doped cobalt ferrite (BSCF) is another alternative material recently investigated for SOFC cathodes [29, 30]. Recent advances and challenges in SOFC cathode materials have been extensively reviewed elsewhere [31–34].

Thin films of LSCO [35–39], LSCF [40–42], and BSCF [43, 44] materials have also fabricated for SOFC cathodes to improve their electrical, structural, and chemical properties. LSCO perovskite thin films with thickness of tens to hundreds of nanometers have been fabricated by sputtering [35, 36] and pulsed laser deposition (PLD) [37–39] on YSZ substrates. Furthermore, the LSCO thin films grown by PLD were highly (001)-oriented and showed significant enhancement in the surface oxygen kinetics [37]. The mechanism and kinetics of oxygen reduction in thin dense two-dimensional LSCF cathodes have been studied theoretically and experimentally by Prestat and the group [41, 42]. LSCF films with thicknesses ranging from 16 to 766 nm were grown on gadolinia-doped ceria substrates by PLD, which confirmed the enhanced electrochemical performance of LSCF ultrathin films [41]. A reaction model for oxygen reduction of these thin films was also developed, which showed parallel bulk and surface pathway reaction mechanisms [42]. PLD-grown BSCF thin film cathodes with ~100 nm thickness showed extremely low surface exchange resistance, which is significantly higher than that of LSCF [43, 44]. However, the research on thin film cathodes mostly devoted to model cathodes to study the reaction mechanisms [13].

Similar to cathode materials, anode materials should also have high electrical conductivity, structural and thermal compatibility, and chemical compatibility with the electrolyte. Ni/YSZ is the typical anode material used for SOFC due to its excellent chemical properties and low cost. Ni acts as an excellent electrocatalyst for electrochemical oxidation of hydrogen with high electronic conductivity, while YSZ helps to inhibit the coarsening of Ni powders [45]. YSZ also exhibits significant ionic conductivity and plays an important role in the creation of additional reaction sites in the cathode [31]. However, sulfur poisoning and carbon deposition are two major drawbacks of Ni/YSZ materials [46]. Thus, new anode materials with high tolerance toward sulfur poisoning and carbon deposition need to be developed and this effort is mainly focused on perovskite-based oxide materials [47–51]. The various strategies have also been attempted to improve the electrochemical properties of Ni/YSZ anode including nano-structured materials [31, 46] and thin films [13, 14, 16]. However, thin film anode materials have mostly been used to study the reaction mechanisms as model systems [52, 53] and hardly used in current-state SOFCs [16].

The oxygen ionic conductivity of SOFC electrolytes can be enhanced at lower temperatures by either reducing the electrolyte thickness of existing YSZ electrolytes or developing alternative electrolytes with high ionic conductivity. In order to maintain the power density of high temperature SOFCs at intermediate and lower temperatures, the thickness of electrolyte must be reduced from a couple of hundred microns to several microns by synthesizing them as thin films. For an example, in order to lower the operating temperature from 950 to 500 °C of a SOFC with open circuit voltage of 1 V and power density of 1 W cm^{-2} while maintaining the same area-specific resistance, the YSZ electrolyte thickness needs to be reduced from 150 μm to <1 μm [54, 55]. Thus, thin film electrolytes of doped zirconia and ceria-based materials have been fabricated by various deposition techniques for SOFC applications and fundamental understanding of oxygen ion conductance [13, 16]. The oxygen ionic conductivity of thin films can be further enhanced by optimizing the dopant concentration and structural properties. Thus, single crystal thin film electrolytes with optimum dopant concentration have shown significant enhancement in the ionic conductivity compared to polycrystalline thin films [56–58]. The enhanced ionic conductivity in single crystal thin films was attributed to the fast oxygen ion diffusion through ordered oxygen vacancies. The ionic conductivity of thin film electrolytes was significantly improved by the fabrication of multilayer hetero-structures with engineered nano-scale interfaces [17, 59, 60]. The enhanced interfacial conductivity of these multilayers due to the strained layers and extended defects at the interfaces increased the total ionic conductivity. Therefore, multilayer thin film electrolytes have shown the maximum ionic conductivity reported up to date [60–62]. In the following sections of this chapter, the evolution of thin film SOFC electrolytes will be discussed in detail including both experimental and theoretical aspects. Those sections will be devoted to doped zirconia, doped ceria, and multilayer hetero-structures-based thin film electrolytes.

6.2 Zirconia-Based Thin Film Electrolytes

Zirconia (ZrO_2)-based materials have been extensively studied as electrolytes in SOFC's, due to their high ionic conductivity at elevated temperatures, low electronic conductivity, chemical stability, and high mechanical strength [63, 64]. ZrO_2 exists in three different phases, namely monoclinic, tetragonal, and cubic. Monoclinic zirconia is the stable form at room temperature and up to 1,000 °C; however, it has low electrical conductivity [65]. The phase transformations from monoclinic to tetragonal and tetragonal to cubic occur at 1,170 and 2,370 °C, respectively [63]. However, high temperature cubic and tetragonal zirconia can be stabilized at low temperatures, with the addition of metal oxides like CaO, MgO, Sc_2O_3, and Y_2O_3 [64]. Various types of rare-earth oxides have been also introduced to stabilize the cubic ZrO_2 structure. The choice of dopant depends on the cation radius and structure [66], where those metal oxides should have a cubic structure with the cation radius similar to that of Zr^{4+} ion. The substitution of Zr^{4+} ions by these metal cations led to oxygen vacancy formation in the cubic ZrO_2 lattice, increasing the ionic conductivity. It is reported that the ionic conductivity of doped zirconia materials strongly depends on the dopant concentration. The dopant Sc^{3+} exhibits the highest ionic conductivity, which has the closest ionic radius to that of the Zr^{4+} [67]. The dopant concentration needed to fully stabilize the cubic zirconia structure is 7–9 mol% for Y_2O_3 and Sc_2O_3, 12–13 mol% for CaO, and 8–12 mol% for other rare-earth oxides [64]. The tetragonal phase of ZrO_2 can be stabilized with low dopant concentration (2.5–3 mol% of Y_2O_3 and Sc_2O_3) compared to cubic structure and has high mechanical strength and toughness.

Among stabilized cubic zirconia-based systems, YSZ is the most widely used and extensively studied SOFC electrolyte due to its high ionic conductivity, compatibility with electrodes, and chemical stability in both oxidizing and reducing atmospheres. However, the ionic conductivity of YSZ decreases at lower temperatures [63, 68]. Also, the most commonly used 8 mol% yttria-doped zirconia (8YSZ) electrolyte exhibits degradation in conductivity after high temperature annealing [69, 70]. We can overcome this issue by increasing the yttrium concentration [63], but there is an optimum yttrium concentration for the highest conductivity of YSZ. The conductivity of YSZ increases with the yttrium content up to 7–8 mol% Y_2O_3 and decreases with higher Y_2O_3 concentration [68, 71]. This optimum concentration value is also the minimum Y_2O_3 concentration needed to stabilize cubic zirconia [63, 64]. At higher dopant concentrations, association of point defects leads to a reduction in defect mobility, which in turn reduces the conductivity [72]. The ionic conductivity of YSZ also depends on the fabrication method [73], sintering temperature [74], and grain size of the polycrystalline materials [75]. The oxygen ionic conductivity of YSZ can be further improved by growing as single crystal or nanocrystalline materials [76–78].

Enhancement in the ionic conductivity is reported for scandia-stabilized zirconia (ScSZ), but the phase changes at high temperatures are a challenge for using them in SOFC. On the other hand, ScSZ has chemical and thermo-mechanical properties very similar to 8YSZ. Enhanced ionic conductivity of ScSZ is about twice that of YSZ [79], which is due to the smaller mismatch of ionic radii between Zr^{4+} and Sc^{3+} with respect to that of Zr^{4+} and Y^{3+}, which leads to increased defect mobility [80–82]. The phase diagram of ScSZ indicates seven different phases including monoclinic, tetragonal, cubic, metastable tetragonal-rich t', and three rhombohedral phases. ScSZ contains monoclinic phase below 3 mol% Sc_2O_3. It shows tetragonal structure from 3 to 9 mol% of Sc_2O_3 concentration and tetragonality decreases with the increase in Sc_2O_3 concentration [83]. In this region, monoclinic and cubic phases were also detected at lower and higher ends, respectively. Above 9 mol% of Sc_2O_3, it is stable with cubic structure up to 11 mol% of Sc_2O_3. However, an ordered beta-phase ($Sc_2Zr_7O_{17}$) has been recorded for Sc_2O_3 concentration of 11–13 mol% [83, 84].

The highest conductivity of ScSZ was observed at about 8–9 mol% of Sc_2O_3 [67, 83, 85–88]. However, there is an inconsistency in optimum concentration values of Sc_2O_3 for maximum conductivity and phase changes, reported by various researchers. It can be attributed to the difference in sample preparation methods and deposition conditions. As reported by Badwal et al., ScSZ shows the highest conductivity and stability at 9.3 mol% of Sc_2O_3, which was stable with single phase cubic structure [85]. In this study, they have investigated the conductivity of ScSZ by increasing the Sc_2O_3 concentration from 7 to 11 mol%. For the specimens with or below 9.5 mol% of Sc_2O_3, conductivity increases with the temperature; however, both 10 and 11 mol% Sc_2O_3-stabilized zirconia samples showed abrupt decrease in the conductivity below 500–600 °C. It might be associated with the presence of substantial amounts of the beta-phase in these compositions. According to Lee et al., 8 mol% scandia-stabilized zirconia (8 ScSZ) shows the maximum conductivity and the conductivity decreases with the Sc_2O_3 content beyond that limit [87]. However, they have observed both tetragonal and cubic phases of 8ScSZ in contrast to previous studies, where it contained only cubic phase [67, 83, 88]. Recent studies by Huang et al. also confirmed the 8ScSZ as the optimum composition for the maximum ionic conductivity of ScSZ with the cubic structure [86]. They have also identified the phase changes from monoclinic to rhombohedral with the increase in Sc_2O_3 content using X-ray diffraction. These studies suggested that 8 ScSZ is a strong candidate to replace 8YSZ as a SOFC electrolyte.

6.2.1 *Yttria-Stabilized Zirconia*

Even though the YSZ with optimum dopant concentration showed required oxygen ionic conductivity for SOFC only at temperatures higher than 800 °C, still it is the most commonly used SOFC electrolyte due to excellent chemical and mechanical properties. In order to achieve the same area-specific resistance (ASR) obtained for 150 μm thick YSZ electrolyte at 950 °C at lower temperatures (500 °C), the thickness of electrolyte needs to be reduced below 1 μm [55]. Thus, the researchers have employed various techniques to deposit YSZ thin films and enhance their ionic conductivity at low temperatures [77, 89–94].

Kosacki et al. have studied the nanoscale effects on the ionic conductivity of YSZ electrolyte by varying the film thickness from 2 μm to 15 nm [77, 92]. They deposited highly (100)-textured YSZ thin films on MgO(100) substrates by pulsed laser deposition (PLD). The films with thickness ranging from 2 μm to 60 nm showed similar conductivity values in the temperature range of 400–800 °C as shown in Fig. 6.5 [77]. Furthermore, those conductivities were comparable with that of single crystal YSZ. The film with thickness of 29 nm showed about one order

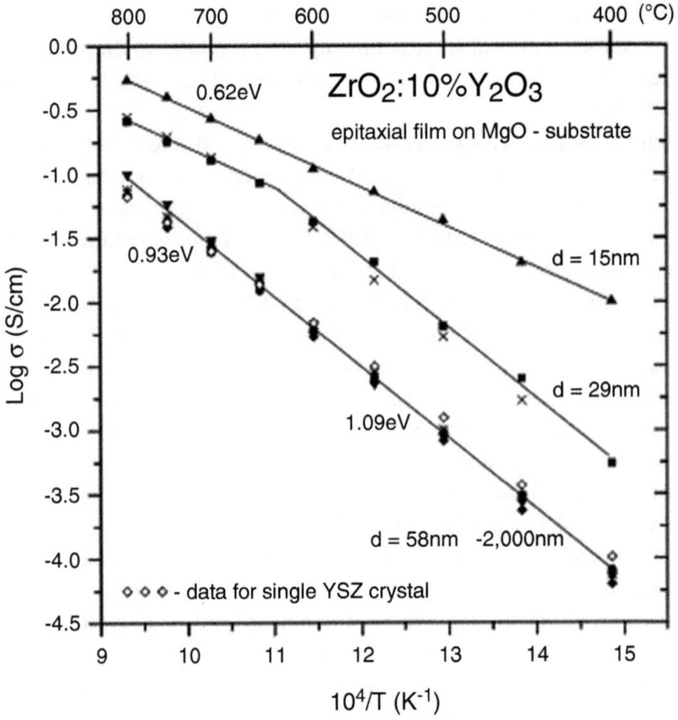

Fig. 6.5 The Arrhenius plots of nanoscale YSZ thin films with thickness ranging from 2,000 to 15 nm. Reprinted with permission from [77], Igor Kosacki et al., *Solid State Ionics*,**176** (2005). © 2005, Elsevier

of magnitude enhancement in the ionic conductivity compared to that of a 60 nm thick film below 650 °C. The 15 nm thick film showed the maximum ionic conductivity of 0.6 S cm^{-1} at 800 °C with an activation energy of ~0.62 eV [77, 92]. The enhancement in the ionic conductivity of these ultra-thin YSZ films was attributed to the increased interfacial conductivity at the MgO/YSZ interface. However, there is a need to address the substrate contribution to the total resistance of the sample when the film thickness is in nanometer scale, especially in the in-plane conductivity measurement configuration [77, 95]. Guo et al. has also studied the nanoscale effects and role of the film/substrate interface on the ionic conductivity of polycrystalline 8 mol% yttria-doped zirconia thin films [90]. They have deposited YSZ films with thicknesses of 12 and 25 nm on MgO(100) substrates using PLD and the transmission electron microscopy (TEM) images confirmed the homogeneity and polycrystalline nature of the films. The high resolution TEM (HRTEM) images also revealed the misfit dislocations at the film/substrate interface, which was atomically non-flat and free of any secondary phases. The average grain size of the films was in the range of ~76 to 88 nm. However, the total conductivity of these polycrystalline nanoscale YSZ thin films was lower than that of 8YSZ bulk ceramic by about a factor of 4 in contrast to epitaxial YSZ thin films.

It was also observed that the conductivity is independent of the film thickness for the tested values. Heiroth et al. have also observed no significant enhancement in the ionic conductivity of polycrystalline YSZ thin films grown by PLD [91]. These observations further confirmed that the epitaxial films in the nanoscale can further enhance the ionic conductivity in contrast to polycrystalline thin films.

Sillassen and co-workers have reported superionic conductivities in the order of 1 S cm^{-1} at 500 °C for ~58 nm thick epitaxial YSZ films deposited on MgO by DC magnetron sputtering [94]. These conductivity values were about one order of magnitude higher than those obtained by Kosacki et al. for 15 nm thick YSZ films. They deposited YSZ thin films on MgO and STO substrates by varying the thickness and observed three orders of magnitude enhancement in the oxygen ionic conductivity at low temperatures (<350 °C) compared to bulk YSZ. At the temperatures above 350 °C, the conductivities of these thin films were similar to that of bulk YSZ. The HRTEM images showed a disordered YSZ/MgO interface in contrast to continuous and flat YSZ/STO interface. The misfit dislocations were also observed at the YSZ/MgO interface, which relaxed the lattice stress induced by large lattice mismatch between YSZ and MgO. The ionic transport along the dislocation lines in the semi-coherent YSZ/MgO interface is much faster than bulk diffusion, which resulted in superionic conductivities at low temperatures. This study further showed that the in-plane ionic conductivity of epitaxial YSZ films in the nanoscale can be tailored by tuning the film thickness and lattice mismatch between the film and substrate.

YSZ thin films in nanometer scale have also been grown by atomic layer deposition (ALD) [89, 93]. Shim et al. have deposited ~50 to 150 nm thick YSZ films on Si_3N_4-buffered Si(100) substrates [93]. The X-ray photoelectron spectroscopy (XPS) depth profile confirmed the uniform elemental depth distributions of the films with 7 atom% yttrium oxide. According to the X-ray diffraction (XRD) measurements, these YSZ thin films are polycrystalline. They also tested the performance of fuel cells with a ~60 nm thick YSZ layer and two porous platinum electrodes. The maximum power density of the best performing cell was found to be 270 mW cm^{-2} at 350 °C. The fuel cells with ALD-grown YSZ electrolytes showed significant increase in the maximum power density compared to that of fuel cells with RF-sputtered ~50 nm thick YSZ electrolytes, which showed a power density of 200 mW cm^{-2} at 350 °C [93].

Realistic computer simulations are needed to augment experimental observations regarding the dopant concentration-dependent ionic conductivity and aging effects of YSZ electrolyte [96]. Thus, computer simulation studies have been carried out to understand the defect structure, oxygen vacancy distribution, and activation energy for oxygen ion diffusion in stabilized zirconia [96–100]. Devanathan et al. have carried out molecular dynamics (MD) simulations to study the defect energetics and oxygen ion diffusion in YSZ electrolytes [96]. They investigated the effects of yttria dopant content on the lattice parameter and coordination number of Y and Zr in YSZ. The simulated lattice parameters agreed well with the experimental values showing increase in the lattice parameter with the increasing Y_2O_3 concentration. The coordination number of Y is higher than that of

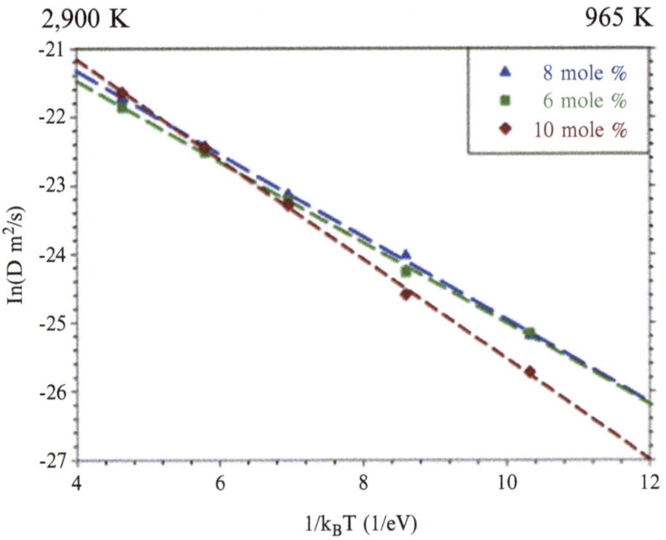

Fig. 6.6 Arrhenius plot of the diffusion coefficient for oxygen ion migration (*D*) as a function of reciprocal temperature (1/*T*) for 6, 8, and 10 mol% YSZ. Reprinted with permission from [96], R. Devanathan et al., *Solid State Ionics,* **177** (2006). © 2006, Elsevier

Zr. Thus, the first neighbor environment is different for Zr and Y in YSZ and oxygen vacancies prefer to be occupied the second nearest neighbor position to Y. The Arrhenius plots of diffusion coefficient for oxygen ion migration as a function of reciprocal temperature for 6, 8, and 10 mol% YSZ are shown in Fig. 6.6 [96]. The highest diffusion coefficient was observed for 8 mol% YSZ, which agreed with the experimentally observed optimum dopant concentration. The activation energy for oxygen ion diffusion was also calculated as a function of yttria concentration and it showed an increase with the increasing dopant concentration from 6 to 15 mol% yttria. The activation energy of the 8 mol% YSZ was found to be 0.6 eV. Devanathan and co-workers further explored the energetics of dopant–dopant and dopant–oxygen vacancy interactions as a function of separation to understand these interactions. They have examined a neutral complex consisted with two dopants which are first neighbors and an oxygen vacancy in the common second neighbor position. It will reduce free oxygen vacancies available for diffusion. Thus, it is believed that this stable complex may play an important role in the degradation of ionic conductivity upon aging of YSZ electrolyte [96].

6.2.2 Scandia-Stabilized Zirconia

Even though ScSZ has shown the maximum oxygen ionic conductivity among all zirconia-based electrolytes, there are limited applications and studies on ScSZ thin film electrolytes. It is also related to the phase stability issues discussed earlier and limited availability and high cost of scandium. Thus, we will discuss the studies reported in the literature for <10 μm thick ScSZ films. These thin film electrolytes have been grown by chemical vapor deposition [101], wet powder spraying [102], electrostatic spray deposition [103], sol–gel deposition [104], dip coating [105], drop-coating [106], pulsed laser deposition [107], and molecular beam epitaxy [58] techniques. The results of these studies will be summarized in this section including theoretical simulation studies on SCSZ.

The ScSZ thin film electrolytes with 8 mol% Sc_2O_3 (8ScSZ) have been grown by metal organic chemical vapor deposition (MOCVD) on Si(100), Al_2O_3, and NiO/YSZ cermet substrates to study their ionic transport properties [101]. The thickness of the electrolyte was kept at 1 and 3.5–4 μm for Al_2O_3 and NiO/YSZ cermet substrates, respectively, to align with the different porosities of the electrodes. The crystalline structure of the films was studied by XRD, which confirms the polycrystalline nature of the films grown on Si(100). However, 8ScSZ thin films were not fully stabilized in the cubic structure, which have a minor tetragonal phase according to the XRD patterns. The cross-sectional scanning electron microscopy (SEM) micrographs of the 8ScSZ thin films confirmed crack-free, thick, and dense coatings adherent to Al_2O_3 and NiO/YSZ porous substrates [101]. The electrical properties of 8ScSZ thin films on Al_2O_3 substrates were investigated and the results showed activation energy of 1.09 eV for oxygen ion diffusion and ionic conductivity of 0.002 S cm^{-1} at 600 °C. The ionic conductivity of 8ScSZ at 600 °C is twice that of 13 mol% yttria-stabilized zirconia (13YSZ) prepared in the same manner [101], which is consistent with the ionic conductivity enhancement observed for ScSZ compared to YSZ with the different synthesis routes [79].

Shi et al. have fabricated a dense ScSZ thin film electrolyte with ~4 μm thickness on NiO/ScSZ porous anode substrate by the wet powder spraying technique [102]. The cross-sectional SEM micrograph showed a dense ScSZ electrolyte layer with no cracks, which is well-adhered to the NiO/ScSZ anode substrate. Following the material synthesis and characterization, two different fuel calls were fabricated using ScSZ electrolyte on NiO/ScSZ anode with $La_{0.8}Sr_{0.2}Sc_{0.1}Mn_{0.9}O_3$ (LSScM) and $Ba_{0.5}Sr_{0.5}Co_{0.8}Fe_{0.2}O_{3-\delta}$ (BSCF) cathodes [102]. Then the electrochemical performance of these fuel cells was tested under the same conditions. The fuel cell with LSScM cathode showed a peak power density of 788 mW cm^{-2} with an ohmic resistance of 0.23 Ω cm^2 at 800 °C. However, the electrode polarization resistance loss of this fuel cell was significant, which degraded the electrochemical performance. Thus, BSCF cathode with excellent electrochemical properties was used to further improve the cell performance. However, BSCF can react with ScSZ electrolyte at elevated temperatures, which significantly increases the interfacial polarization resistance. To avoid the interfacial reactions between BSCF and ScSZ, a samarium-doped ceria (SDC) interlayer was introduced in the system, which improved the cell performance. However, it also introduced a new ScSZ/SDC interface where the inter-diffusion of cations could occur at elevated temperatures, which could increase the ohmic resistance of the fuel cell. Thus, Shi et al. have also discussed the importance of optimizing the properties of doped ceria interlayer to avoid these implications [102]. The NiO/ScSZ anode supported fuel

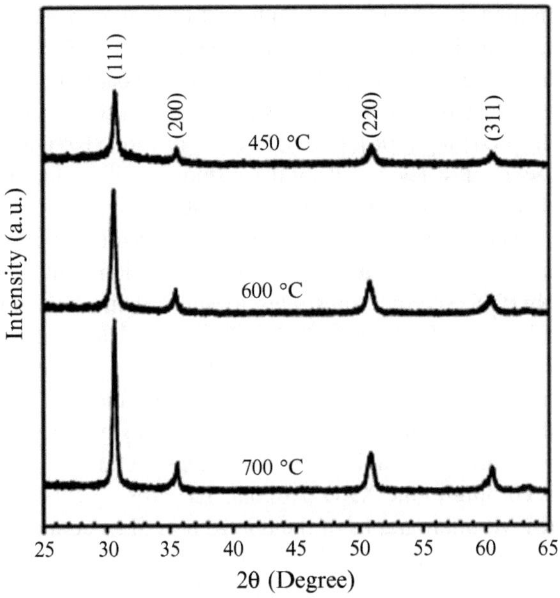

Fig. 6.7 The XRD patterns of as-grown (450 °C) and heat-treated (at 600 and 700 °C) ScSZ thin films grown by electrostatic spray deposition. Reprinted with permission from [103], Nae-Eung Lee et al., *Thin Solid films*, **181** (2010). © 2010, Elsevier

cell with ScSZ electrolyte and BSCF/SDC cathode showed a high power density of 1,760 mW cm^{-2} at 800 °C.

The 10 mol% Sc_2O_3 thin films were deposited on stainless steel substrates using electrostatic spray deposition (ESD) by Kaspabutr and co-workers [103]. The average ScSZ film thickness deposited using three different solvents was about 200 nm. SEM micrographs showed dense electrolytes with crack-free surfaces synthesized using all solvents. However, ScSZ film synthesized using Ethanol/ Butyl carbitol mixed solvent showed the smoothest surface with a root mean square (rms) roughness of 11.82 nm as calculated using atomic force microscopy (AFM) images. The dopant concentration of ScSZ thin films was also confirmed by energy dispersive spectroscopy (EDS) analysis. The XRD patterns of the as-deposited and heat-treated samples indicated the presence of single phase zirconia stabilized in cubic fluorite structure as shown in Fig. 6.7 [103]. Furthermore, the heat-treated samples were highly (111)-oriented. The maximum ionic conductivity of the heat-treated films was found to be 0.33 S cm^{-1} at 800 °C and no phase transformation was observed at this temperature according to XRD patterns. The activation energy for oxygen ion diffusion calculated using Arrhenius equation was 1.09 eV at 800 °C. Therefore, these thin films showed promising characteristics as IT-SOFC electrolytes [103].

The 10 mol% ScSZ (10ScSZ) thin films with a thickness of ~400 nm were deposited on Al_2O_3(0001) substrates by PLD [107]. The 10ScSZ thin films were also co-doped with 1 mol% of yttria and ceria to prevent the phase transformation at high temperatures. The XRD patterns revealed the (111) preferential orientation of all films; however, the (111) peak in the 10ScSZ can be associated with cubic or rhombohedral phase. This peak shifted towards the right side of the XRD pattern with the addition of ceria or yttria and ceria-doped 10ScSZ exhibited a strong second phase peak. The surface and cross-sectional SEM micrographs showed columnar grain growth for all films. A substantial decrease in the ionic conductivity was observed for 10ScSZ film near 550 °C due to the cubic-rhombohedral phase transformation. The conductivity of yttria and ceria-doped 10ScSZ was linear with reciprocal temperature at all temperatures; however, the conductivity was lower than that of 10ScSZ. Thus, authors concluded that these thin films are not useful for SOFC applications unless the conductivity can be enhanced by controlling the dopant concentration [107].

The oxygen ionic conductivity of ScSZ can be further enhanced by depositing high-quality epitaxial thin films in the nanometer scale. Yu et al. have grown epitaxial ScSZ thin films on Al_2O_3(0001) substrates by molecular beam epitaxy (MBE) and investigated the influence of dopant concentration and crystal structure on the oxygen ionic conductivity [58]. The film thickness was also reduced down to ~150 nm as compared to a few micron thick films discussed above. They have observed that ScSZ is stable in the cubic structure over a much wider dopant concentration range (4.7–17.6 mol% Sc_2O_3) than that reported in the previous studies for polycrystalline ScSZ ceramics as confirmed by the XRD measurements. The XRD patterns shown in Fig. 6.8 further suggest the growth of highly

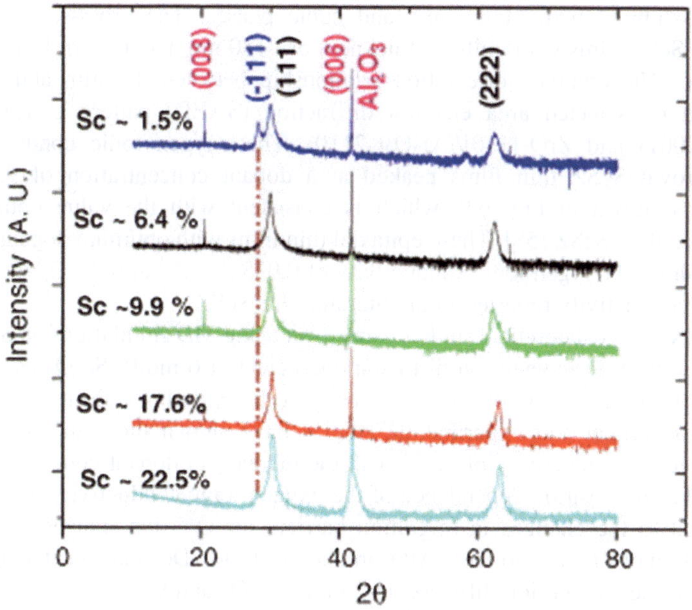

Fig. 6.8 The XRD patterns of as-grown ScSZ thin films with different dopant concentrations grown by molecular beam epitaxy indicating highly oriented cubic (111) and monoclinic (−111) phases. Reprinted with permission from [58], Z.Q. Yu et al., *Solid State Ionics,* **181** (2010). © 2010, Elsevier

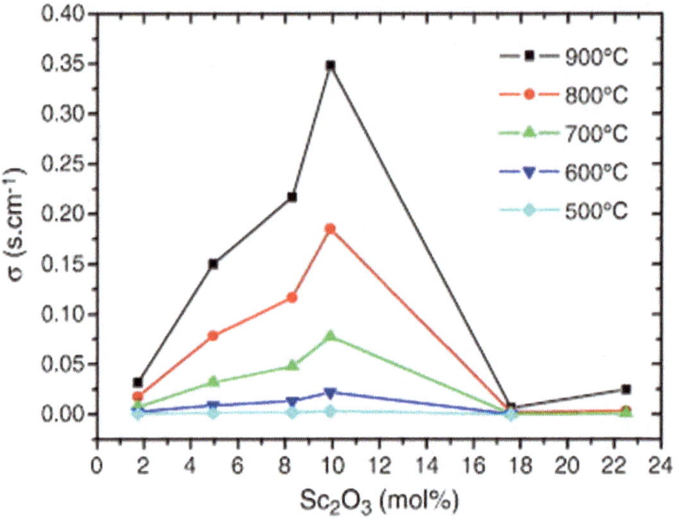

Fig. 6.9 The ionic conductivity of ScSZ thin films as a function of scandia concentration at different temperatures indicating the maximum conductivity at 10 mol% scandia. Reprinted with permission from [58], Z.Q. Yu et al., *Solid State Ionics,* **181** (2010). © 2010, Elsevier

(111)-oriented thin films on Al_2O_3(0001) [58]. The films with 1.5 and 22.5 mol% Sc_2O_3 exhibited both monoclinic and cubic phases. The cross-sectional TEM showed ScSZ films with uniform thickness of ~150 nm and a sharp film/substrate interface. The epitaxial orientation relationship between the film and substrate obtained by selected area electron diffraction (SAED) pattern is ZrO_2(111)//Al_2O_3(0001) and ZrO_2[110]//Al_2O_3(-2110). The oxygen ionic conductivity of MBE-grown ScSZ thin films peaked at a dopant concentration of 9.9 mol% Sc_2O_3 as shown in Fig. 6.9, which is consistent with the value obtained for polycrystalline ScSZ [58]. These epitaxial thin films with optimum dopant concentration showed oxygen ionic conductivity of 0.035 S cm^{-1} at 600 °C, which is the highest conductivity reported in the literature for ScSZ.

The extensive theoretical studies carried out using MD simulations on ScSZ also showed a peak in oxygen ion diffusion coefficient at 6 mol% Sc_2O_3 as shown in Fig. 6.10 [108]. In spite of the discrepancy in the optimum dopant concentration value, theoretical and experimental studies have shown the same trend in the oxygen ionic conductivity of ScSZ with the increase in dopant concentration [58, 108]. The composition dependence of the oxygen ionic conductivity of ScSZ was attributed to the changes in migration barriers for O^{2-} ion transport with Sc^{3+} substitution of Zr^{4+} as shown by MD simulations [108]. Devanathan et al. have also simulated the oxygen ion diffusion in 2 mol% Y_2O_3 and 9 mol% Sc_2O_3 co-doped zirconia and YSZ in addition to ScSZ for the comparison. The simulated lattice parameters of ScSZ decreased with the increase in Sc_2O_3 concentration in contrast to that of YSZ, which agreed with the experimental results [96, 108]. The highest oxygen diffusion coefficient obtained for 6 mol% ScSZ is greater than that of the

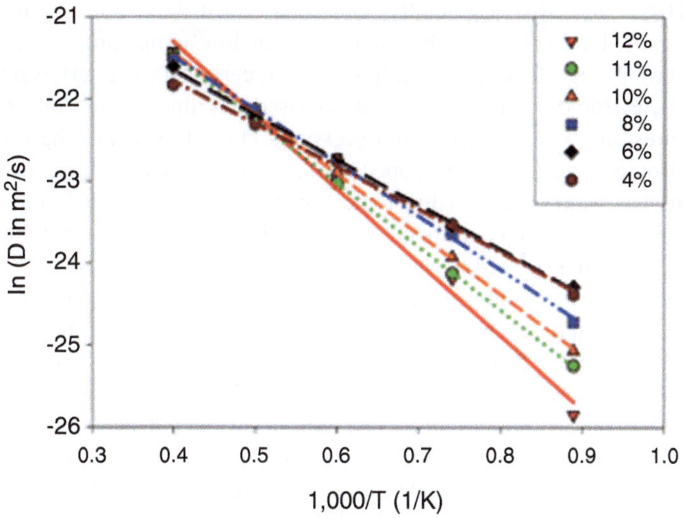

Fig. 6.10 Oxygen diffusion coefficient (*D*) in ScSZ as a function of reciprocal temperature (1/*T*) for various Sc_2O_3 contents (mol%). Reprinted with permission from [108], R. Devanathan et al., *Physical Chemistry Chemical Physics* **11** (2009). © 2009, Royal Society of Chemistry

8 mol% YSZ, which confirmed the enhancement in the ionic conductivity of ScSZ compared to YSZ. The co-doped zirconia showed smaller diffusion coefficient compared to optimized ScSZ and YSZ. The activation energy for oxygen ion diffusion in ScSZ increased with the increase in dopant concentration up to the maximum value tested (17 mol% Sc_2O_3), which agreed with the experimental values obtained by impedance measurements. The Zr–O and Sc–O bond lengths in ScSZ were similar to each other. The coordination numbers of Zr and Sc cations in ScSZ were also similar, which indicated that oxygen vacancy does not have a preference for Zr or Sc as first neighbor. The association of dopants and oxygen vacancies was further studied by examining binding energy of a neutral cluster with the oxygen vacancy being a common first neighbor to the dopants. The formation of these defect clusters can bind oxygen vacancies and degrade the ionic conductivity of ScSZ electrolytes during prolonged operation of SOFC [108].

6.3 Ceria-Based Thin Film Electrolytes

Doped cerium oxide materials have attracted the attention of researchers during last two decades as a SOFC electrolyte, because of their high oxygen ionic conductivity at low temperatures and better chemical compatibility with the electrode materials. Thus, they have been extensively studied as electrolytes for intermediate temperature solid oxide fuel cells (IT-SOFC) [4, 72, 109–112]. The ionic conductivity of doped ceria materials strongly depends on the radius and the concentration of the dopant [109, 113, 114]. Especially, gadolinia-doped ceria (GDC) and samaria-doped ceria (SDC) exhibit high conductivity at low temperatures, compared to zirconia-based electrolytes [115–122]. However, ceria-based materials have shown increased electronic conductivity, when exposed to reducing environment at high temperatures and low oxygen partial pressures [113, 123–127]. The mixed ionic and electronic conductivity of doped ceria materials under different operating conditions has been studied using ionic transference number. It is reported that, 20 mol% $SmO_{1.5}$ and $GdO_{1.5}$-doped ceria have ionic transference numbers of 0.77 and 0.88, respectively [128]. The electronic conductivity of doped ceria can be suppressed to some extent by adding co-dopants [129, 130]. Also, ultra-thin layer of YSZ has been used as an electronic barrier in bilayer thin films to block the electronic conduction [131–133]. In addition to gadolinia and samaria, dopants like yttria, calcia, and strontia have been investigated and used in ceria-based systems [134–138]. Co-doping was also used to further enhance the ionic conductivity of ceria-based systems [139–150]. Unlike zirconia-based systems, the aging behavior of ceria-based electrolytes is not significant, due to the absence of any phase transitions at high temperatures [151]. However, it is reported that, high temperature aging led to an increase in the total conductivity of GDC at lower dopant concentrations ($\leq$20 mol% $GdO_{1.5}$), while at higher dopant concentrations it had a detrimental effect on the conductivity [152]. A similar behavior was observed for yttria-doped ceria electrolytes [153].

GDC is the most extensively studied material among ceria-based electrolytes, due to the enhanced electrical conductivity at intermediate temperatures [119–122]. Below 600 °C, it shows significant enhancement in conductivity compared to zirconia-based materials. The conductivity of GDC depends on the gadolinia concentration and shows linear temperature dependency. It is reported that, GDC exhibits the maximum electrical conductivity around 15–25 mol% $GdO_{1.5}$ [119, 120, 122, 127]. Kudo and Obayashi have observed the optimum conductivity of GDC at 20 mol% $GdO_{1.5}$, wherein the conductivity at 700 °C almost equals to the value of calcia-stabilized zirconia at 1,000 °C [119]. Tianshu et al. have also obtained the maximum conductivity of GDC at 20 mol% $GdO_{1.5}$ for the temperature range of 400–800 °C [121]. However, according to Steele and D. J. Seo et al. the electrical conductivity is maximized at 25 mol% $GdO_{1.5}$ [120, 127]. In contrast to these studies, Zha et al. have reported the highest conductivity of GDC at a dopant concentration of 15 mol% $GdO_{1.5}$ [122]. In addition, pressure dependence of the conductivity of 10 and 20 mol% gadolinia-doped ceria (10GDC and 20GDC) has been studied to understand the increase in electronic conductivity at lower oxygen partial pressure values [154, 155]. Wang et al. have concluded that, the increase in thedopant concentration from 10 to 20 mol% $GdO_{1.5}$ increases the ionic conductivity, but reduces the electronic conductivity [155]. However, 10GDC has better stability at reduced oxygen pressures than 20GDC, wherein the incorporation of Pr improves the stability [156]. Co-doping of GDC was also used to enhance the conductivity. The addition of cobalt oxide into GDC increases the conductivity, which is attributed to the electronic conduction of grain boundary phase rich with CoO [157]. Co-doping of GDC by small amount of TiO_2 also reduces the grain-boundary resistance [158]. The most widely studied 10GDC and 20GDC electrolytes [159–162] have been tested in fuel cell environment and reported to result in a good open circuit voltage [127, 163, 164].

Among ceria-based electrolytes, SDC exhibits the highest ionic conductivity [165–167]. It is reported that, the ionic conductivity of SDC is even 1.5–2 times higher than that of GDC [165]. Recently, enormous attempts were made to optimize the ionic conductivity of SDC in terms of different growth methods, dopant concentrations, co-doping, and geometry of the electrolyte. The optimum $SmO_{1.5}$ concentration values for the highest conductivity of SDC show some discrepancy due to the difference in growth techniques and texture of the samples [115–118, 168]. However, most of the researchers have observed the 20 mol% $SmO_{1.5}$ (20SDC) as the optimum concentration, especially for polycrystalline SDC thin films [115, 116, 118, 165, 169]. For highly oriented SDC thin films grown by MBE, the optimum dopant concentration was found to be 12–15 mol% $SmO_{1.5}$ [168, 170]. In addition, Zha et al. have observed the similar optimum $SmO_{1.5}$ concentration (15 mol%) for polycrystalline SDC, synthesized using co-precipitation process [122]. SDC thin films prepared by ion beam-assisted electron beam evaporation have also exhibited the highest conductivity at 12.3 mol% $SmO_{1.5}$, which is consistent with the value obtained for epitaxial thin films [171]. The maximum conductivity recorded for 20SDC is 5.6×10^{-2} S cm^{-1} at 800 °C [118]. The increase in the conductivity with the $SmO_{1.5}$ concentration up to optimum limit is due to the introduction of oxygen ion vacancies, while defect association and ordering at higher dopant concentration leads to decrease in the conductivity [115, 168, 170].

6.3.1 Gadolinia-Doped Ceria

The thickness of the GDC electrolyte has also been reduced below a few microns in order to optimize its oxygen ionic conductivity [163, 172–178]. These GDC thin film electrolytes have been fabricated by different deposition techniques including MBE [172], PLD [173, 177, 179, 180], magnetron sputtering [133, 175, 181, 182], chemical vapor deposition (CVD) [183–185], electron beam vapor deposition [186], spray techniques [174, 187], solid state reaction [176], tape casting [163], spin coating [178], and sol–gel process [188]. Some of the selected GDC electrolytes fabricated by these techniques will be discussed here. The thicknesses of these thin films are ranging from a few hundred nanometers to a few microns.

Bera et al. have deposited highly oriented GDC thin films using MBE by reducing the film thickness to ~150 nm and investigated the structural, chemical, and oxygen ion-conducting properties [172]. These thin films were grown on pure and zirconia-buffered $Al_2O_3(0001)$ substrates to obtain high-quality epitaxial thin films. The Gd concentration was varied from 6 to 24 atom% to optimize the ionic conductivity in terms of dopant concentration and XPS was used to confirm the film composition. XPS depth profile also showed uniform Ce and O concentration

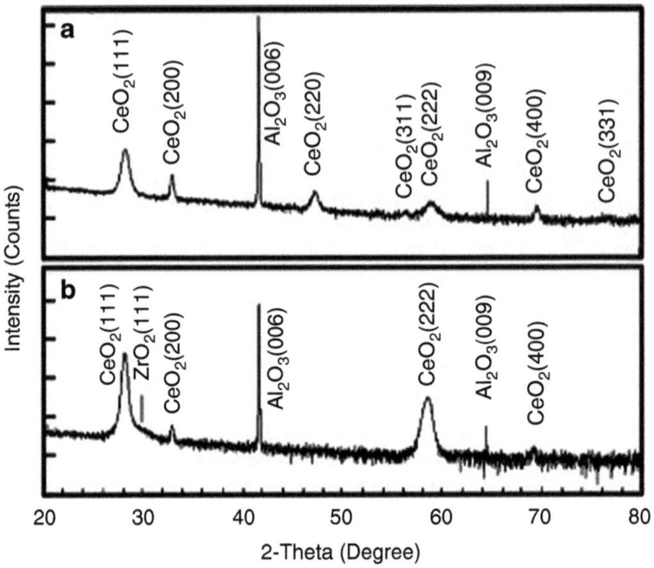

Fig. 6.11 XRD patterns of gadolinia-doped ceria thin films grown by molecular beam epitaxy on (**a**) $Al_2O_3(0001)$ and (**b**) zirconia-buffered $Al_2O_3(0001)$ substrates. Reprinted with permission from [172], Debasis Bera et al., *Thin Solid Films* **516** (2008). © 2008, Elsevier

throughout the film and no inter-diffusion of metal atoms at the film/substrate interface. The high resolution TEM and EDS data further confirmed the sharp film/substrate interface with no inter-diffusion of atoms. Quantitative analysis of EDS spectra also suggested the uniformity of Gd distribution along the thickness of the film. XRD patterns indicated the polycrystalline nature of GDC films grown on pure $Al_2O_3(0001)$ substrates; however, the films grown with ~3 nm zirconia buffer layer were highly (111)-oriented as shown in Fig. 6.11 [172]. The zirconia buffer layer reduced the lattice mismatch between GDC and $Al_2O_3(0001)$, which improved the epitaxial quality. The oxygen ionic conductivity of these thin films was measured using four probe van der Pauw technique. The Arrhenius plots of 5.3, 8.4, and 19.2 atom% Gd doped ceria thin films are shown in Fig. 6.12, which showed the maximum ionic conductivity for 19.2 atom% Gd doped ceria [172]. The Arrhenius plots for single and polycrystalline YSZ and polycrystalline GDC are also shown for the comparison, which indicated the enhanced ionic conductivity of highly oriented, ~150 nm thick GDC films.

The PLD is the most commonly used deposition technique to fabricate GDC thin films [173, 177, 179, 180]. Joo and Choi have grown 0.5–1.8 μm thick GDC thin films ($Ce_{0.79}Gd_{0.21}O_{2-\delta}$) on Pt and $Al_2O_3(0001)$ substrates [180]. XRD patterns indicated the polycrystalline and highly (111)-oriented growth of GDC on Pt(111) and $Al_2O_3(0001)$, respectively. SEM showed average grain size of ~40 nm for the film grown on $Al_2O_3(0001)$. These thin films exhibited about one order of magnitude enhancement in the ionic conductivity compared to the YSZ film in the

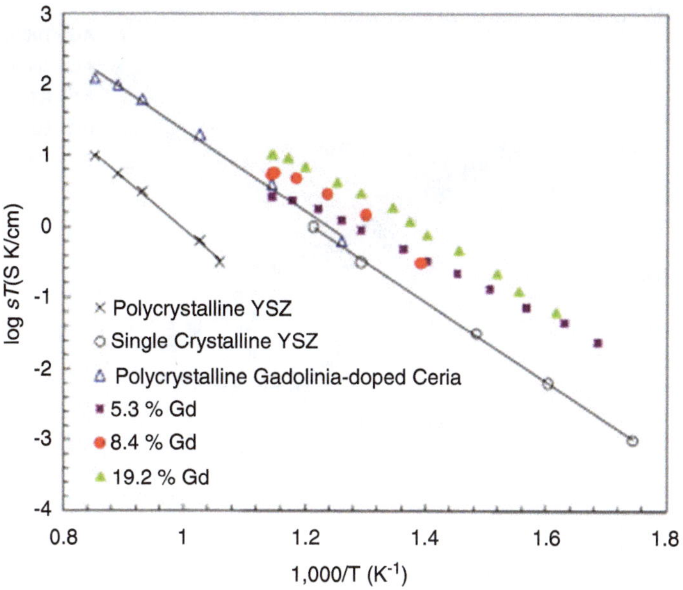

Fig. 6.12 Arrhenius plots of the MBE grown gadolinia-doped ceria thin films with different dopant concentrations. The data from polycrystalline YSZ and GDC and single crystal YSZ are also shown for the comparison. Reprinted with permission from [172], Debasis Bera et al., *Thin Solid Films* **516** (2008). © 2008, Elsevier

temperature range of 300–450 °C. However, the total conductivity of GDC thin films tested in this study was slightly lower than that of bulk GDC and the influence of substrate on the total conductivity was not significant [180]. Cho and Yoon have also deposited GDC (10 mol% of gadolinia) thin films in the range of 500–700 nm thickness on MgO(001) and YSZ(001) substrates using PLD [173]. They were able to tailor the grain size and crystallinity of the films by tuning the growth temperature. The thin films deposited on MgO(001) and YSZ(001) at temperatures above 500 °C showed highly (001)-textured growth as confirmed by XRD measurements. The grain size of the films increases with increasing growth temperature and the samples were very dense and free of cracks as observed in the cross-sectional TEM images. The grain size dependence on the growth temperature was also confirmed by AFM images. The GDC films grown on MgO(001) substrates were used for the electrical conductivity measurements due to the insulating nature of the substrate. The thin films grown at higher temperatures with larger grain size and improved crystallinity showed higher conductivity and lower activation energy as shown in the Arrhenius plots (Fig. 6.13) [173]. It further confirmed the enhanced ionic conductivity observed in nano-scale epitaxial thin film electrolytes. Infrotuna et al. also studied the effects of PLD growth temperature and ambient pressure on the microstructure and electrical conductivity of GDC thin films [179]. To suppress the electronic conductivity of PLD grown 1 μm thick GDC electrolytes under reducing conditions, a YSZ blocking layer has also been used [177]. In this study,

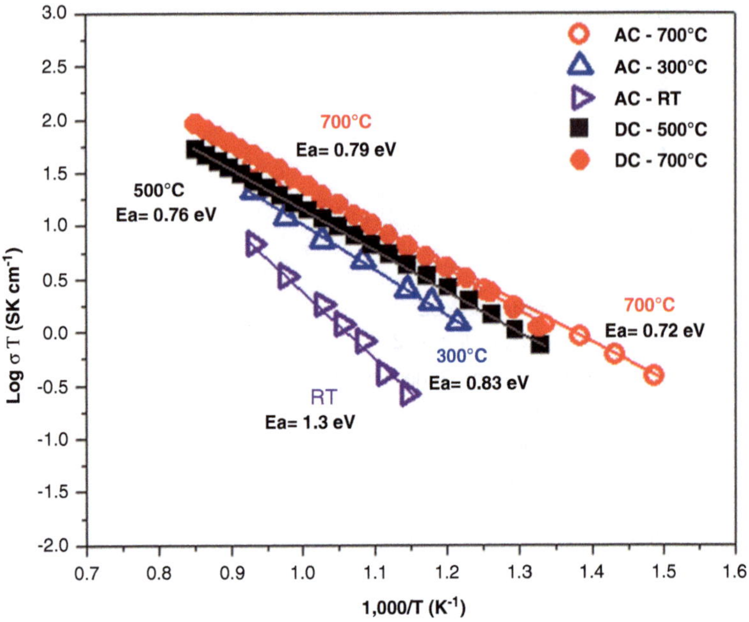

Fig. 6.13 Arrhenius plots of total conductivity of PLD grown gadolinia-doped ceria thin films as a function of deposition temperature. Reprinted with permission from [173], Sungmee Cho et al., *Journal of Material Research* **26** (2011). © 2011, Cambridge University Press

they fabricated a NiO-YSZ anode-supported fuel cell with GDC/YSZ thin film electrolyte and lanthanum strontium cobaltite (LSC) cathode. They measured the open cell voltage (OCV) and power density of this SOFC by varying the thickness of PLD grown YSZ blocking layer from 0 to 200 nm. The SOFC with a YSZ blocking layer exhibited OCV of 1 V compared to OCV of 0.6 V of the cell without the blocking layer. The power density was also increased from 337 to 1,000 mW cm^{-2} at 600 °C with the insertion of YSZ blocking layers. Due to the improved performance of SOFC, GDC/YSZ bi-layer electrolyte has also been investigated by the other researchers [189–192].

The GDC thin film electrolytes have also been fabricated by magnetron sputtering [133, 175, 182]. Lin et al. have deposited 20 mol% Gd-doped ceria thin films on polycrystalline Al_2O_3 substrates by RF magnetron sputtering using both oxide and metallic targets [182]. XRD confirmed the crystalline phase of $Gd_{0.2}Ce_{0.8}O_{1.9}$ after the heat treatment at 900 °C for 2 h. The heat-treated samples were dense with a ~650 nm film thickness. They exhibited a total conductivity of 0.001 S cm^{-1} at 650 °C, which is higher than that of bulk YSZ. GDC thin film electrolytes with a thickness of 1–5 μm have been deposited by electron beam vapor deposition [186]. The 3 μm thick electrolyte exhibited an ionic conductivity of 2.58×10^{-3} S cm^{-1} at 700 °C with activation energy of 0.34 eV. Chourashiya and Jadhav have synthesized GDC thin films on NiO-GDC anode substrate using spray pyrolysis technique (SPT) [187]. However, they observed that GDC thin films were non-adherent to ceramic substrate. Thus, they modified the substrate surface by heating in reducing atmosphere before electrolyte deposition. The heat treatment transformed NiO-GDC substrate to Ni-GDC and produced a rough and porous substrate surface, which may improve the adhesion of GDC to the substrate. However, SPT thin film deposition on metallic substrates was quite difficult, therefore the GDC synthesis method was modified [187]. The thickness of the films grown by this method was found to be 13 μm. The grain interior (GI) conductivity of the half-cell fabricated in this study was 0.1 S cm^{-1} at 500 °C, which is well-desired for IT-SOFC [187].

Lee et al. have carried out an experimental study combined with theoretical simulations to understand the effect of mechanical strain on the electrical properties of GDC thin films [193]. They deposited epitaxial GDC(111) thin films on Al_2O_3(0001) substrates by varying the film thickness from 50 to 400 nm. The lattice strain in the GDC films with different thicknesses was calculated using XRD data. The lattice was relaxed with the increase in film thickness and 99 % of strain was released within a thickness of 26 nm. The Arrhenius plots and activation energy as a function of film thickness of the samples are shown in Fig. 6.14 [193]. The oxygen partial pressure measurements confirmed that the conductivities of these samples are pure ionic. The ionic conductivity of ~400 nm thick film was similar to that of bulk GDC. However, the ionic conductivity and its temperature dependence changed significantly with the decrease in film thickness as shown in Fig. 6.14a. The activation energy for oxygen ion diffusion also increased with the decreasing thickness (Fig. 6.14b), which is comparable with

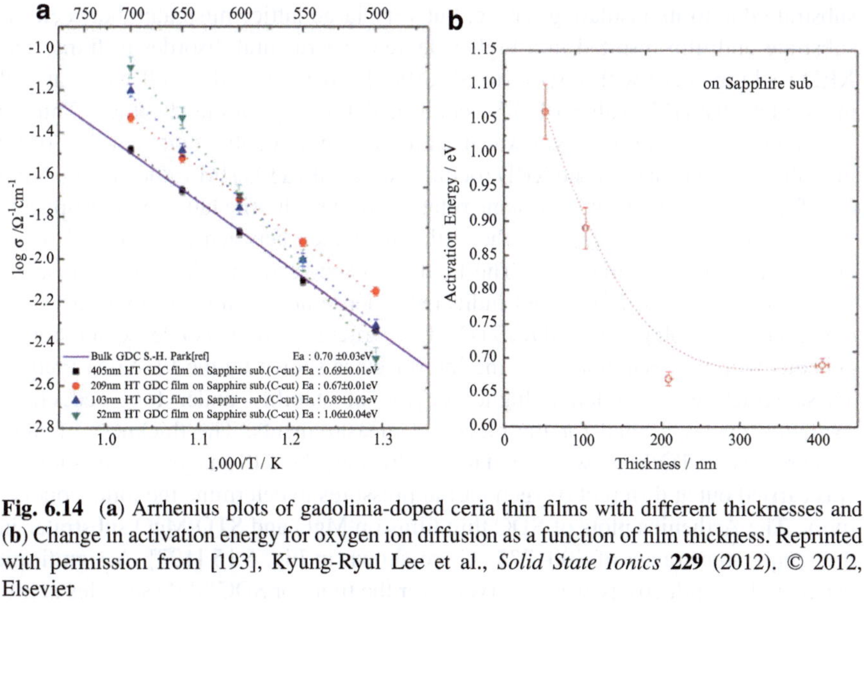

Fig. 6.14 (**a**) Arrhenius plots of gadolinia-doped ceria thin films with different thicknesses and (**b**) Change in activation energy for oxygen ion diffusion as a function of film thickness. Reprinted with permission from [193], Kyung-Ryul Lee et al., *Solid State Ionics* **229** (2012). © 2012, Elsevier

the other studies. These changes were mainly attributed to the lattice strain associated with film thickness. The strain state of this system is rather complex with compressive strain in the lateral direction and tensile stress in the vertical direction, thus a new atomistic migration model was proposed to explain the strain effects on the electrical properties of GDC thin films [193]. From ab initio calculation results, higher activation energy for oxygen ion migration in stressed GDC thin films was expected by considering the consequential atomic positions, bonding lengths, and angles in the stressed ceria lattice. The calculated activation energies at different dilative strain values in the lateral direction were agreed with the experimental results [193].

6.3.2 Samaria-Doped Ceria

Since Sm was considered as one of the most suitable dopant for ceria electrolytes [194], SDC thin film electrolytes have been fabricated to further enhance the ionic conductivity. SDC thin films with different thicknesses, crystalline orientation and quality, and microstructures have been extensively studied and tested for SOFC applications. The deposition techniques such as e-beam evaporation [171, 195], MBE [57, 170, 196], PLD [197], spin coating [198–200], and magnetron sputtering [201] were employed to fabricate these thin films. The fabrication and electrical properties of SDC thin film electrolytes only with <1 μm thickness will be discussed here.

Sanna et al. have deposited 20 mol% samaria-doped ceria thin films on $SrTiO_3$ (STO) buffered MgO(001) substrates by PLD [197]. MgO(001) was selected as the substrate due to its insulating nature, but very large lattice mismatch between the substrate and film resulted in a higher degree of structural disorder in films. The XRD rocking curve with respect to SDC(002) reflection with a full width at half maximum (FWHM) value of 3.5° confirmed this crystalline disorder. Thus, a ~50 nm thick STO buffer layer was used to grow high-quality epitaxial SDC(001) thin films as confirmed by the XRD rocking curve with a FWHM value of 0.3°. XRD data further confirmed a 45° in-plane rotation of SDC fluorite lattice with respect to STO orientation, which reduced the lattice mismatch between SDC and STO and lattice stress in SDC thin films. The field-emission scanning electron microscopy (FE-SEM) images of SDC/MgO indicated a dense and homogeneous film with a columnar morphology. The SDC/STO/MgO structures were also dense and homogeneous with no delamination at the interfaces as shown by the FE-SEM images. These structures exhibited a higher degree of stability under long reduction-oxidation cycles needed for the electrical measurements. The thickness of SDC film grown on STO/MgO was found to be ~260 nm. The EIS analysis of the samples was carried out at different oxygen partial pressures to determine the ionic conductivity. The Arrhenius plots of SDC thin films on MgO and STO/MgO substrates in the temperature range of 400–775 °C are shown in Fig. 6.15 [197]. A significant decay in the conductivity was observed over the time for SDC/MgO samples at low

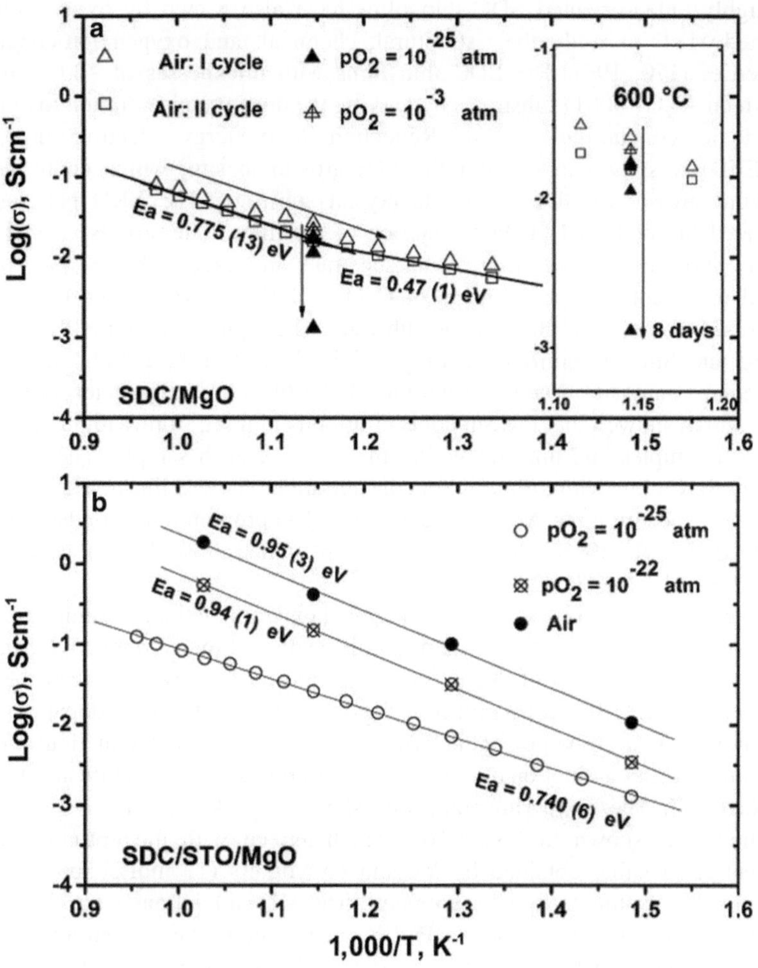

Fig. 6.15 Arrhenius plots of samaria-doped ceria thin films grown on (**a**) MgO (**b**) STO/MgO substrates measured in the temperature range of 400–775 °C and at different oxygen partial pressures. Reprinted with permission from [197], Simone Sanna et al., *Advanced Functional Materials* **19** (2009). © 2009, John Wiley and Sons

oxygen partial pressures and this degradation was irreversible even after re-exposing the samples to air. Thus, it was attributed to the microstructural instability of SDC/MgO samples observed in FE-SEM images. In contrast to SDC/MgO samples, SDC/STO/MgO samples were stable during all EIS measurements and the data was reproducible. The SDC epitaxial film on STO/MgO exhibited a conductivity of about 0.07 S cm^{-1} in air at 700 °C with activation energy of 0.74 eV. The oxygen partial pressure EIS measurements confirmed that the total conductivity is predominantly ionic for $pO_2 > 10^{-17}$ atm [197].

Highly (111)-oriented SDC thin films have also grown by oxygen plasma-assisted MBE to study their structural, chemical, and oxygen ion-conducting properties [170, 196]. The SDC thin films with thicknesses of ~200 nm were grown on $Al_2O_3(0001)$ substrates by varying the dopant concentration to optimize the ionic conductivity [196]. Reflection high-energy electron diffraction (RHEED) was used to monitor the film growth in situ, which confirmed the epitaxial growth of SDC on single crystal substrate. The XRD patterns also indicated the preferred (111) orientation of the films. The cross-sectional TEM image showed a very sharp film/substrate interface. The SAED pattern revealed the epitaxial relationship of $CeO_2(111)//Al_2O_3(0001)$ and $CeO_2[110]//$ $Al_2O_3(-2110)$ between the film and substrate. The high resolution XPS spectra of Ce 3d and Sm 3d confirmed a fully oxidized ceria surface with Sm_2O_3. The elemental composition and distribution of the films was also determined using XPS, which showed that the Sm concentration is ranging from 1 to 33 atom% for different samples and uniform within the depth of each sample. The Arrhenius plots of SDC films with different Sm concentrations showed linear relationship for all samples according to Arrhenius equation. The optimum ionic conductivity was obtained for 5 atom% Sm doped ceria thin film, which showed a total conductivity of 0.04 S cm^{-1} at 600 °C [170].

Sanghavi et al. have carried out integrated experimental and atomistic simulation study to understand the oxygen ion diffusion and ionic conductivity in SDC thin films as a function of dopant concentration [57]. They optimized the ionic conductivity of MBE grown epitaxial SDC thin films in terms of dopant concentration using two probe dc conductivity measurements. The conductivity plots of SDC thin films as a function of dopant concentration and temperature are shown in Fig. 6.16 [57]. These experimental results showed a peak in conductivity at 15 mol % $SmO_{1.5}$ as shown in Fig. 6.16a, which agreed with the optimum dopant concentration value obtained by Yu and co-workers (13 mol% $SmO_{1.5}$) using four probe ac conductivity measurements. The MD simulations were also carried out on the SDC system using DL_POLY code to study the oxygen ion diffusion coefficient (D) as a function of temperature (T) and dopant concentration. The simulations also showed a peak in the oxygen ion diffusion coefficient and ionic conductivity at about 13 mol% $SmO_{1.5}$ concentration as shown in Fig. 6.17 [57], which is very similar to the experimental optimum dopant concentration observed in the ionic conductivity of SDC thin films. The Arrhenius plots of D versus 1/T also showed an increase in diffusion coefficient with the increasing temperature (Fig. 6.17a). The ionic conductivities of SDC obtained from MD simulations as a function dopant concentration at different temperatures are also shown in Fig. 6.17b. The activation energy for oxygen ion diffusion in SDC increased with increasing dopant concentration. These values were found to be in the range of 0.76–1.1 eV for the dopant concentrations of 0–30 mol% samaria. The simulated activation energy of optimized SDC was in the range of 0.8–1.0 eV, which is comparable with experimental activation energy of optimized SDC (0.6–0.9 eV) calculated by several research groups [57]. Sanghavi and co-workers also studied the defect and dopant interactions in SDC using radial distribution

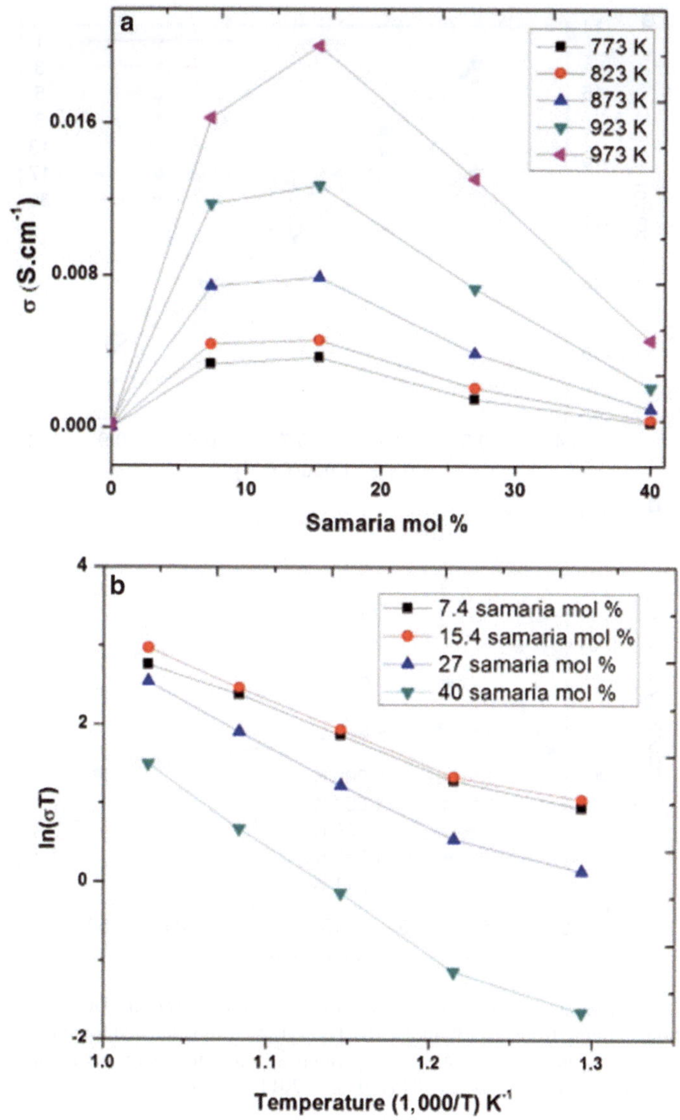

Fig. 6.16 (**a**) The total conductivity of MBE grown samaria-doped ceria thin films as a function of samaria concentration at various temperatures. (**b**) Arrhenius plots of ionic conductivity for SDC films with various dopant concentrations. Reprinted with permission from [57], R. Sanghavi et al., *Solid State Ionics* **204** (2011). © 2011, Elsevier

functions and coordination numbers [57]. The first neighbor distributions for Ce–O and Sm–O were different; however, the second and subsequent distributions were similar. The coordination numbers of Ce and Sm cations were less than the typical value of eight for cubic fluorite lattice due to the oxygen vacancies and

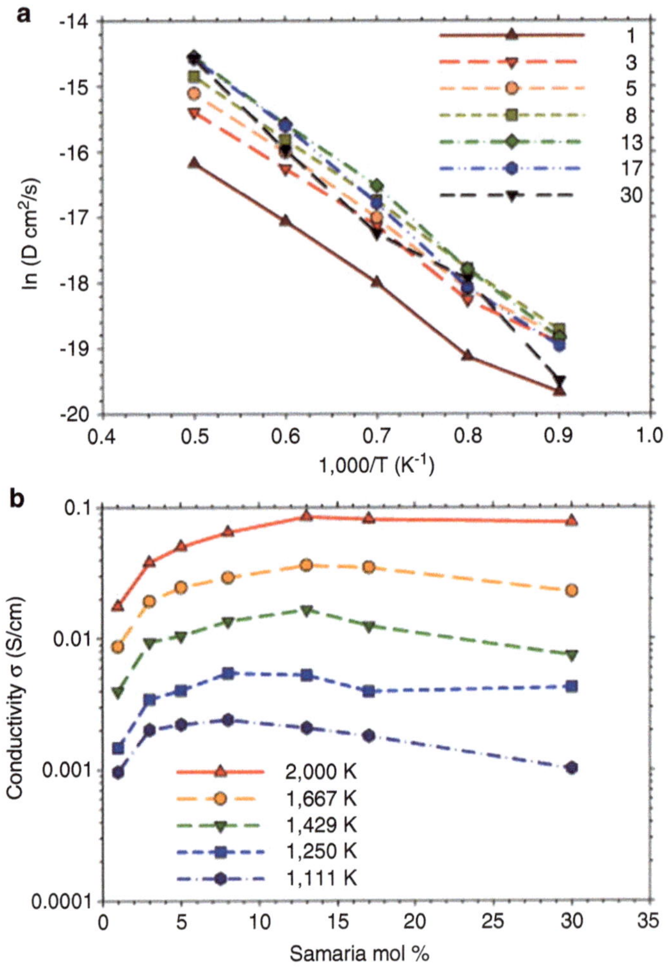

Fig. 6.17 (**a**) Arrhenius plots of oxygen ion diffusion coefficient (*D*) in SDC for various $SmO_{1.5}$ concentrations ranging from 1 to 30 mol% (**b**) The variation in simulated ionic conductivity as a function of $SmO_{1.5}$ mol% at different temperatures. Reprinted with permission from [57], R. Sanghavi et al., *Solid State Ionics* **204** (2011). © 2011, Elsevier

decreased with increasing dopant concentration as shown in Fig. 6.18 [57]. Furthermore, the coordination number of Ce is higher than that of Sm indicating the oxygen vacancy preference of Sm as the nearest neighbor. It was further suggested that the oxygen vacancy-dopant association in SDC is the origin of the decrease in ionic conductivity beyond the optimum dopant concentration [57].

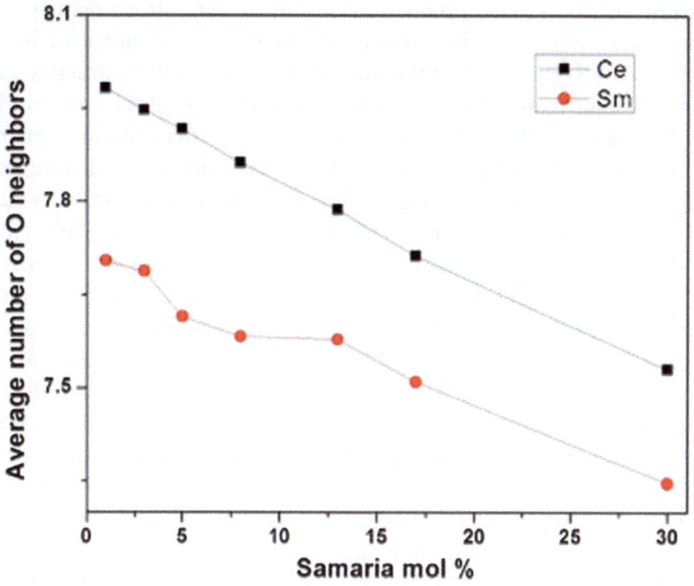

Fig. 6.18 Average coordination number of Ce and Sm cations in SDC as a function of dopant concentration at 1,111 K. Reprinted with permission from [57], R. Sanghavi et al., *Solid State Ionics* **204** (2011). © 2011, Elsevier

6.4 Multilayer Hetero-Structured Thin Film Electrolytes

Nanoscale materials often display properties very different from the bulk materials due to the enhanced surface to volume ratio. The ionic conductivity of nanoscale oxides is greatly influenced by the defect interaction between the dopant and the vacancies that facilitate oxygen ion conduction. Also, at the nanoscale level, increased contribution of the interfacial conductivity leads to an enhancement in the total conductivity [77]. Thus, the ionic conductivity of nanoscale materials can be further enhanced by developing multilayer thin films with nanoscale interfaces [202–204]. It has been observed that, nanoscale hetero-structures of CaF_2 and BaF_2 exhibit significantly higher ionic conductivity along the interfacial direction at moderate temperatures due to the redistribution of mobile fluoride ions in the overlapped space-charge regions of CaF_2 and BaF_2 [18, 205, 206]. These results paved the way to develop multilayer thin film electrolytes to enhance the oxygen ionic conductivity by introducing hetero-structured interfaces [59, 61, 207–214].

The first oxygen ion-conducting multilayer thin film electrolyte reported in the literature is gadolinia-doped ceria/zirconia, which was developed by Azad and co-workers [59]. They have observed a significant increase in ionic conductivity by increasing the number of layers in these multilayer thin films and their results will be discussed in detail in the following section. Gadolinia-doped ceria and zirconia multilayer thin films have also been grown by sputter deposition to check the feasibility of mass scale production of these hetero-structures [214]. Garcia-Barriocanal have observed eight orders of magnitude enhancement in the ionic conductivity of YSZ/$SrTiO_3$(STO) epitaxial multilayer hetero-structures at 357 K [61]. They have attributed it to the atomic reconstruction at the coherent interface between YSZ and STO, which causes high carrier concentration and mobility. However, these results were questioned due to the inability of distinguishing the electronic and ionic contributions and the contribution from STO substrate to the total conductivity of the system [215]. These arguments were strengthened by Cavallaro and co-workers, who claimed that the enhancement obtained in the conductivity of YSZ/STO multilayer system is related to electronic rather than ionic conductivity [208]. Thus, Pennycook et al. have carried out extensive microscopy studies [216] and density functional theory (DFT) calculations [217] to explain the origin of colossal ionic conductivity in YSZ/STO multilayers. The annular dark field scanning transmission electron microscopy (STEM) images revealed the existence of coherent and incoherent interfaces of two different YSZ/STO multilayer samples grown under similar growth conditions indicating even the slight differences in growth environment can result coherent or incoherent interfaces [216]. They observed a disordered oxygen sub-lattice in the strained YSZ layer of coherent YSZ/STO multilayers, which can be considered as the origin of high ionic conductivity of these hetero-structures. They have also shown that the strained regions exist around dislocation cores at the incoherent YSZ/STO interfaces, which can further increase the ionic conductivity [216]. It was further confirmed by density functional theory (DFT) calculations, which explained the colossal ionic conductivity of YSZ/STO multilayers in terms of lattice strain and extreme disordering of oxygen sub-lattice in YSZ [217]. The electronic conductivity in YSZ/STO multilayers observed by Cavallaro et al. [208] was attributed to incoherent disconnected YSZ islands embedded in STO, which caused n-type doping of STO/YSZ nanocomposites [216].

Janek's group has developed a theoretical model to explain the fast ionic transport and enhanced oxygen ionic conductivity along the nano-scale interfaces in multilayer hetero-structures in terms of elastic strain and density of misfit dislocations at the interfaces [211, 218]. According to this model, with the increasing lattice mismatch between two phases, the oxygen ionic conductivity along the interfaces increases due to the decrease in activation energy for oxygen ion diffusion. If the lattice mismatch is very small, a coherent interface will be formed between two phases and the interfacial conductivity only depends on the elastic strain. If the lattice mismatch cannot be compensated only by the strain, a semi-coherent interface will be formed with misfit dislocations. The oxygen ionic conductivity along these semi-coherent interfaces is linearly dependent on the dislocation density. In the case of completely incoherent interfaces, the increase in the interfacial ionic conductivity can be more pronounced [210]. This qualitative model has been utilized to explain the enhanced oxygen ionic conductivity observed in various multilayer systems [209–211, 218]. Here, the oxygen ion-conducting properties of GDC/GSZ and SDC/YSZ multilayer systems will be discussed in detail.

6.4.1 Gadolinia-Doped Ceria and Zirconia

The multilayer thin films of gadolinia-doped ceria and zirconia (GDC/GSZ) were grown using molecular beam epitaxy and magnetron sputtering techniques to understand the influence of nano-scale interfaces on the oxygen ionic conductivity [59, 214]. Azad et al. have grown ~150 nm thick GDC/GSZ multilayer thin films on Al_2O_3(0001) using MBE by varying the number of layers from 2 to 16 [59]. The XRD measurements confirmed the growth of highly (111)-oriented ceria and zirconia layers with minor GSZ(100) phase [56]. Rutherford backscattering spectrometry (RBS) and XPS depth profiles confirmed uniform dopant distributions in both layers with average Gd concentration of 12 at.%. HRTEM images indicated the presence of structural domains within each layer and internal and interfacial misfit dislocations in these multilayer thin films regardless of the number of layers [59]. The misfit dislocation density was also similar at the interfaces between film and substrate and subsequent layers. These multilayer hetero-structures showed a significant increase in ionic conductivity compared to YSZ and polycrystalline GDC as shown in Fig. 6.19 [59]. The increase in the conductivity of these multilayer thin films cannot be attributed to the space charge effects near the interfaces, since the space charge region is very small. Thus, it was attributed to extended defects and lattice strain near

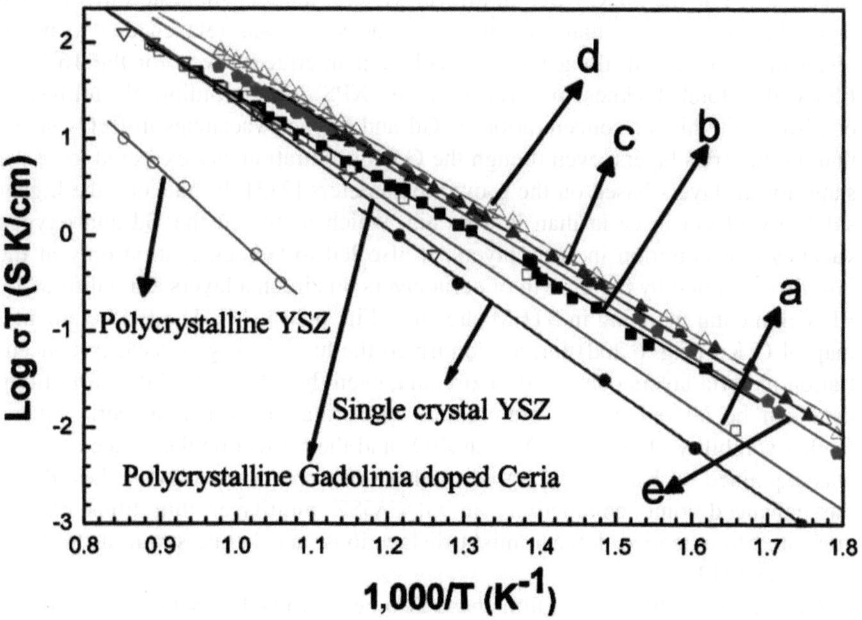

Fig. 6.19 Arrhenius plots of oxygen ionic conductivity of (**a**) two, (**b**) four, (**c**) eight, (**d**) ten, and (**e**) sixteen layer gadolinia-doped ceria-zirconia multilayer thin films [59]. The data from polycrystalline YSZ and GDC and single crystal YSZ are also shown for the comparison. Reprinted with permission from [59], S. Azad et al., *Applied Physics Letters*, **86** (2005). © 2005, AIP Publishing LLC

the layer interfaces, which may increase the solubility of Gd and therefore the density of oxygen vacancies in ceria. The conductivity of GDC/GSZ multilayers increased with the increase in number of layers; however, 10-layer film showed the maximum ionic conductivity. The decrease in the ionic conductivity below 15 nm of individual layer thickness was attributed to the strain effects at these smaller thicknesses [59, 60].

Wang and co-workers have carried out extensive HRTEM, STEM, EDS, and XPS multi-technique characterization of MBE-grown GDC/GSZ hetero-structures to understand their oxygen ion-conducting properties in terms of microstructure and oxygen vacancy and dopant distributions [213]. The cross-sectional TEM micrographs and SAED patterns of GDC/GSZ thin films revealed the layered structure and well-defined epitaxial relationship between the layers. The separation of the higher order diffraction spots in the SAED patterns confirmed the single crystal nature of ceria and zirconia layers with different lattice parameters. The strain induced at the layer interfaces due to the lattice mismatch between ceria and zirconia layers was released by the misfit dislocations, which are distributed along the interfaces. These interface dislocations are 7 nm apart from each other as shown by Fourier filtered HRTEM images [213]. It is theoretically expected that a lattice is strained and forms coherent interfaces between layers below a critical layer thickness [219, 220], which was also experimentally evidenced [221–223]. For the GDC/GSZ multilayer system, authors suggested this critical layer thickness is less than 10 nm, since the strain was relaxed due to misfit dislocations appeared along the semi-coherent interfaces, even for the 16-layer film with a total thickness of ~155 nm. The XPS depth profiling of multilayers confirmed the higher concentration of Gd and oxygen vacancies in CeO_2 layers than in the ZrO_2 layers, even though the Gd concentration was expected to be the same for all layers based on the growth parameters [213]. It indicated the higher solubility of Gd in ceria than in zirconia, which increased the Gd and oxygen vacancy concentration in ceria layers. It also led to Gd segregation only at the interfaces formed by the growth of ceria layers on zirconia layers as confirmed by EDS elemental mapping in STEM shown in Fig. 6.20 [213]. The EDS elemental map of O Kα (Fig. 6.20d) further confirmed the higher oxygen vacancy concentration in ceria layers compared to zirconia, even though the XPS depth profiling data can be altered by the preferential sputtering of oxygen in ceria layers. Higher solubility of Gd in CeO_2 than ZrO_2 and their effect on the lattice constant was explained by Kim [224]. Based on these observations, authors believed that the enhanced ionic conductivity in GDC/GSZ multilayer thin films can be attributed to extended defects, misfit dislocations, and lattice strain at the layer interfaces [213].

GDC/GSZ multilayer thin films have also been grown by sputter deposition on Al_2O_3(0001) substrates with relatively larger total film thicknesses compared to MBE grown films [214]. XRD confirmed the preferred growth of ceria and zirconia layers in (111) orientation. A reduced lattice mismatch of 4.4 % was observed between ceria and zirconia layers using XRD data in contrast to the expected value of 5.5 %, which was attributed to the different lattice expansions caused by Gd

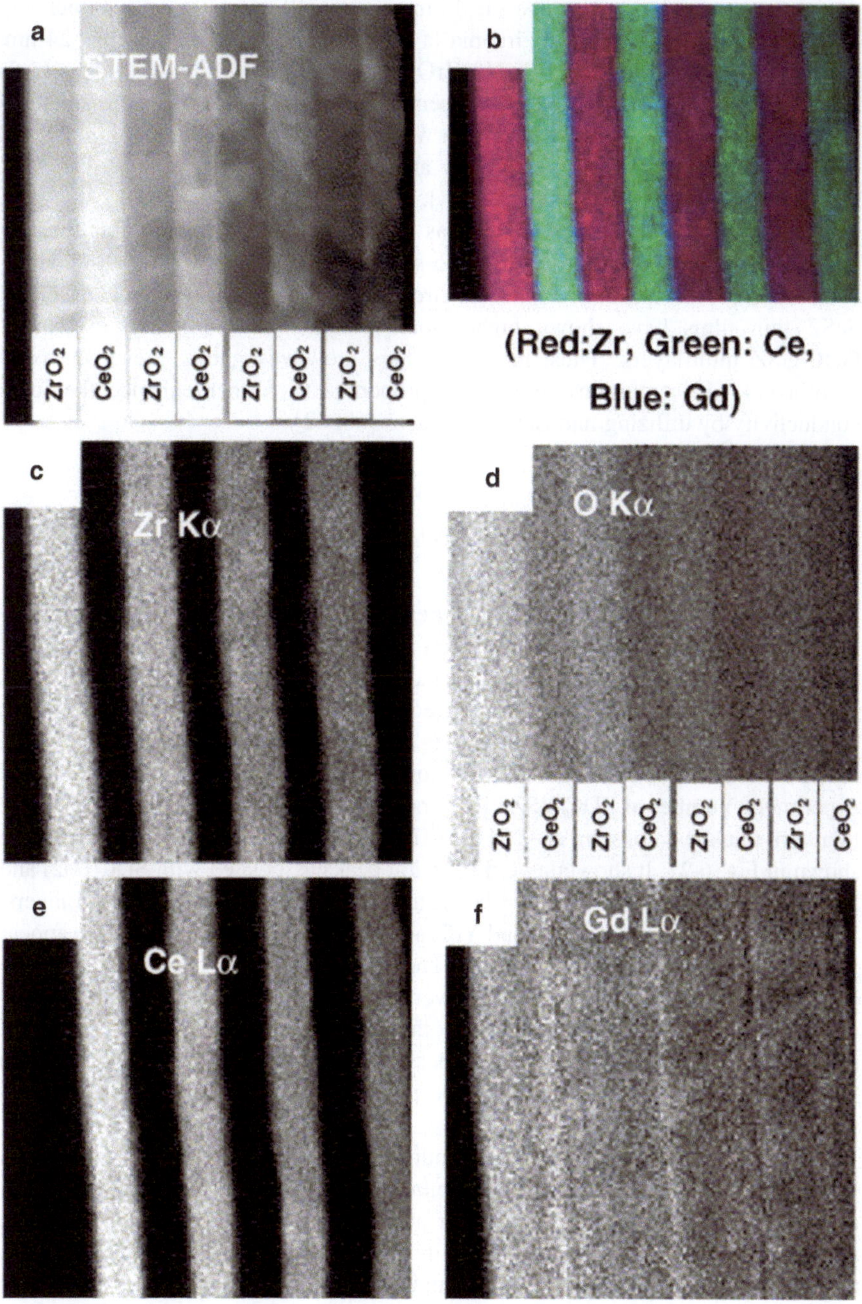

Fig. 6.20 EDS elemental mapping of the 8-layer Gd doped ceria-zirconia thin film with (**a**) STEM-ADF image (**b**) RGB color overlay of the film, (**c**) Zr map, (**d**) O map (**e**) Ce map, and (**f**) Gd map. Reprinted with permission from [213], C.M. Wang et al., *Solid State Ionics*, **177** (2006). © 2010, Elsevier

doping in ceria and zirconia. The TEM image of the 10-layer film with a thickness of ~240 nm showed ceria and zirconia layers with uniform thickness of ~24 nm. The epitaxial relationships of (111)ZrO_2//(111)CeO_2//(0001)Al_2O_3 and (11-2) ZrO_2//(11-2)CeO_2//(-2110)Al_2O_3 between the film and substrate were obtained using SAED patterns. The 50-layer film with nominal thickness of ~2 μm showed enhanced ionic conductivity, especially at lower temperatures compared to YSZ thin films and polycrystalline GDC [214].

The GDC/GSZ multilayer system has attracted a greater attention as the first oxygen ion conducting multilayer hetero-structure reported in the literature. However, Gd is not the optimum dopant for zirconia and single-layer epitaxial SDC and ScSZ thin films have shown higher ionic conductivity compared to that of GDC/GSZ multilayers. Thus, the researchers are investigating the other oxide multilayers with optimum single-layer properties to further enhance the ionic conductivity by utilizing nanoscale interfaces [60, 62].

6.4.2 *Samaria-Doped Ceria/Yttria-Stabilized Zirconia*

The multilayer heteroepitaxial structures of samaria-doped ceria and yttria-stabilized zirconia (SDC/YSZ) have been grown on STO (~10 nm thick) buffered MgO(001) substrates by PLD to understand their oxygen ion-conducting properties [60]. Sanna and co-workers used optimized 20 mol% SDC and 8 mol% YSZ layers to fabricate these multilayers by keeping the total film thickness and number of layers constant for two different sets of experiments, respectively. For the first set of experiment, number of SDC/YSZ bi-layers was varied from 1 to 20 by keeping the total film thickness constant at ~154 nm. The XRD patterns of these multilayers are shown in Fig. 6.21. It showed two diffraction peaks associated with SDC(002) and YSZ(002) for the film with one bi-layer, indicating the epitaxial growth of layers. The interference effects of SDC and YSZ and superlattice features started to appear with the increase in number of bilayers. The well-defined superlattice features were observed for 20-bilayer film with an average structure peak and two sharp first-order satellite peaks, which indicated the presence of high-quality SDC/YSZ interfaces. The individual bilayer thickness of the 20-bilayer film was found to be 7.7 nm, which was calculated using the positions of satellite peaks in the XRD pattern.

For the other set of experiments, the individual bilayer thickness was varied from 7.7 to 28 nm by keeping the number of bilayers constant at 20. The Arrhenius plots of both sets of samples collected in air at temperatures from 400 to 800 °C are shown in Fig. 6.22. These multilayer thin films exhibited enhanced ionic conductivity compared to single-layer SDC and YSZ thin films grown on STO/MgO substrates and bulk SDC. The ionic conductivity of SDC/YSZ multilayer films increased with the increase in number of bilayers, indicating a significant contribution from the interfacial conductivity. Furthermore, the interfacial conductivity increased with the decrease in individual layer thickness. The enhancement in the

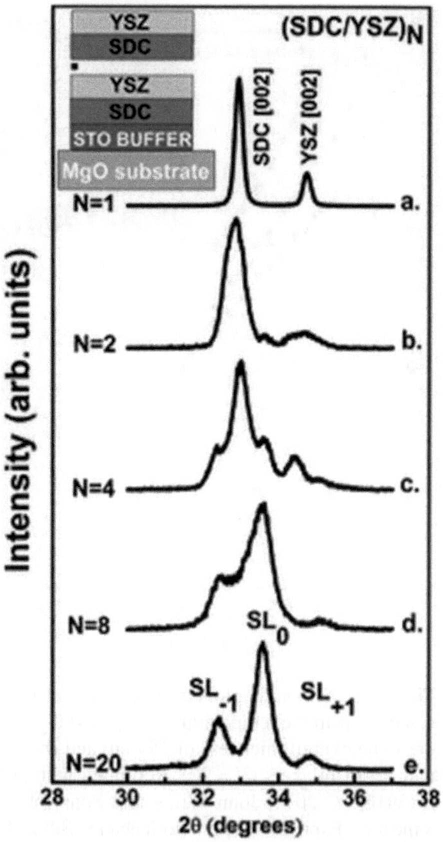

Fig. 6.21 XRD patterns of SDC/YSZ multilayer hetero-structures with different number of bilayers (N) ranging from 1 to 20 with the same overall thickness. A sketch of the hetero-structure is shown in the inset. Reprinted with permission from [60], Simone Sanna et al., *Small*, **6** (2010). © 2010, John Wiley and Sons

ionic conductivity of 20-bilayer sample with a bilayer thickness of 7.7 nm is about one order of magnitude compared to SDC and YSZ single-layer thin films. In contrast to GDC/GSZ multilayers and YSZ nanoscale thin films, the ionic conductivity did not decrease below a critical individual layer thickness down to the smallest tested value of ~4 nm. The EIS measurements of SDC/YSZ thin films at different oxygen partial pressures confirmed the predominant ionic conductivity at $pO_2 > 10^{-15}$ atm [60]. At low oxygen partial pressures ($<10^{-15}$ atm), the electronic contribution dominated as confirmed by the increase in total conductivity with decreasing oxygen partial pressure.

The XRD reciprocal space maps were also used to calculate in-plane and out-of-plane lattice parameters. For the sample with one bilayer, it showed two different lattice parameters for SDC and YSZ, which agree with the standard values of 0.544 and 0.514 nm, respectively. With the decrease in layer thickness, YSZ in-plane lattice parameter approached that of SDC, which resulted in highly strained YSZ layer and fully relaxed SDC layer. They calculated the theoretical value for ratio between interfacial conductivity (σ_{int}) and bulk conductivity (σ_{bulk}) using 6 % epitaxial strain of YSZ layer, which was well-agreed with the

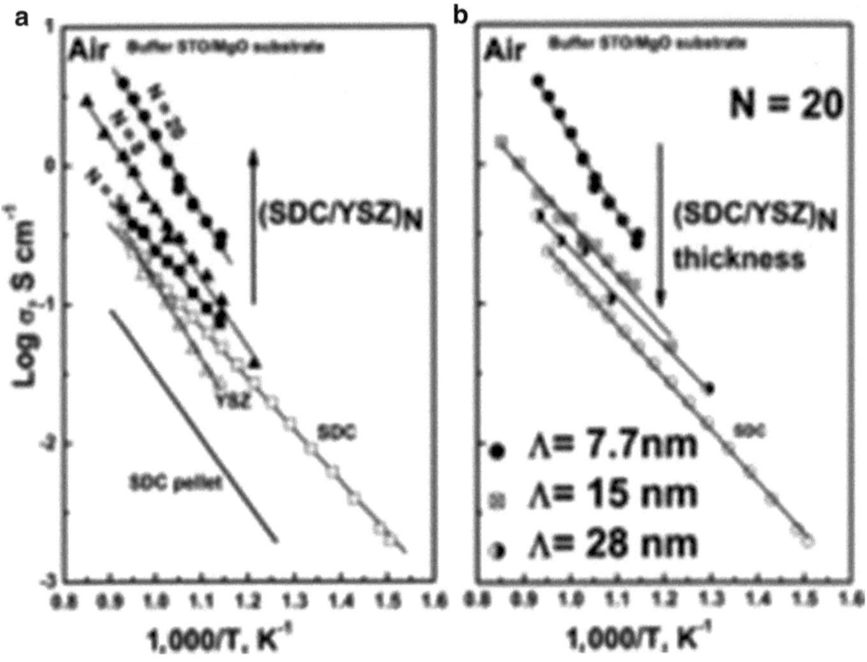

Fig. 6.22 Arrhenius plots of ionic conductivity of SDC/YSZ multilayer hetero-structures measured in air in the temperature range of 400–800 °C with (**a**) different number of bilayers (N) and the same overall thickness of 154 nm and (**b**) different individual layer thicknesses and the same number of bilayers ($N = 20$). Reprinted with permission from [60], Simone Sanna et al., *Small*, 6 (**2010**). © 2010, John Wiley and Sons [60] (Reproduced from Small **6**(17), Simone Sanna, Vincenzo Esposito, Antonello Tebano, Silvia Licoccia, Enrico Traversa, Giuseppe Balestrino: Enhancement of Ionic Conductivity in Sm-Doped Ceria/Yttria-Stabilized Zirconia Heteroepitaxial Structures, 1863-1867 (2010) with permission from John Wiley and Sons)

experimental value of 2.4 [60]. Thus, the enhancement in the oxygen ionic conductivity of SDC/YSZ multilayers was attributed to tensile strain effects as predicted by the theoretical studies.

6.5 Summary

One of the main focus areas in current clean energy technologies is associated with the development of SOFC for clean energy production. High efficiency, minimal pollution, and wide choice of fuels made the SOFC device as a promising energy technology. Currently, significant efforts are taken to address several scientific issues associated with many different components of these devices. Here we have extensively discussed the current status of SOFC thin film electrolytes in both fundamental and device application point of view. We have also briefly discussed the role of thin films in advanced SOFC anode and cathode materials.

There has been a quest for viable replacements to YSZ as electrolyte to reduce the current operating temperature of fuel cells, thereby improving the overall economics of the SOFCs. The prevalent approach to increase the ionic conductivity at lower temperatures is based on the reduction of electrolyte thickness and incorporation of different dopants to ceria- and zirconia-based electrolytes. Nanostructured membrane materials including epitaxial and multilayer thin films also have tremendous potential to revolutionize SOFC technology through the increase of ionic conduction at lower temperatures. Among ceria- and zirconia-based electrolytes, SDC and ScSZ epitaxial thin films have shown the highest oxygen ionic conductivity reported up to date. Hetero-structured multilayer thin films have also shown significant enhancement in the ionic conductivity, which is even higher than that of single crystal thin films. Even though a number of research efforts have been focused on studying the properties of thin film electrolytes (TFE), there were limited efforts to apply TFE in SOFC devices due to the issues associated with their fabrication on porous electrodes, structural and chemical stability, chemical compatibility with the other SOFC components, and mass-scale production. One of the grand challenges in incorporating nanoscale TFE in electrode-supported SOFC devices is fabricating them on porous electrode materials. However, a few micron thick electrolytes have been successfully fabricated on anode supports in some of the examples we have discussed in this chapter, even though the long-term stability of these devices needs to be extensively studied. We believe that the developing next-generation SOFCs with thin film electrolytes is a trade-off between the enhanced ionic conductivity at nanoscale and successfully fabricating these materials on porous electrodes. Thus, more research efforts and device testing need to be carried out to optimize the properties of TFE fabricated on porous electrodes. Another major future perspective in the development of SOFCs is the mass scale production of thin film electrolytes. Even though many different techniques are available for thin film fabrication, process complexity and cost, choice of materials, and fabrication size need to be considered in industrial fabrication of TFE. The sputter and chemical vapor depositions are two major techniques which can be employed in the industrial production of TFE.

The Sr-doped lanthanum manganite-based composites are still the most widely used SOFC cathode materials due to their high catalytic activity for oxygen reduction, enhanced electronic conductivity, and long-term stability. The mixed ionic and electronic conductors have also attracted attention as cathode materials due to the extension of active sites for oxygen reduction reactions beyond the triple phase boundaries and hence the improved performance. However, the cathode polarization loss is still a major challenge for development of cathode materials especially for low temperature SOFCs. The chromium poisoning of cathode is another issue in the SOFCs with Cr-containing interconnect materials. To overcome these challenges, various approaches have been employed including thin film coatings on cathode surface and deposition of nanoparticles using infiltration process. The future perspectives are focused on the development of new materials with high catalytic activity, reduced polarization, and enhanced stability at low cost. The understanding of oxygen reduction reaction mechanisms using model systems and thin films is another growing field in SOFC cathodes. Ni-YSZ cermet is the most commonly used anode material in SOFCs with excellent performance in clean hydrogen or reformed fuels. However, the deactivation of the anode due to the re-oxidation of Ni, carbon deposition in hydrocarbon fuels, and sulfur poisoning are major drawbacks in Ni-YSZ anode material. Thus, Ni-free anode materials have also been studied; however, they have structural and thermal compatibility issues with the electrolyte. Another approach of improving the tolerance for carbon deposition and sulfur contamination is replacing YSZ with the other oxygen ion conductors. The future perspectives in the development of SOFC anode materials include the fabrication of nano-structured anodes using infiltration method and investigation of reaction mechanisms using thin film model systems.

In general, it is a challenge to understand the reaction mechanisms and interactions between different SOFC components with the development of new materials for low temperature operation. These reaction mechanisms include the oxygen ion conduction of electrolytes, oxygen reduction in cathode, and hydrogen oxidation in anode. Furthermore, the structural, chemical, and thermal compatibility of different SOFC components needs to be investigated with the inclusion of new materials. Thus, the fundamental understanding of these processes is very important to improve the performance of SOFCs. In this regard, thin film structures play a significant role in understanding the physical, chemical, and electrical properties of new SOFC materials. Recently, micro-SOFCs have emerged as the next-generation portable clean energy devices due to their high energy density. The main objective in these efforts is to develop smaller (micro size) devices compared to conventional SOFCs. In order to develop new micro-SOFCs, the fabrication of thin film SOFC components is necessary. In summary, the application of thin film technologies to SOFC had a greater influence in developing new materials and understanding their properties for different SOFC components. However, it will be a grand challenge to incorporate these thin film structures in SOFC devices in the future.

Acknowledgements A portion of this research was performed using EMSL, a national scientific user facility sponsored by the Department of Energy's Office of Biological and Environmental Research and located at Pacific Northwest National Laboratory (PNNL). PNNL is operated for the U.S. DOE by Battelle Memorial Institute under contract number DE-AC05-76RL01830.

References

1. U.S. Energy Information Administration. Ann. Energy Rev. (2011)
2. S.P.S. Badwal, K. Foger, Solid oxide electrolyte fuel cell review. Ceram. Int. **22**(3), 257–265 (1996)
3. S.C. Singhal, Advances in solid oxide fuel cell technology. Solid State Ion. **135**(1–4), 305–313 (2000)
4. O. Yamamoto, Solid oxide fuel cells: fundamental aspects and prospects. Electrochim. Acta **45**(15–16), 2423–2435 (2000)

5. S.C. Singhal, Solid oxide fuel cells for stationary, mobile, and military applications. Solid State Ion. **152–153**, 405–410 (2002)
6. N.Q. Minh, Solid oxide fuel cell technology–features and applications. Solid State Ion. **174** (1–4), 271–277 (2004)
7. S.C. Singhal, Solid oxide fuel cells. Electrochem. Soc. Interface **16**(4), 41 (2007)
8. E.D. Wachsman, K.T. Lee, Lowering the temperature of solid oxide fuel cells. Science **334** (6058), 935–939 (2011)
9. H. Tu, U. Stimming, Advances, aging mechanisms and lifetime in solid-oxide fuel cells. J. Power Sources **127**(1–2), 284–293 (2004)
10. B.C.H. Steele, A. Heinzel, Materials for fuel-cell technologies. Nature **414**(6861), 345–352 (2001)
11. C. Ding, T. Hashida, High performance anode-supported solid oxide fuel cell based on thin-film electrolyte and nanostructured cathode. Energy Environ. Sci. **3**(11), 1729–1731 (2010)
12. S.P.S. Badwal, Stability of solid oxide fuel cell components. Solid State Ion. **143**(1), 39–46 (2001)
13. D. Beckel et al., Thin films for micro solid oxide fuel cells. J. Power Sources **173**(1), 325–345 (2007)
14. S.J. Litzelman et al., Opportunities and challenges in materials development for thin film solid oxide fuel cells. Fuel Cells **8**(5), 294–302 (2008)
15. H.L. Tuller, S.J. Litzelman, W. Jung, Micro-ionics: next generation power sources. Phys. Chem. Chem. Phys. **11**(17), 3023–3034 (2009)
16. J. Santiso, M. Burriel, Deposition and characterisation of epitaxial oxide thin films for SOFCs. J. Solid State Electrochem. **15**(5), 985–1006 (2011)
17. F. Emiliana, P. Daniele, T. Enrico, Ionic conductivity in oxide heterostructures: the role of interfaces. Sci. Tech. Adv. Mater. **11**(5), 054503 (2010)
18. X. Guo, J. Maier, Ionically conducting two-dimensional heterostructures. Adv. Mater. **21** (25–26), 2619–2631 (2009)
19. J.M. Vohs, R.J. Gorte, High-performance SOFC cathodes prepared by infiltration. Adv. Mater. **21**(9), 943–956 (2009)
20. S.P. Jiang, A comparison of O2 reduction reactions on porous (La, Sr)MnO3 and (La, Sr)(Co, Fe)O3 electrodes. Solid State Ion. **146**(1–2), 1–22 (2002)
21. E.P. Murray, T. Tsai, S.A. Barnett, Oxygen transfer processes in (La, Sr)MnO3/Y2O3-stabilized ZrO2 cathodes: an impedance spectroscopy study. Solid State Ion. **110**(3–4), 235–243 (1998)
22. T. Horita et al., Oxygen reduction mechanism at porous $La_{1-x}Sr_xCoO_{3-d}$ cathodes/$La_{0.8}Sr_{0.2}Ga_{0.8}Mg_{0.2}O_{2.8}$ electrolyte interface for solid oxide fuel cells. Electrochim. Acta **46**(12), 1837–1845 (2001)
23. A. Mineshige et al., Metal–insulator transition and crystal structure of $La_{1-x}Sr_xCoO_3$ as functions of Sr-content, temperature, and oxygen partial pressure. J. Solid State Chem. **142** (2), 374–381 (1999)
24. H. Uchida, S. Arisaka, M. Watanabe, High performance electrodes for medium-temperature solid oxide fuel cells: activation of $La(Sr)CoO_3$ cathode with highly dispersed Pt metal electrocatalysts. Solid State Ion. **135**(1–4), 347–351 (2000)
25. Y. Huang, J.M. Vohs, R.J. Gorte, Fabrication of Sr-doped $LaFeO_3$ YSZ composite cathodes. J. Electrochem. Soc. **151**(4), A646–A651 (2004)
26. J.M. Ralph, C. Rossignol, R. Kumar, Cathode materials for reduced-temperature SOFCs. J. Electrochem. Soc. **150**(11), A1518–A1522 (2003)
27. S.P. Simner et al., Performance variability of La (Sr) FeO3 SOFC cathode with Pt, Ag, and Au current collectors. J. Electrochem. Soc. **152**(9), A1851–A1859 (2005)
28. A. Esquirol et al., Electrochemical characterization of La0.6Sr0.4Co0.2Fe0.8 O3 cathodes for intermediate-temperature SOFCs. J. Electrochem. Soc. **151**(11), A1847–A1855 (2004)
29. Z. Shao, S.M. Haile, A high-performance cathode for the next generation of solid-oxide fuel cells. Nature **431**(7005), 170–173 (2004)

30. B. Wei et al., Thermal and electrical properties of new cathode material Ba0.5Sr0.5Co0.8Fe0.2O3 – δ for solid oxide fuel cells. Electrochem. Solid State Lett. **8**(8), A428–A431 (2005)
31. S.P. Jiang, Nanoscale and nano-structured electrodes of solid oxide fuel cells by infiltration: advances and challenges. Int. J. Hydrogen Energy **37**(1), 449–470 (2012)
32. J. Richter et al., Materials design for perovskite SOFC cathodes. Monatsh. Chem. **140**(9), 985–999 (2009)
33. S.J. Skinner, Recent advances in perovskite-type materials for solid oxide fuel cell cathodes. Int. J. Inorg. Mater. **3**(2), 113–121 (2001)
34. A. Tarancon et al., Advances in layered oxide cathodes for intermediate temperature solid oxide fuel cells. J. Mater. Chem. **20**(19), 3799–3813 (2010)
35. A. Bieberle-Hütter, M. Søgaard, H.L. Tuller, Electrical and electrochemical characterization of microstructured thin film $La_{1-x}Sr_xCoO_3$ electrodes. Solid State Ion. **177**(19–25), 1969–1975 (2006)
36. A. Bieberle-Hütter, H. Tuller, Fabrication and structural characterization of interdigitated thin film $La_{1-x}Sr_xCoO_3$ (LSCO) electrodes. J. Electroceram. **16**(2), 151–157 (2006)
37. E.J. Crumlin et al., Oxygen electrocatalysis on epitaxial $La_{0.6}Sr_{0.4}CoO_{3-\delta}$ perovskite thin films for solid oxide fuel cells. J. Electrochem. Soc. **159**(7), F219–F225 (2012)
38. G.J. La O' et al., Catalytic activity enhancement for oxygen reduction on epitaxial perovskite thin films for solid-oxide fuel cells. Angew. Chem. Int. Ed. **49**(31), 5344–5347 (2010)
39. E. Mutoro et al., Enhanced oxygen reduction activity on surface-decorated perovskite thin films for solid oxide fuel cells. Energy Environ. Sci. **4**(9), 3689–3696 (2011)
40. A.C. Johnson et al., Fabrication and electrochemical performance of thin-film solid oxide fuel cells with large area nanostructured membranes. J. Power Sources **195**(4), 1149–1155 (2010)
41. M. Prestat et al., Oxygen reduction at thin dense $La_{0.5}2Sr_{0.48}Co_{0.18}Fe_{0.82}O_{3-\delta}$ electrodes. J. Electroceram. **18**(1–2), 111–120 (2007)
42. M. Prestat, J.-F. Koenig, L. Gauckler, Oxygen reduction at thin dense $La_{0.52}Sr_{0.48}Co_{0.18}Fe_{0.82}O_{3-\delta}$ electrodes. J. Electroceram. **18**(1–2), 87–101 (2007)
43. F.S. Baumann et al., $Ba_{0.5}Sr_{0.5}Co_{0.8}Fe_{0.2}O_{3-\delta}$ thin film microelectrodes investigated by impedance spectroscopy. Solid State Ion. **177**(35–36), 3187–3191 (2006)
44. F.S. Baumann, J. Maier, J. Fleig, The polarization resistance of mixed conducting SOFC cathodes: a comparative study using thin film model electrodes. Solid State Ion. **179**(21–26), 1198–1204 (2008)
45. W.Z. Zhu, S.C. Deevi, A review on the status of anode materials for solid oxide fuel cells. Mater. Sci. Eng. A **362**(1–2), 228–239 (2003)
46. R.J. Gorte, J.M. Vohs, Nanostructured anodes for solid oxide fuel cells. Curr. Opin. Colloid Interface Sci. **14**(4), 236–244 (2009)
47. Y.-H. Huang et al., Double perovskites as anode materials for solid-oxide fuel cells. Science **312**(5771), 254–257 (2006)
48. O.A. Marina, N.L. Canfield, J.W. Stevenson, Thermal, electrical, and electrocatalytical properties of lanthanum-doped strontium titanate. Solid State Ion. **149**(1–2), 21–28 (2002)
49. P.R. Slater, J.T.S. Irvine, Niobium based tetragonal tungsten bronzes as potential anodes for solid oxide fuel cells: synthesis and electrical characterisation. Solid State Ion. **120**(1–4), 125–134 (1999)
50. S. Tao, J.T.S. Irvine, Optimization of mixed conducting properties of Y_2O_3–ZrO_2–TiO_2 and Sc_2O_3–Y_2O_3–ZrO_2–TiO_2 solid solutions as potential SOFC anode materials. J. Solid State Chem. **165**(1), 12–18 (2002)
51. S. Tao, J.T.S. Irvine, A redox-stable efficient anode for solid-oxide fuel cells. Nat. Mater. **2** (5), 320–323 (2003)
52. A. Bieberle, L.P. Meier, L.J. Gauckler, The electrochemistry of Ni pattern anodes used as solid oxide fuel cell model electrodes. J. Electrochem. Soc. **148**(6), A646–A656 (2001)
53. J. Mizusaki et al., Kinetic studies of the reaction at the nickel pattern electrode on YSZ in H_2/ H_2O atmospheres. Solid State Ion. **70–71**(Part 1(0)), 52–58 (1994)

54. B.C.H. Steele, Oxygen transport and exchange in oxide ceramics. J. Power Sources **49**(1–3), 1–14 (1994)
55. B.C.H. Steele, Interfacial reactions associated with ceramic ion transport membranes. Solid State Ion. **75**, 157–165 (1995)
56. M.I. Nandasiri et al., Nanoscale thin film electrolytes for clean energy applications. Nanosci. Nanotechnol. Lett. **4**(2), 124–131 (2012)
57. R. Sanghavi et al., Integrated experimental and modeling study of the ionic conductivity of samaria-doped ceria thin films. Solid State Ion. **204–205**, 13–19 (2011)
58. Z.Q. Yu et al., Integrated experimental and modeling study of ionic conductivity of scandia-stabilized zirconia thin films. Solid State Ion. **181**(8–10), 367–371 (2010)
59. S. Azad et al., Nanoscale effects on ion conductance of layer-by-layer structures of gadolinia-doped ceria and zirconia. Appl. Phys. Lett. **86**(13), 131906-3 (2005)
60. S. Sanna et al., Enhancement of ionic conductivity in Sm-doped ceria/yttria-stabilized zirconia heteroepitaxial structures. Small **6**(17), 1863–1867 (2010)
61. J. Garcia-Barriocanal et al., Colossal ionic conductivity at interfaces of epitaxial ZrO_2:Y_2O_3/$SrTiO_3$ heterostructures. Science **321**(5889), 676–680 (2008)
62. M.I Nandasiri, Engineered interfaces and nano-scale thin films for solid oxide fuel cell electrolytes (2013)
63. S.P.S. Badwal, Zirconia-based solid electrolytes: microstructure, stability and ionic conductivity. Solid State Ion. **52**(1–3), 23–32 (1992)
64. T.H. Etsell, S.N. Flengas, Electrical properties of solid oxide electrolytes. Chem. Rev. **70**(3), 339–376 (1970)
65. M.M. Nasrallah, D.L. Douglass, Ionic and electronic conductivity in Y 2 O 3-doped monoclinic ZrO2. J. Electrochem. Soc. **121**(2), 255–262 (1974)
66. R.J. Stafford, S.J. Rothman, J.L. Routbort, Effect of dopant size on the ionic conductivity of cubic stabilised ZrO_2. Solid State Ion. **37**(1), 67–72 (1989)
67. Y. Arachi et al., Electrical conductivity of the ZrO_2-Ln_2O_3 (Ln = lanthanides) system. Solid State Ion. **121**(1–4), 133–139 (1999)
68. J.M. Dixon et al., Electrical resistivity of stabilized zirconia at elevated temperatures. J. Electrochem. Soc. **110**(4), 276–280 (1963)
69. C. Haering, A. Roosen, H. Schichl, Degradation of the electrical conductivity in stabilised zirconia systems. Part I: yttria-stabilised zirconia. Solid State Ion. **176**(3–4), 253–259 (2005)
70. J. Kondoh et al., Effect of aging on yttria-stabilized zirconia. J. Electrochem. Soc. **145**(5), 1527–1536 (1998)
71. A.I. Ioffe, D.S. Rutman, S.V. Karpachov, On the nature of the conductivity maximum in zirconia-based solid electrolytes. Electrochim. Acta **23**(2), 141–142 (1978)
72. J.W. Fergus, Electrolytes for solid oxide fuel cells. J. Power Sources **162**(1), 30–40 (2006)
73. J. Herle, A.J. McEvoy, K.R. Thampi, Conductivity measurements of various yttria-stabilized zirconia samples. J. Mater. Sci. **29**(14), 3691–3701 (1994)
74. X.J. Chen et al., Influence of microstructure on the ionic conductivity of yttria-stabilized zirconia electrolyte. Mater. Sci. Eng. A **335**(1–2), 246–252 (2002)
75. M. Han et al., Fabrication, microstructure and properties of a YSZ electrolyte for SOFCs. J. Power Sources **165**(2), 757–763 (2007)
76. S. Ikeda et al., Electrical conductivity of yttria-stabilized zirconia single crystals. J. Mater. Sci. **20**(12), 4593–4600 (1985)
77. I. Kosacki et al., Nanoscale effects on the ionic conductivity in highly textured YSZ thin films. Solid State Ion. **176**(13–14), 1319–1326 (2005)
78. A. Rivera, J. Santamaria, C. Leon, Electrical conductivity relaxation in thin-film yttria-stabilized zirconia. Appl. Phys. Lett. **78**(5), 610–612 (2001)
79. Y. Mizutani et al., From rare earth doped zirconia to 1 kW solid oxide fuel cell system. J. Alloys Compd. **408–412**, 518–524 (2006)
80. J.W. Fergus, Doping and defect association in oxides for use in oxygen sensors. J. Mater. Sci. **38**(21), 4259–4270 (2003)

81. V.V. Kharton, F.M.B. Marques, A. Atkinson, Transport properties of solid oxide electrolyte ceramics: a brief review. Solid State Ion. **174**(1–4), 135–149 (2004)
82. K. Nomura et al., Aging and Raman scattering study of scandia and yttria doped zirconia. Solid State Ion. **132**(3–4), 235–239 (2000)
83. O. Yamamoto et al., Electrical conductivity of stabilized zirconia with ytterbia and scandia. Solid State Ion. **79**, 137–142 (1995)
84. S.P.S. Badwal, J. Drennan, Microstructure/conductivity relationship in the scandia-zirconia system. Solid State Ion. **53–56**(Part 2), 769–776 (1992)
85. S.P.S. Badwal, F.T. Ciacchi, D. Milosevic, Scandia-zirconia electrolytes for intermediate temperature solid oxide fuel cell operation. Solid State Ion. **136–137**, 91–99 (2000)
86. H. Huang et al., Structure, local environment, and ionic conduction in scandia stabilized zirconia. Solid State Ion. **179**(27–32), 1442–1445 (2008)
87. D. Lee et al., Characterization of scandia stabilized zirconia prepared by glycine nitrate process and its performance as the electrolyte for IT-SOFC. Solid State Ion. **176**(11–12), 1021–1025 (2005)
88. Y. Mizutani et al., Development of high-performance electrolyte in SOFC. Solid State Ion. **72** (Part 2), 271–275 (1994)
89. C.N. Ginestra et al., Atomic layer deposition of Y2O3 / ZrO2 nanolaminates: a route to ultrathin solid-state electrolyte membranes. Electrochem. Solid State Lett. **10**(10), B161–B165 (2007)
90. X. Guo et al., Ionic conduction in zirconia films of nanometer thickness. Acta Mater. **53**(19), 5161–5166 (2005)
91. S. Heiroth et al., Microstructure and electrical conductivity of YSZ thin films prepared by pulsed laser deposition. Appl. Phys. A Mater. Sci. Proc. **93**(3), 639–643 (2008)
92. I. Kosacki et al., Surface/interface-related conductivity in nanometer thick YSZ films. Electrochem. Solid State Lett. **7**(12), A459–A461 (2004)
93. J.H. Shim et al., Atomic layer deposition of yttria-stabilized zirconia for solid oxide fuel cells. Chem. Mater. **19**(15), 3850–3854 (2007)
94. M. Sillassen et al., Low-temperature superionic conductivity in strained yttria-stabilized zirconia. Adv. Funct. Mater. **20**(13), 2071–2076 (2010)
95. H.-R. Kim et al., 'Illusional' nano-size effect due to artifacts of in-plane conductivity measurements of ultra-thin films. Phys. Chem. Chem. Phys. **13**(13), 6133–6137 (2011)
96. R. Devanathan et al., Computer simulation of defects and oxygen transport in yttria-stabilized zirconia. Solid State Ion. **177**(15–16), 1251–1258 (2006)
97. A. Bogicevic, C. Wolverton, Nature and strength of defect interactions in cubic stabilized zirconia. Phys. Rev. B **67**(2), 024106 (2003)
98. F. Pietrucci et al., Vacancy-vacancy interaction and oxygen diffusion in stabilized cubic ZrO_2 from first principles. Phys. Rev. B **78**(9), 094301 (2008)
99. Y. Yamamura, S. Kawasaki, H. Sakai, Molecular dynamics analysis of ionic conduction mechanism in yttria-stabilized zirconia. Solid State Ion. **126**(1–2), 181–189 (1999)
100. M.O. Zacate et al., Defect cluster formation in M_2O_3-doped cubic ZrO_2. Solid State Ion. **128** (1–4), 243–254 (2000)
101. N. El Habra et al., A study on Sc_2O_3-stabilized zirconia obtained by MOCVD as a potential electrolyte for solid oxide fuel cells. Chem. Vapor Depos. **18**(10–12), 289–294 (2012)
102. H. Shi, R. Ran, Z. Shao, Wet powder spraying fabrication and performance optimization of IT-SOFCs with thin-film ScSZ electrolyte. Int. J. Hydrogen Energy **37**(1), 1125–1132 (2012)
103. B. Ksapabutr et al., Fabrication of scandium stabilized zirconia thin film by electrostatic spray deposition technique for solid oxide fuel cell electrolyte. Thin Solid Films **518**(22), 6518–6521 (2010)
104. Y. Zhang et al., Annealing effects on the phase and microstructure transformations of nanocrystalline $(ZrO_2)_{1-x}(Sc_2O_3)_x$ ($x = 0.02$-0.16) thin films deposited by sol-gel method. Solid State Commun. **122**(7–8), 439–444 (2002)

105. C. Yang et al., Intermediate temperature micro-tubular SOFCs with enhanced performance and thermal stability. Electrochem. Commun. **34**, 231–234 (2013)
106. B. He et al., Ni-Sm_2O_3 cermet anodes for intermediate-temperature solid oxide fuel cells with stabilized zirconia electrolytes. Int. J. Hydrogen Energy **36**(9), 5589–5594 (2011)
107. J.H. Joo, G.M. Choi, Electrical conductivity of scandia-stabilized zirconia thin film. Solid State Ion. **179**(21–26), 1209–1213 (2008)
108. R. Devanathan, S. Thevuthasan, J.D. Gale, Defect interactions and ionic transport in scandia stabilized zirconia. Phys. Chem. Chem. Phys. **11**(26), 5506–5511 (2009)
109. H. Inaba, H. Tagawa, Ceria-based solid electrolytes. Solid State Ion. **83**(1–2), 1–16 (1996)
110. Z. Bi et al., Electrochemical evaluation of $La_{0.6}Sr_{0.4}CoO_3$-$La_{0.45}Ce_{0.55}O_2$ composite cathodes for anode-supported $La_{0.45}Ce_{0.55}O_2$-$La_{0.9}Sr_{0.1}Ga_{0.8}Mg_{0.2}O_{2.85}$ bilayer electrolyte solid oxide fuel cells. Solid State Ion. **176**(7–8), 655–661 (2005)
111. W.G. Wang, M. Mogensen, High-performance lanthanum-ferrite-based cathode for SOFC. Solid State Ion. **176**(5–6), 457–462 (2005)
112. X. Xu et al., Fabrication and performance of functionally graded cathodes for IT-SOFCs based on doped ceria electrolytes. Solid State Ion. **176**(17–18), 1513–1520 (2005)
113. M. Mogensen, N.M. Sammes, G.A. Tompsett, Physical, chemical and electrochemical properties of pure and doped ceria. Solid State Ion. **129**(1–4), 63–94 (2000)
114. H. Yoshida et al., Investigation of the relationship between the ionic conductivity and the local structures of singly and doubly doped ceria compounds using EXAFS measurement. Solid State Ion. **140**(3–4), 191–199 (2001)
115. K. Eguchi, Ceramic materials containing rare earth oxides for solid oxide fuel cell. J. Alloys Compd. **250**(1–2), 486–491 (1997)
116. Y.-P. Fu, S.-B. Wen, C.-H. Lu, Preparation and characterization of Samaria-doped ceria electrolyte materials for solid oxide fuel cells. J. Am. Ceram. Soc. **91**(1), 127–131 (2008)
117. W. Huang, P. Shuk, M. Greenblatt, Properties of sol-gel prepared $Ce_{1-x}Sm_xO_{2-x/2}$ solid electrolytes. Solid State Ion. **100**(1–2), 23–27 (1997)
118. G.-B. Jung, T.-J. Huang, C.-L. Chang, Effect of temperature and dopant concentration on the conductivity of samaria-doped ceria electrolyte. J. Solid State Electrochem. **6**(4), 225–230 (2002)
119. T. Kudo, H. Obayashi, Mixed electrical conduction in the fluorite-type $Ce_{1-x}Gd_xO_{2-x/2}$. J. Electrochem. Soc. **123**(3), 415–419 (1976)
120. D.J. Seo et al., Synthesis and properties of $Ce_{1-x}Gd_xO_{2-x/2}$ solid solution prepared by flame spray pyrolysis. Mater. Res. Bull. **41**(2), 359–366 (2006)
121. Z. Tianshu et al., Ionic conductivity in the CeO_2-Gd_2O_3 system ($0.05 <= Gd/Ce <= 0.4$) prepared by oxalate coprecipitation. Solid State Ion. **148**(3–4), 567–573 (2002)
122. S. Zha, C. Xia, G. Meng, Effect of Gd (Sm) doping on properties of ceria electrolyte for solid oxide fuel cells. J. Power Sources **115**(1), 44–48 (2003)
123. R.N. Blumenthal, F.S. Brugner, J.E. Garnier, The electrical conductivity of CaO-doped nonstoichiometric cerium dioxide from 700° to 1500°C. J. Electrochem. Soc. **120**(9), 1230–1237 (1973)
124. H. Yahiro, K. Eguchi, H. Arai, Electrical properties and reducibilities of ceria-rare earth oxide systems and their application to solid oxide fuel cell. Solid State Ion. **36**(1–2), 71–75 (1989)
125. I. Riess, M. Gödickemeier, L.J. Gauckler, Characterization of solid oxide fuel cells based on solid electrolytes or mixed ionic electronic conductors. Solid State Ion. **90**(1–4), 91–104 (1996)
126. M. Godickemeier, L.J. Gauckler, Engineering of solid oxide fuel cells with ceria-based electrolytes. J. Electrochem. Soc. **145**(2), 414–421 (1998)
127. B.C.H. Steele, Appraisal of $Ce_{1-y}Gd_yO_{2-y/2}$ electrolytes for IT-SOFC operation at 500 °C. Solid State Ion. **129**(1–4), 95–110 (2000)

128. C. Milliken, S. Guruswamy, A. Khandkar, Properties and performance of cation-doped ceria electrolyte materials in solid oxide fuel cell applications. J. Am. Ceram. Soc. **85**(10), 2479–2486 (2002)
129. N. Kim, B.-H. Kim, D. Lee, Effect of co-dopant addition on properties of gadolinia-doped ceria electrolyte. J. Power Sources **90**(2), 139–143 (2000)
130. D.L. Maricle, T.E. Swarr, S. Karavolis, Enhanced ceria – a low-temperature SOFC electrolyte. Solid State Ion. **52**(1–3), 173–182 (1992)
131. A. Tsoga et al., Performance of a double-layer CGO/YSZ electrolyte for solid oxide fuel cells. Ionics **4**(3), 234–240 (1998)
132. S.-G. Kim et al., Fabrication and characterization of a YSZ/YDC composite electrolyte by a sol-gel coating method. J. Power Sources **110**(1), 222–228 (2002)
133. C. Brahim et al., Electrical properties of thin bilayered YSZ/GDC SOFC electrolyte elaborated by sputtering. J. Power Sources **156**(1), 45–49 (2006)
134. M.D. Hurley, D.K. Hohnke, Mixed electrical conduction in $Ce_{1-x}Ca_xO_{2-x}$. J. Phys. Chem. Solid **41**(12), 1349–1353 (1980)
135. D.Y. Wang et al., Oxygen-ion conductivity and defect interactions in yttria-doped ceria. Solid State Ion. **2**(2), 95–105 (1981)
136. H. Yahiro et al., Electrical properties and microstructure in the system ceria-alkaline earth oxide. J. Mater. Sci. **23**(3), 1036–1041 (1988)
137. Y. Gu et al., Sintering and electrical properties of coprecipitation prepared $Ce_{0.8}Y_{0.2}O_{1.9}$ ceramics. Mater. Res. Bull. **35**(2), 297–304 (2000)
138. B. Zhu et al., Cost-effective yttrium doped ceria-based composite ceramic materials for intermediate temperature solid oxide fuel cell applications. J. Mater. Sci. Lett. **20**(7), 591–594 (2001)
139. F.-Y. Wang, S. Chen, S. Cheng, Gd^{3+} and Sm^{3+} co-doped ceria based electrolytes for intermediate temperature solid oxide fuel cells. Electrochem. Commun. **6**(8), 743–746 (2004)
140. F.-Y. Wang et al., Study on Gd and Mg co-doped ceria electrolyte for intermediate temperature solid oxide fuel cells. Catal. Today **97**(2–3), 189–194 (2004)
141. S. Omar, E.D. Wachsman, J.C. Nino, A co-doping approach towards enhanced ionic conductivity in fluorite-based electrolytes. Solid State Ion. **177**(35–36), 3199–3203 (2006)
142. S. Omar, E.D. Wachsman, J.C. Nino, Higher ionic conductive ceria-based electrolytes for solid oxide fuel cells. Appl. Phys. Lett. **91**(14), 144106-3 (2007)
143. X. Sha et al., Study on La and Y co-doped ceria-based electrolyte materials. J. Alloys Compd. **428**(1–2), 59–64 (2007)
144. Y. Tsung-Her, C. Chen-Chia, Ionic conductivity investigation in samarium and strontium co-doped ceria system. Phys. Scripta **2007**(T129), 303 (2007)
145. P. Devi, S. Banerjee, Search for new oxide-ion conducting materials in the ceria family of oxides. Ionics **14**(1), 73–78 (2008)
146. X. Guan et al., High performance Gd^{3+} and Y^{3+} co-doped ceria-based electrolytes for intermediate temperature solid oxide fuel cells. Mater. Res. Bull. **43**(4), 1046–1054 (2008)
147. S. Omar, E.D. Wachsman, J.C. Nino, Higher conductivity Sm^{3+} and Nd^{3+} co-doped ceria-based electrolyte materials. Solid State Ion. **178**(37–38), 1890–1897 (2008)
148. Y. Zheng et al., La and Ca co-doped ceria-based electrolyte materials for IT-SOFCs. Mater. Res. Bull. **44**(8), 1717–1721 (2009)
149. Y. Zheng et al., The effect of Sr on the properties of Y-doped ceria electrolyte for IT-SOFCs. J. Alloys Compd. **486**(1–2), 586–589 (2009)
150. B. Li, X. Wei, W. Pan, Improved electrical conductivity of $Ce_{0.9}Gd_{0.1}O_{1.95}$ and $Ce_{0.9}Sm_{0.1}O_{1.95}$ by co-doping. Int. J. Hydrogen Energy **35**(7), 3018–3022 (2010)
151. T.S. Zhang et al., Aging behavior and ionic conductivity of ceria-based ceramics: a comparative study. Solid State Ion. **170**(3–4), 209–217 (2004)
152. T.S. Zhang et al., High-temperature aging behavior of Gd-doped ceria. Electrochem. Solid State Lett. **7**(6), J13–J15 (2004)

153. T.S. Zhang et al., Effects of dopant concentration and aging on the electrical properties of Y-doped ceria electrolytes. Solid State Sci. **5**(11–12), 1505–1511 (2003)
154. S. Wang et al., Nonstoichiometry of $Ce_{0.9}Gd_{0.1}O_{1.95-x}$. Solid State Ion. **107**(1–2), 73–79 (1998)
155. S. Wang et al., Electrical and ionic conductivity of Gd-doped ceria. J. Electrochem. Soc. **147** (10), 3606–3609 (2000)
156. V.V. Kharton et al., Ceria-based materials for solid oxide fuel cells. J. Mater. Sci. **36**(5), 1105–1117 (2001)
157. E. Jud, L. Gauckler, The effect of cobalt oxide addition on the conductivity of $Ce_{0.9}Gd_{0.1}O_{1.95}$. J. Electroceram. **15**(2), 159–166 (2005)
158. J.R. Jurado, Present several items on ceria-based ceramic electrolytes: synthesis, additive effects, reactivity and electrochemical behaviour. J. Mater. Sci. **36**(5), 1133–1139 (2001)
159. A. Sin et al., Preparation and sintering of $Ce_{1-x}Gd_xO_{2-x/2}$ nanopowders and their electrochemical and EPR characterization. Solid State Ion. **175**(1–4), 361–366 (2004)
160. H.S. Kang et al., The characteristics of nano-sized Gd-doped CeO_2 particles prepared by spray pyrolysis. J. Alloys Compd. **398**(1–2), 240–244 (2005)
161. P. Datta, P. Majewski, F. Aldinger, Study of gadolinia-doped ceria solid electrolyte surface by XPS. Mater Charact **60**(2), 138–143 (2009)
162. Z. He et al., Densification and grain growth during early-stage sintering of $Ce_{0.9}Gd_{0.1}O_{1.95-\delta}$ in a reducing atmosphere. Acta Mater. **58**(11), 3860–3866 (2010)
163. M. Sahibzada et al., Development of solid oxide fuel cells based on a $Ce(Gd)O_{2-x}$ electrolyte film for intermediate temperature operation. Catal. Today **38**(4), 459–466 (1997)
164. B. Dalslet et al., Assessment of doped ceria as electrolyte. J. Solid State Electrochem. **10**(8), 547–561 (2006)
165. H. Yahiro et al., Oxygen ion conductivity of the ceria-samarium oxide system with fluorite structure. J. Appl. Electrochem. **18**(4), 527–531 (1988)
166. T. Inoue et al., Study of a solid oxide fuel cell with a ceria-based solid electrolyte. Solid State Ion. **35**(3–4), 285–291 (1989)
167. K. Eguchi et al., Electrical properties of ceria-based oxides and their application to solid oxide fuel cells. Solid State Ion. **52**(1–3), 165–172 (1992)
168. S. Gupta et al., Influence of samaria doping on the resistance of ceria thin films and its implications to the planar oxygen sensing devices. Sens. Actuators B Chem. **139**(2), 380–386 (2009)
169. Z. Zhan et al., AC impedance investigation of samarium-doped ceria. J. Electrochem. Soc. **148**(5), A427–A432 (2001)
170. Z.Q. Yu et al., Conductivity of oriented Samaria-doped ceria thin films grown by oxygen-plasma-assisted molecular beam epitaxy. Electrochem. Solid State Lett. **11**(5), B76–B78 (2008)
171. C. Mansilla et al., Microstructure and transport properties of ceria and samaria doped ceria thin films prepared by EBE-IBAD. Surf. Coat. Technol. **202**(4–7), 1256–1261 (2007)
172. D. Bera et al., Growth and characterization of highly oriented gadolinia-doped ceria(111) thin films on zirconia(111)/sapphire(0001) substrates. Thin Solid Films **516**(18), 6088–6094 (2008)
173. S. Cho et al., Microstructural and electrical properties of $Ce_{0.9}Gd_{0.1}O_{1.95}$ thin-film electrolyte in solid-oxide fuel cells. J. Mater. Res. **26**(07), 854–859 (2011)
174. C. Ding et al., A simple, rapid spray method for preparing anode-supported solid oxide fuel cells with GDC electrolyte thin films. J. Membr. Sci. **350**(1–2), 1–4 (2010)
175. E. Gourba et al., Characterisation of thin films of ceria-based electrolytes for intermediate temperature — solid oxide fuel cells (IT-SOFC). Ionics **9**(1), 15–20 (2003)
176. Y.J. Leng et al., Low-temperature SOFC with thin film GDC electrolyte prepared in situ by solid-state reaction. Solid State Ion. **170**(1–2), 9–15 (2004)

177. D.-H. Myung et al., The effect of an ultra-thin zirconia blocking layer on the performance of a 1-μm-thick gadolinia-doped ceria electrolyte solid-oxide fuel cell. J. Power Sources **206**, 91–96 (2012)
178. T. Suzuki, I. Kosacki, H.U. Anderson, Microstructure-electrical conductivity relationships in nanocrystalline ceria thin films. Solid State Ion. **151**(1–4), 111–121 (2002)
179. A. Infortuna, A.S. Harvey, L.J. Gauckler, Microstructures of CGO and YSZ thin films by pulsed laser deposition. Adv. Funct. Mater. **18**(1), 127–135 (2008)
180. J.H. Joo, G.M. Choi, Electrical conductivity of thin film ceria grown by pulsed laser deposition. J. Eur. Ceram. Soc. **27**(13–15), 4273–4277 (2007)
181. Y.-L. Kuo et al., Gadolinia-doped ceria films deposited by RF reactive magnetron sputtering. Solid State Ion. **180**(26–27), 1421–1428 (2009)
182. S.-E. Lin et al., Characterization of electrolyte films deposited by using RF magnetron sputtering a 20 mol% gadolinia-doped ceria target. Thin Solid Films **518**(24), 7229–7232 (2010)
183. G. Meng et al., Application of novel aerosol-assisted chemical vapor deposition techniques for SOFC thin films. Solid State Ion. **175**(1–4), 29–34 (2004)
184. M.V.F. Schlupp et al., Gadolinia doped ceria thin films prepared by aerosol assisted chemical vapor deposition and applications in intermediate-temperature solid oxide fuel cells. Fuel Cells **13**(5), 658–665 (2013)
185. H.Z. Song et al., Aerosol-assisted MOCVD growth of Gd_2O_3-doped CeO_2 thin SOFC electrolyte film on anode substrate. Solid State Ion. **156**(3–4), 249–254 (2003)
186. Y.S. Hong et al., Fabrication and characterization GDC electrolyte thin films by e-beam technique for IT-SOFC. Curr. Appl. Phys. **11**(5 Suppl), S163–S168 (2011)
187. M.G. Chourashiya, L.D. Jadhav, Synthesis and characterization of 10%Gd doped ceria (GDC) deposited on NiO-GDC anode-grade-ceramic substrate as half cell for IT-SOFC. Int. J. Hydrogen Energy **36**(22), 14984–14995 (2011)
188. Y.-G. Choi et al., Ceria-based electrolyte reinforced by sol–gel technique for intermediate-temperature solid oxide fuel cells. Int. J. Hydrogen Energy **38**(23), 9867–9872 (2013)
189. S. Cho et al., High power density thin film SOFCs with YSZ/GDC bilayer electrolyte. Electrochim. Acta **56**(16), 5472–5477 (2011)
190. H. Choi, G. Cho, S.-W. Cha, Fabrication and characterization of anode supported YSZ/GDC bilayer electrolyte SOFC using dry press process. Int. J. Prec. Eng. Manuf. Green Tech. **1**(2), 95–99 (2014)
191. T. Mukai et al., Fabrication of Y_2O_3-doped zirconia/gadolinia-doped ceria bilayer electrolyte thin film SOFC cells of SOFCs by single-pulsed laser deposition processing. J. Fuel Cell Sci. Tech. **10**(6), 061006 (2013)
192. H.-S. Noh et al., The potential and challenges of thin-film electrolyte and nanostructured electrode for yttria-stabilized zirconia-base anode-supported solid oxide fuel cells. J. Power Sources **247**, 105–111 (2014)
193. K.-R. Lee et al., Lattice distortion effect on electrical properties of GDC thin films: experimental evidence and computational simulation. Solid State Ion. **229**, 45–53 (2012)
194. D.A. Andersson et al., Optimization of ionic conductivity in doped ceria. Proc. Natl. Acad. Sci. U. S. A. **103**(10), 3518–3521 (2006)
195. G. Laukaitis, J. Dudonis, D. Virbukas, Samarium doped cerium oxide thin films deposited by e-beam technique. Surf. Coat. Technol. **204**(12–13), 2028–2031 (2010)
196. Z.Q. Yu et al., Growth and structure of epitaxial $Ce_{0.8}Sm_{0.2}O_{1.9}$ by oxygen-plasma-assisted molecular beam epitaxy. J. Cryst. Growth **310**(10), 2450–2456 (2008)
197. S. Sanna et al., Fabrication and electrochemical properties of epitaxial samarium-doped ceria films on $SrTiO_3$-buffered MgO substrates. Adv. Funct. Mater. **19**(11), 1713–1719 (2009)
198. N. Ai et al., Low temperature solid oxide fuel cells based on $Sm_{0.2}Ce_{0.8}O_{1.9}$ films fabricated by slurry spin coating. J. Power Sources **159**(1), 637–640 (2006)
199. R. Hui et al., Fabrication of ceramic films for solid oxide fuel cells via slurry spin coating technique. J. Power Sources **172**(2), 840–844 (2007)

200. R.-J. Yang et al., Fabrication and characterization of a $Sm_{0.2}Ce_{0.8}O_{1.9}$ electrolyte film by the spin-coating method for a low-temperature anode-supported solid oxide fuel cells. J. Power Sources **206**, 111–118 (2012)
201. Y. Yoo, Fabrication and characterization of thin film electrolytes deposited by RF magnetron sputtering for low temperature solid oxide fuel cells. J. Power Sources **160**(1), 202–206 (2006)
202. J. Maier, Defect chemistry and conductivity effects in heterogeneous solid electrolytes. J. Electrochem. Soc. **134**(6), 1524–1535 (1987)
203. J. Maier, Ionic conduction in space charge regions. Prog. Solid State Chem. **23**(3), 171–263 (1995)
204. S. Ramanathan, Interface-mediated ultrafast carrier conduction in oxide thin films and superlattices for energy. J. Vac. Sci. Technol. A **27**(5), 1126–1134 (2009)
205. X. Guo, J. Maier, Comprehensive modeling of ion conduction of nanosized CaF_2/BaF_2 multilayer heterostructures. Adv. Funct. Mater. **19**(1), 96–101 (2009)
206. N. Sata et al., Mesoscopic fast ion conduction in nanometre-scale planar heterostructures. Nature **408**(6815), 946–949 (2000)
207. S. Azad et al., *Growth and Characterization of Single-Crystal Multilayer Nanostructures for Fast Ion Conduction*, in Nanotechnology and the Environment (American Chemical Society, Washington, DC, 2004), pp. 133–141
208. A. Cavallaro et al., Electronic nature of the enhanced conductivity in YSZ-STO multilayers deposited by PLD. Solid State Ion. **181**(13–14), 592–601 (2010)
209. C. Korte et al., Ionic conductivity and activation energy for oxygen ion transport in superlattices-the semicoherent multilayer system YSZ (ZrO_2 + 9.5 mol% Y_2O_3)/Y_2O_3. Phys. Chem. Chem. Phys. **10**(31), 4623–4635 (2008)
210. A. Peters et al., Ionic conductivity and activation energy for oxygen ion transport in superlattices – the multilayer system CSZ (ZrO2 + CaO) / Al2O3. Solid State Ion. **178** (1–2), 67–76 (2007)
211. N. Schichtel et al., Elastic strain at interfaces and its influence on ionic conductivity in nanoscaled solid electrolyte thin films-theoretical considerations and experimental studies. Phys. Chem. Chem. Phys. **11**(17), 3043–3048 (2009)
212. C.M. Wang et al., Microstructure of ZrO_2-CeO_2 hetero-multi-layer films grown on YSZ substrate. Acta Mater. **53**(7), 1921–1929 (2005)
213. C.M. Wang et al., Distribution of oxygen vacancies and gadolinium dopants in ZrO_2-CeO_2 multi-layer films grown on [alpha]-Al_2O_3. Solid State Ion. **177**(15–16), 1299–1306 (2006)
214. Y. Wang et al., Microstructure and ionic conductivity of alternating-multilayer structured Gd-doped ceria and zirconia thin films. J. Mater. Sci. **44**(8), 2021–2026 (2009)
215. X. Guo, Comment on "colossal ionic conductivity at interfaces of epitaxial ZrO_2:Y_2O_3/$SrTiO_3$ heterostructures". Science **324**(5926), 465 (2009)
216. T.J. Pennycook et al., Seeing oxygen disorder in YSZ/SrTiO3 colossal ionic conductor 1518 heterostructures using EELS. Eur. Phys. J. Appl. Phys. **54**(3), 33507 (2011)
217. T.J. Pennycook et al., Origin of colossal ionic conductivity in oxide multilayers: interface induced sublattice disorder. Phys. Rev. Lett. **104**(11), 115901 (2010)
218. C. Korte et al., Influence of interface structure on mass transport in phase boundaries between different ionic materials. Monatsh. Chem. **140**(9), 1069–1080 (2009)
219. R. Hull, J.C. Bean, Misfit dislocations in lattice-mismatched epitaxial films. Crit. Rev. Solid State Mater. Sci. **17**(6), 507–546 (1992)
220. J.R. Willis, S.C. Jain, R. Bullough, The energy of an array of dislocations: implications for strain relaxation in semiconductor heterostructures. Philos. Mag. A **62**(1), 115–129 (1990)
221. G. Gutekunst, J. Mayer, M. Rühle, Atomic structure of epitaxial Nb-Al_2O_3 interfaces I. Coherent regions. Philos. Mag. A **75**(5), 1329–1355 (1997)
222. G. Gutekunst et al., Atomic structure of epitaxial Nb-Al_2O_3 interfaces II. Misfit dislocations. Philos. Mag. A **75**(5), 1357–1382 (1997)

223. S.I. Yi et al., Morphological and structural investigation of the early stages of epitaxial growth of α-Fe_2O_3 (0001) on α-Al_2O_3 (0001) by oxygen-plasma-assisted MBE. Surf. Sci. **443** (3), 212–220 (1999)
224. D.-J. Kim, Lattice parameters, ionic conductivities, and solubility limits in fluorite-structure MO_2 oxide [$M = Hf^{4+}$, Zr^{4+}, Ce^{4+}, Th^{4+}, U^{4+}] solid solutions. J. Am. Ceram. Soc. **72**(8), 1415–1421 (1989)

Chapter 7
Thin Film Thermoelectric Materials for Sensor Applications: An Overview

Ashoka Bali, Raju Chetty, and Ramesh Chandra Mallik

Abstract The modern world depends completely on electricity, the demand for which is increasing day by day. However, carbon-based fuels for power generation and transportation being currently used, e.g., coal, petroleum, etc., are expected to ultimately deplete, due to which there is an urgent need for non-conventional energy sources. At the same time, a large (~60 %) amount of heat is wasted from these industries and vehicles, which will increase further with economic growth and industrialization. Other renewable sources have their own limitations and many are region-specific. Thermoelectric (TE) materials are a means of utilization of waste heat on a small scale by directly converting it into electricity. The generated voltage also finds application in several types of sensors. In addition, these materials can be used for cooling, which is equivalent to direct conversion of electricity into heat, without using any harmful chlorofluorocarbon (CFC) gases. This chapter is intended to give an overview of thin film thermoelectric materials with a focus on their application as different sensors.

The chapter is divided into six sections. Section 7.1 introduces the phenomenon of thermoelectricity—Seebeck, Peltier, and Thomson effects, and the thermoelectric figure of merit (zT). Section 7.2 deals with the basics of transport properties, while in Sect. 7.3 the different strategies to improve zT are briefly summarized. Section 7.4 describes the need of going towards lower dimensions, i.e., the thin film state, for improvement of zT. Different fabrication methods are summarized in Sect. 7.5. Finally, in Sect. 7.6, application of thermoelectric thin films as different types of sensors is described.

A. Bali • R. Chetty • R.C. Mallik (✉)
Thermoelectric Materials and Devices Laboratory, Department of Physics,
Indian Institute of Science, Bangalore 560 012, Karnataka, India
e-mail: rcmallik@physics.iisc.ernet.in

S. Babu Krishna Moorthy (ed.), *Thin Film Structures in Energy Applications*,
DOI 10.1007/978-3-319-14774-1_7

7.1 Introduction to Thermoelectric Effects

Thermoelectricity basically involves the conversion of thermal energy into electrical energy, and vice-versa. It deals mainly with three effects—Seebeck effect, Peltier effect and Thomson effect, which are discussed below.

7.1.1 *Seebeck Effect*

It is one of the most important fundamental thermoelectric phenomena, which was discovered by Thomas Johann Seebeck in 1821. When the two junctions of dissimilar materials are joined together and maintained at different temperatures, a potential difference can be produced. This phenomenon is known as Seebeck effect (Fig. 7.1).

The potential difference (dV) that develops at the junctions is proportional to the temperature difference (dT) applied to it. The proportionality constant, which is an intrinsic property of material, is called Seebeck coefficient (S) or thermopower and is defined as the ratio of voltage produced to the temperature difference.

$$S = \frac{\mathrm{d}V}{\mathrm{d}T}. \tag{7.1}$$

Seebeck effect occurs microscopically when temperature difference is applied across a material and can be explained as follows. The carriers from hot side of the material diffuse towards the cold side, leaving the positive charge at hot end and resulting in negative charge buildup at cold side. This happens because the carriers gain more kinetic energy at hot side than the cold side. The diffusion of carriers from hot end to the cold end continues until it develops the opposite potential, which stops any further diffusion of carriers when the system is in equilibrium.

7.1.2 *Peltier Effect*

The reverse of Seebeck effect was observed by Jean Charles Athanase Peltier in 1834, which is known as Peltier effect. This effect states that when an electrical

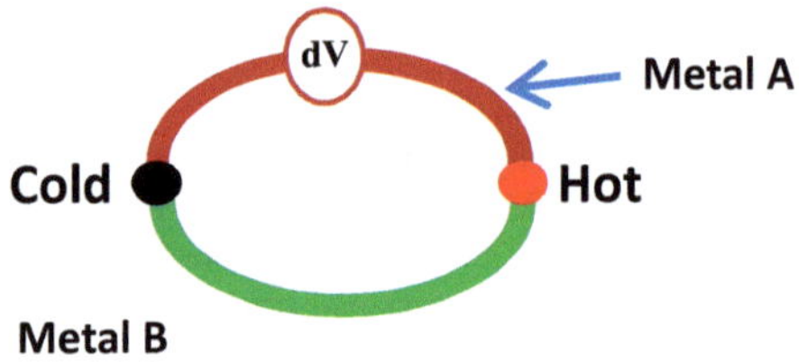

Fig. 7.1 A schematic diagram showing the Seebeck effect

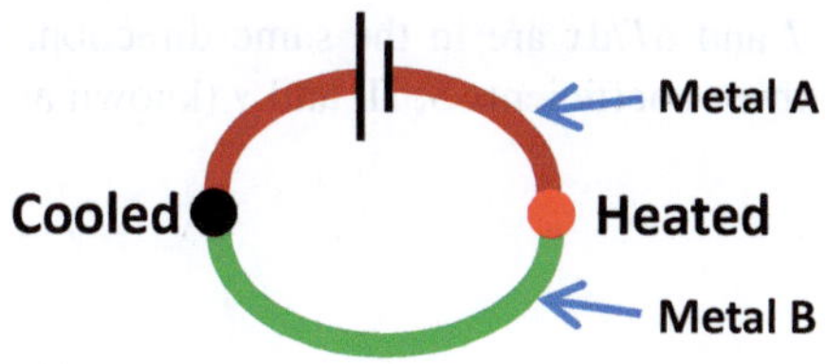

Fig. 7.2 A schematic diagram showing the Peltier effect

current passes across the junctions of two dissimilar materials, heat will be evolved at one end of the junction and absorbed at other (Fig. 7.2).

The heat absorbed or evolved is directly proportional to the electrical current passed through the junction. The proportionality constant (Π) is known as the Peltier coefficient and is defined as the ratio of amount heat energy absorbed or liberated per unit time ($\dot{Q}$) to the electrical current (I) passed across the junction

$$\dot{Q} = (\Pi_{\mathrm{A}} - \Pi_{\mathrm{B}})I. \tag{7.2}$$

In the above equation, Π_{A} and Π_{B} are the Peltier coefficients of materials A and B, respectively.

Peltier heat is dependent on the direction of flow of the current passed across junction. It is reversible, which means that if heat is evolved when current passes through metal A to metal B, the same amount of heat is absorbed when the current direction is reversed. Peltier heat can be easily distinguished from Joule heat, which is independent of current direction. The reversible heat mainly depends on the intrinsic properties of metals (absolute Peltier coefficient), not on the contact between them. Two metals are necessary to explain Seebeck and Peltier effects, while the absolute coefficients of a single material can be explained by Thomson effect.

7.1.3 Thomson Effect

Thomson effect was observed by William Thomson, who was also known as Lord Kelvin, in 1854. Thomson effect states that the liberation or absorption of heat takes place in a homogeneous current carrying conductor when the temperature gradient is applied across it. The amount of net heat produced per unit volume ($\dot{q}$) of a conductor is given by

$$\dot{q} = \rho J^2 - \mu J \frac{\mathrm{d}T}{\mathrm{d}x}. \tag{7.3}$$

Here ρ is the resistivity, J the current density, and μ is the Thomson coefficient. The first term in above equation is irreversible Joule heating and the second term is the reversible Thomson heat. Thomson coefficient depends on the temperature gradient and current density. μ is positive if heat is absorbed from the surroundings when

J and $\mathrm{d}T/\mathrm{d}x$ are in the same direction. Thomson derived the relationship among three coefficients S, Π, and γ (known as Kelvin relations), which are given as:

$$\gamma = T\frac{\mathrm{d}S}{\mathrm{d}T}, \tag{7.4}$$

$$\Pi = TS. \tag{7.5}$$

Using these relations, the absolute Seebeck coefficient and Peltier coefficient can be calculated if temperature-dependent Thomson coefficient is known.

7.1.4 *Thermoelectric Figure of Merit* (*z*T)

Edmund Altenkirch proposed a theory for the efficiency of thermoelectric generator and the coefficient of performance of the refrigerator, according to which a material should possess large Seebeck coefficient (S) (to generate large electrical power), low electrical resistivity (ρ) (to minimize the Joule heating effect), and low thermal conductivity (λ) (to maintain the large temperature difference across the junction of material). The S, ρ, and λ are related to each other and defined by the parameter called the dimensionless thermoelectric figure of merit (zT)

$$z\mathrm{T} = \frac{S^2}{\rho\lambda}T, \tag{7.6}$$

where S is the Seebeck coefficient, ρ the electrical resistivity, and λ is the total thermal conductivity $\lambda = \lambda_\mathrm{C} + \lambda_\mathrm{L}$, where λ_C is the carrier and λ_L is the lattice contribution.

Thermoelectric figure of merit (zT) for a single material will only not serve the purpose of the research, because a series of TE couples are required to be used in module or device. Generally, thermoelectric modules or devices consist of two types of semiconductors with similar zT, which are connected electrically in series and thermally in parallel. Thermoelectric figure of merit of the device, denoted by ZT, is given by [1, 2]:

$$ZT = \frac{(S_\mathrm{p} - S_\mathrm{n})^2 T}{\left[(\rho_\mathrm{p}\lambda_\mathrm{p})^{1/2} + (\rho_\mathrm{n}\lambda_\mathrm{n})^{1/2}\right]^2}. \tag{7.7}$$

7.2 Transport Properties

This section describes briefly the three transport parameters—Seebeck coefficient, electrical resistivity, and thermal conductivity.

7.2.1 Seebeck Coefficient

The Seebeck coefficient or thermopower of a material is a voltage induced by the temperature gradient across it. It is also defined as the entropy per carrier and is measured in terms of μV/K. The Seebeck coefficient depends on the type of the carriers, i.e., Seebeck coefficient of n-type (S_n) carriers is negative and that of p-type (S_p) carriers is positive. The expression for degenerate semiconductors is given below [3]:

$$S = -\frac{\pi^2 k_B^2 T}{3eE_F}\left[\left(\frac{\mathrm{d}\ln n(E)}{\mathrm{d}\ln E}\right) + \frac{\mathrm{d}\ln v^2(E)}{\mathrm{d}E} + \frac{\mathrm{d}\ln \tau(E)}{\mathrm{d}E}\right], \tag{7.8}$$

where n(E) is the density of states, $v(E)$ the average velocity of charge carrier, τ(E) the relaxation time of charge carriers, E_F the Fermi energy, e the electron charge, T the absolute temperature, and k_B is the Boltzmann constant. Three parameters n(E), v(E) and τ(E) are functions of the charge carrier energy.

For non-degenerate semiconductors, Seebeck coefficient can be expressed as [3]:

$$S = \left(\frac{k_B}{e}\right)\left[\frac{5}{2} + \left(\frac{\mathrm{d}\ln \tau(E)}{\mathrm{d}\ln E}\right) - \left(\frac{E_C - E_F}{k_B T}\right)\right] \quad \text{for n-type material,} \tag{7.9a}$$

and

$$S = \left(\frac{k_B}{e}\right)\left[\frac{5}{2} + \left(\frac{\mathrm{d}\ln \tau(E)}{\mathrm{d}\ln E}\right) - \left(\frac{E_F - E_V}{k_B T}\right)\right] \quad \text{for p-type material,} \tag{7.9b}$$

where E_V and E_C are the energies of valence band maximum and conduction band minimum, respectively.

The Fermi energy can be expressed as [4]

$$E_F = k_B T \ln\left(\frac{1}{2} N h^3 (2\pi m^* k_B T)^{-3/2}\right). \tag{7.10}$$

Here, N is the density of available free charge carriers, which is a function of temperature and effective mass (m*) of carriers, and h is Planck's constant.

When two types of charge carriers (say electrons and holes) are present in a material and participate in diffusion under temperature gradient, the Seebeck coefficient is reduced. In such a case, it is given by the following expression [4]

$$S = \frac{\sigma_n S_n + \sigma_p S_p}{\sigma_n + \sigma_p}. \tag{7.11}$$

Here, σ_n and σ_p are the electrical conductivities for electron and holes, respectively, while S_n and S_p are the Seebeck coefficients for electrons and holes, respectively.

7.2.2 Electrical Resistivity

Electrical resistivity is one of the influencing parameters in the improvement of *z*T. ρ must be as low as possible to optimize the *z*T. Electrical resistivity is an intrinsic property of a material, which is an indicator of opposition to the movement of electric charge. Electrical resistivity of a material is dependent on the carrier concentration (n) and mobility (μ) and can be expressed as

$$\rho = \frac{1}{ne\mu}, \tag{7.12}$$

where e is the electric charge ($e = 1.6 \times 10^{-19}$ C). Electrical resistivity varies with temperature. Electrical resistivity of metals increases with increase in temperature due to decrease in carrier mobility by carrier-phonon scattering, whereas the carrier concentration is almost independent of temperature. For semiconductors, ρ varies exponentially with temperature because of increase in carrier concentration. Energy gap (E_g) of semiconductor can be calculated using the temperature-dependent electrical resistivity and is given by

$$\rho = \rho_o \exp\left(\frac{E_g}{2k_B T}\right), \tag{7.13}$$

where $\rho_o = \text{constant} \times \frac{1}{e(\mu_e + \mu_h)} T^{-3/2}$ μ_e is electron mobility, μ_h hole mobility, and T is the temperature.

7.2.3 Thermal Conductivity

Thermal conductivity of a material should be as low as possible to achieve high thermoelectric figure of merit. It is an ability of a material to transport the heat energy. Thermal conductivity can be expressed by using the Fourier's law as [5]

$$\lambda = \frac{\vec{Q}}{\nabla T}, \tag{7.14}$$

where $\vec{Q}$ is heat flow rate vector, ∇T temperature gradient, and λ is thermal conductivity.

Generally, thermal conductivity of a solid can be expressed as

$$\lambda_T = \lambda_C + \lambda_L, \tag{7.15}$$

where λ_T is total thermal conductivity, λ_C carrier contribution, and λ_L is due to the lattice contribution. In metals, thermal conductivity is mainly dominated by the free carrier contribution and lattice contribution is negligible. The electrical resistivity and thermal conductivity are related by the Wiedemann–Franz relation, expressed as [6]

$$\lambda_C = \frac{LT}{\rho}, \tag{7.16}$$

where

$$L = \frac{\pi^2}{3}\left(\frac{k_B}{e}\right)^2 = 2.44 \times 10^{-8}\ \text{W}\Omega\text{K}^{-2}. \tag{7.17}$$

Hence, the reduction in carrier thermal conductivity is not a good choice for thermoelectric materials, because it leads to an increase of electrical resistivity. In case of semiconductors, thermal conductivity is contributed by both carriers and phonons. The lattice part of thermal conductivity can be expressed as

$$\lambda_L = \frac{1}{3} C_v \bar{v} l, \tag{7.18}$$

where C_v is the specific heat at constant volume, $\bar{v}$ is the average speed of phonons, and l is the mean free path of the phonons. Therefore, tuning of lattice thermal conductivity is possible for the enhancement of zT, while it is difficult to tune λ_C because of dependence of S and ρ on carrier concentration.

When there are sufficient number of both electrons and holes, and their concentrations are unequal, the semiconductor is said to show mixed conduction. When the concentrations of electrons and holes are equal, the semiconductor is said to be intrinsic. In case of bipolar thermal diffusion of carriers, the carrier thermal conductivity can be expressed by [2, 5]

$$\lambda_C = \lambda_e + \lambda_h + \frac{\sigma_n \sigma_p}{\sigma_n + \sigma_p}\left(S_p - S_n\right)^2, \tag{7.19}$$

where λ_e and λ_h are thermal conductivities of electrons and holes, σ_n and σ_p are partial electrical conductivities, and S_n and S_p are Seebeck coefficients of electrons and holes, respectively. Total carrier thermal conductivity is the sum of the partial thermal conductivities due to electrons and holes and bipolar thermal diffusion term. Bipolar thermal diffusion effect is mostly found in semiconductors of narrow band gap or at high temperatures.

7.3 Improvement of Thermoelectric Figure of Merit

As already mentioned, in order to achieve large zT, a material should possess high Seebeck coefficient, low electrical resistivity, and thermal conductivity. But these three parameters are interdependent on each other and dependent on crystal structure of the material, its electronic structure, and carrier concentration [7, 8]. In case of metals, electrical resistivity as well as Seebeck coefficient is very low. In case of insulators, electrical resistivity and Seebeck coefficient are very high, which lead to lowering of the power factor (S^2/ρ). Therefore, metals and insulators are not a choice for thermoelectric materials. In case of semiconductors, electrical resistivity and Seebeck coefficient values are in range that is between metals and insulators, which can lead to optimum power factor (S^2/ρ). This can be achieved by optimizing the carrier concentration in the range of 10^{19} to 10^{21} cm^{-3} (depending on the material system) [8]. The improvement of figure of merit can be achieved by either increasing the power factor or reducing thermal conductivity. Thermoelectric figure of merit for bulk can be improved by reducing thermal conductivity by using the heavy constituent elements, alloying, and using compounds with complex crystal structures. Also, zT can be enhanced by increasing the power factor through resonant energy level doping. In this chapter, the main focus is on the enhancement of zT for low-dimensional materials.

7.4 Enhancement of zT for Low-Dimension Materials

The increase of Seebeck coefficient without disturbing electrical resistivity is highly essential to improve the power factor in thermoelectric research. Thermal conductivity of bulk materials is restricted by the alloying limit. Therefore, the decoupling of transport properties, namely S, ρ, and λ, is an interesting topic in thermoelectric research, which can be addressed by the reducing the dimensions of the materials. Low-dimensional thermoelectric materials are expected to have better thermoelectric performance than the bulk materials due to (1) increase of power factor caused by quantum confinement by large density of states near the Fermi level, and (2) decrease of thermal conductivity by increasing the number of interfaces and surfaces that causes enhancement of phonon scattering [9]. This idea was first introduced by Hicks and Dresselhaus [10, 11] in 1993. The effect of quantum well or two-dimensional structures on thermoelectric properties was studied theoretically by assuming parabolic bands and time-independent relaxation time in one band material [10]. Electrical conductivity of two-dimensional materials [10, 12] is expressed as

$$\sigma_{2D} = \frac{1}{\rho_{2D}} = \frac{1}{2\pi w}\left(m_x^* m_y^*\right)^{1/2} F_0 e\mu_x \left(\frac{2k_B T}{h^2}\right), \tag{7.20}$$

where w is the quantum well width, m_x^* and m_y^* are the effective mass components in the x and y direction (in the plane of the 2D material), respectively, F_0 is the Fermi integral, and μ_x is the mobility along x-direction. From the above equation, it is clear that the reduction in width of quantum well increases the electrical conductivity.

Seebeck coefficient can be expressed as [10, 12]

$$S_{2D} = \frac{-k_B}{e}\left(\frac{2F_1}{F_0} - \chi^*_{2D}\right), \tag{7.21}$$

and

$$\chi^*_{2D} = \frac{1}{k_B T}\left(E_F - \frac{\pi^2 k_B^2 T^2}{12 E_F} - \frac{h^2 \pi^2}{2 m_z^* w^2}\right), \tag{7.22}$$

where m_z^* is the component of the effective mass along the z-direction, $F_i = F_i(\chi^*) = \int_0^\infty \frac{x^i \mathrm{d}x}{\exp(x-\chi^*)+1}$ are Fermi–Dirac functions, and χ^*_{2D} is the reduced chemical potential. From the above equation, increase of Seebeck coefficient is expected by decreasing width of the quantum well [12].

For two-dimensionally structured materials, carrier and lattice thermal conductivity are given by

$$\lambda_{L-2D} = \frac{k_B}{2\pi^2 v}\left(\frac{k_B}{h}\right)^2 T^3 \int_0^{\Theta_D/T} \frac{\tau_c \xi^4 e^\xi}{(e^\xi - 1)^2}\,\mathrm{d}\xi, \tag{7.23}$$

and

$$\lambda_{C-2D} = \frac{\tau h^2}{4\pi w}\left(\frac{2k_B T}{h^2}\right)^2 \left(\frac{m_y^*}{m_x^*}\right)^{1/2} k_B\left(3F_2 - 4\frac{F_1^2}{F_0}\right). \tag{7.24}$$

respectively.

Here, Θ_D is the Debye temperature, $\xi = \frac{h\omega}{k_B T}$ ($h\omega$ being the phonon energy), τ is the electron relaxation time defined as $\tau = \frac{\mu_x m_x}{e}$, where μ_x is the mobility along the x-axis, v is the velocity of sound, and τ_c is the combined phonon relaxation time due to all resistive processes (which include boundary scattering, mass difference scattering, scattering at dislocations, and Umklapp scattering processes), which is governed by the Matthiessen's rule.

It can be seen from Eq. (7.23) that the lattice thermal conductivity is mainly affected by width of quantum well and temperature. From Eq. (7.24) it is clear that the carrier thermal conductivity depends on the effective mass components of the carriers because of quantum confinement. The decrease of quantum well width is expected to increase the carrier thermal conductivity. However, overall thermal conductivity decreases due to the decrease in width of the quantum well [12]. Dresselhaus et al. theoretically proposed that the enhancement of zT can be obtained for the two-dimensional quantum well structures due to increase in density of states near the Fermi level, resulting in the increase of power factor [10].

The experimental results of enhancement in zT were reported for Bi_2Te_3/Sb_2Se_3 superlattices along the cross-plane direction and PbTe/PbTeSe quantum dot superlattices along the film-plane direction [13, 14]. Therefore, improvement in thin films is expected to have a large scope in the increase of thermoelectric efficiency.

The advantages of using thin film materials over bulk materials for thermoelectric applications [15] are that:

1. A large improvement of zT might be possible by using the quantum well concept and multi-layered structures.
2. Miniature applications, e.g., low power generators, on-chip coolers (micro-coolers) and generators, surface and micro-sensors, etc. can be synthesized by thin films.
3. Thermoelectric device technology can be used along with IC technology by which mass production of thermoelectric modules may be possible.

7.5 Fabrication of Thin Film Thermoelectric Materials

There are several methods of fabrication of thin films. In this section, techniques most commonly used for thermoelectric materials are briefly described.

7.5.1 Chemical Vapor Deposition

Chemical vapor deposition (CVD) is a method of thin film fabrication in which the material to be deposited is carried over to a heated substrate in the form of precursor gases that react among themselves or decompose to yield the desired material. The material is deposited over the pre-heated substrate, while the other products and unreacted reactants are flushed outside the reactor. An advantage of this method is that the material can be deposited uniformly at both the top as well as sides of the substrate, which is useful for complex substrate geometries. Secondly, deposition can be done simultaneously on a large number of substrates, which is important for reducing the production cost. Factors that are important for this process are thermodynamics, nature of the reactants, molar ratio, fluid dynamics inside the reactor, and mass transport. The microstructure of the films grown can be controlled by varying the deposition parameters like pressure, temperature, supersaturation, and the choice of precursors [16]. Metal Organic CVD (MOCVD) process [17], in which one of the precursor gases consists of an organic-metal bond, is a popular choice for synthesizing thermoelectric materials. The organic radical usually consists of an alkyl, aryl, or alicyclic hydrocarbon which form a ring, while the metal may belong to group IIA, IIB, IIIB, IVB, VB, or VIB. A list of the most commonly used precursors is given in [18]. The advantage of using MOCVD is lower reaction temperature.

Among the thermoelectric materials, Bi_2Te_3, Sb_2Te_3, and their alloys like $Bi_{0.4}Sb_{1.6}Te_3$ have been most commonly synthesized using MOCVD. Trimethylbismuth/trimethylbismuth, diiso-propyltelluride/di-ethyltellurium, and triethylantimony are usually used as precursors for bismuth, tellurium, and antimony, respectively. The deposition rate is varied depending on the thickness required, e.g. 0.34 μm/h for a final thickness of 0.6 μm. Hydrogen is commonly used as a carrier gas for which constant flow rate of 3 L/min [19] has been reported. Pressure inside the chamber affects the crystallite size and number density of the particles in the films and has been varied from 10 to 760 Torr [20]. Different substrates like Pyrex [21], Sapphire, (001) GaAs substrate, etc. can be used, which impart orientation, affect the morphology and faults of the film. This is important, especially when the material (e.g. Bi_2Te_3) is already known to have anisotropic properties in its bulk form. The reaction proceeds by pyrolysis near the substrate preheated to temperatures in the range of 250–400 °C, or the reactor temperature maintained at 350 °C. MOCVD has also been used to fabricate devices of $Bi_{0.4}Sb_{1.6}Te_3$—Bi_2Te_3, which was shown to produce power of 1.3 μW for a temperature difference of 45 K [22] by using standard microfabrication. Another Peltier module consisting of n-type Bi_2Te_3 and p-type $(Bi_{1-x}Sb_x)_2Te_3$ ($x = 0.77$) has been prepared on polyimide substrates by depositing one material after the other [21]. In such devices, the electrical contacts, which are usually made of metals (e.g. Cu), are deposited by other techniques like sputtering and not by CVD. Thin film superlattices with period in the range of 10–100 Å for Bi_2Te_3/Sb_2Te_3 [23] have also been synthesized by this method. A Peltier module based on an n-type Bi_2Te_3/$Bi_2Te_{2.83}Se_{0.17}$ superlattice has also been integrated into silicon chip in an electronic package [24]. Thick films several tens to hundreds of nanometers were synthesized by a horizontal CVD reactor, which consisted of a graphite mixing room heated by radiofrequency induction [20].

7.5.2 Electrodeposition

This method is used very commonly since it is much cheaper as compared to vapor deposition methods, does not require any vacuum, reaction takes place at lower temperatures (<100 °C), and a large amount of material can be deposited at a time. Through this method, nanomaterials can also be grown easily [25]. In this method, when current is passed through an aqueous, organic, or fused salt electrolyte, a chemical reaction (electrolysis), in which the more noble metal displaces the lesser noble metal, occurs and the density of the film depends on the concentration of the electrolyte. The reaction involves reduction of the cation at the interface of the electrolyte and the electrode, which is a thin region having electric field of the order 10^6 to 10^7 V/cm [25]. The entire procedure involves a series of reactions, including mass transfer between the bulk electrolyte and the transfer of electrons. Broadly, the first step includes the arrival and adsorption of anions at the electrode, while the second is the motion of the adatoms on the substrate surface, discussed in detail in [26]. The method of electrodeposition for thermoelectric materials has been reviewed by Xiao et al. [27]. A few of the examples of application of this method for thermoelectric thin films are briefly mentioned below.

$Bi_{0.5}Sb_{1.5}Te_3$ thin film has been deposited in nitric acid bath [28] where Sb III was found to be reduced with more difficulty than Bi^{3+} and $HTeO^{2+}$ ions. $CuSbSe_2$ films have been deposited [29] in a single-step electrodeposition process, in which a three-electrode cell consisting of SnO_2-coated glass electrode, graphite plate as counter electrode, and standard calomel electrode as a standard reference electrode was used. The electrolytes used were 2 mM copper chloride ($CuCl_2$), 2 mM antimony potassium tartrate ($K_2[Sb_2(C_4H_2O_6)_2]\cdot 3H_2O$), 2.5 mM selenious acid (H_2SeO_3), 100 mM ammonium chloride (NH_4Cl), and the pH was controlled at 2.00 ± 0.01 using concentrated HCl. Thin films of BiSb, BiTe, SbTe, and BiSbTe were deposited using chloride aqueous as well as ionic electrolytes. Bi and Sb ions were obtained by dissolving Bi_2O_3 and Sb_2O_3, respectively, and Te ions by solution of TeO_2 in 1 M NaOH. The electrolyte solution was neutralized by concentrated HCl. The ionic liquid was prepared using choline chloride with malonic acid, and $BiCl_3$, $SbCl_3$, and TeO_2 as precursors for Bi, Sb, and Te ions, respectively.

7.5.3 Physical Vapor Deposition

This method of thin film fabrication consists of deposition of vapors of the material to be deposited on the substrate. The deposition occurs under vacuum conditions when the mean free path of the molecules of the material is large enough to travel directly to the substrate without colliding with the air molecules. First, the vapors are generated from the bulk form of the material by heating by resistive, induction, laser beam, or ion beams [30], followed by their transport to the substrate and deposition by normal or oblique incidence [31]. The different deposition methods and the growth modes have been discussed in detail by Jiang et al. [32] and Das [3].

Since good thermoelectric materials are mostly compounds or alloys, one problem with the PVD method is maintaining stoichiometry, as the molecules can dissociate in vapor state. To overcome this, flash evaporation technique is used, in which the material in finely grinded powder form is made to drop on a boat which is heated above the melting point of the alloy. As soon as the material falls on it, evaporation takes place immediately, so that there is very little time for the molecules to dissociate, resulting in preservation of stoichiometry. This method has been used to deposit films of $(Bi_{0.25}Sb_{0.75})_2Te_3$ with 2 % excess Te [33], $Bi_2(Te_{0.8}Se_{0.2})_3$ [34], and $Bi_{88}Sb_{12}$ [35], where the bulk alloys were initially prepared by melt-quench-anneal method followed by their evaporation under vacuum ~10^{-5} Torr. Flash evaporation method has also been used to fabricate generators of p-type $Bi_{0.4}Te_{3.0}Sb_{1.6}$ and n-type $Bi_{2.0}Te_{2.7}Se_{0.3}$ [36], which was annealed in hydrogen atmosphere. Stoichiometric films of Ga-doped $Bi_{0.5}Sb_{1.5}Te$ containing nanodots of GaSb for which a high power factor of 22 $\mu Wcm^{-1}\,K^{-2}$

was achieved at room temperature [37] and Sn doped $Bi_2Te_{2.7}Se_{0.3}$, i.e. $(Bi_{1-x}Sn_x)_2Te_{2.7}Se_{0.3}$ films, for which power factor ~12 $\mu Wcm^{-1}K^{-2}$ was obtained, have also been synthesized [38].

Co-evaporation [39] method, in which all the constituent materials are evaporated together under vacuum and the stoichiometry, is controlled by varying the evaporation rates of the individual elements and have been used to deposit thin films of Bi_2Te_3 and Sb_2Te_3 and a Peltier module based on them has been fabricated [40].

7.5.4 *Sputtering*

Sputtering is another technique of physical vapor deposition (PVD) in which energetic ion beam bombards the surface of bulk material, displaces the atoms or molecules by energy transfer in the collision, which are then deposited on the substrate [41]. These ions, which ideally should have single charge for all ions, are positively charged and are initially produced from an inert gas. The cathode which contains the material to be sputtered is bombarded by these ions, and the resultant recoil atoms are deposited on the cleaned substrate kept directly near it. The theory, including the sputtering yield, has been calculated in detail by Sigmund [42]. In order to maximize the amount of material reaching the substrate and increase the efficiency, magnetron sputtering [43] has been used, in which a magnetic field is used to focus the ions near the substrate. One pole of the magnet is along the axis of the cathode, while the other pole surrounds it, so that dense plasma is produced near the substrate. To produce this plasma, voltages of the order of a few hundred volts (~ −500 V) and pressure $<10^{-2}$ mbar are typically maintained. If the substrate is insulating (e.g. glass) or the material to be deposited is dielectric, there is a possibility of charge buildup at the cathode, which would gradually restrict further ejection of atoms from it, thereby decreasing the emission rate. To overcome this, alternating current (a.c.) field is applied, which falls in the radio frequency (rf) range. During the first half of the cycle, ions reach the target and emit the atoms, while during the second half, electrons reach the target and remove any positive charge buildup. The plasma can be focussed here also by means of magnetrons.

Mostly, the rf sputtering method has been used for depositing materials like Heusler material Fe_2VAl [44], $CoSb_3$ [45], PbTe [46], etc. For PbTe, the deposition was carried out in argon atmosphere (50 mTorr) with rf power of 50 W and deposition rate of 10 nm/s [46] on substrate heated at 330 °C. Radiofrequency diode system has been used to sputter thin films of Bi_2Te_3 and PbTe on Mylar substrates using argon gas and rf power at 100–400 W [47]. Plasma-assisted co-sputtering method was used for depositing Ag-doped Mg_2Sn [48] on Si (111) substrates, where deposition was done by individual Mg and 2 % Ag-doped Sn sources. Here the initial pressure was kept at 4×10^{-5} Torr after which argon was flown and the pressure was maintained at 1.5 mTorr. The composition was obtained by varying the bias voltages of Mg and Sn sources independently.

Apart from the above methods, Pulsed Laser Deposition, Atomic Layer Deposition, and mechanical exfoliating are also being used lately for thin film deposition.

7.6 Application of Thermoelectric Effect in Sensors: Thermal Sensors

There are various sensors available today, e.g. pyrolectric sensors, enzyme thyristors, sensors utilizing photoacoustic spectroscopy, inverse photoelectric method, thermoelectric effect, and photopyroelectric detection. This section focuses primarily on the physical phenomenon of thermoelectricity, especially of thin films, in sensing action. However, only a few important ones that are commonly being used are described in this section.

Thermal sensors can be classified into two broad categories given by Herwaarden [49]:

1. *Power sensors*—The input signal is the thermal power responsible for producing the electrical signal that is detected.
2. *Thermal conductance sensors*—The input signal changes the thermal conductance in the area between the sensitive chemical and the atmosphere.

7.6.1 Performance Parameters

The performance of thermopiles is characterized by several parameters like sensitivity, responsivity, noise equivalent power (NEP), detectivity, etc. These have been discussed in detail in [50, 51] with the relevant mathematical details.

Sensitivity: The ease with which the output signal changes by varying the input stimulus (signal) is called the sensitivity of the sensor. A good sensor should be able to produce a signal even on a very small change of the quantity to be detected. Mathematically, it can be defined [52] as the slope of a linear fit of the response signal (β) and the external stimuli signal (s):

$$\beta = \mathrm{A} + \mathrm{B}\Lambda. \tag{7.25}$$

For a nonlinear fit between the input and output signal, the sensitivity of the sensor is defined as the first derivative of the transfer function given by [52]

$$b_i(\Lambda_i) = \frac{\mathrm{d}\beta(\Lambda)}{\mathrm{d}\Lambda} \approx \frac{\Delta\beta}{\Delta\Lambda}. \tag{7.26}$$

For thermopiles, the sensitivity would arise how much voltage would be produced per unit change in temperature, which essentially is equal to Seebeck coefficient.

The units are in V/K (or V/°C). In Si-based micro-machined thermopiles, sensitivities upto 45 mV/K have been achieved for thermopiles consisting of 92 thermocouples [53].

Response: The response of flow sensor gives the indication about the rate of change of sensor output with respect to change in flow rate or direction. The response should be as small as possible, usually of the order of a few ms is considered good. The response of the sensor is governed more by the thermometer placed upstream than the one placed downstream. A few of different types of sensor applications are listed in further sections.

7.6.2 Radiation Detectors

These consist of a series of thermocouples (known as thermopiles) joined from end to end. One end of the thermocouples is exposed to the radiation which is absorbed and increases the temperature of that end. The other end is kept at a reference temperature. The absorbed radiation increases the temperature of the thermopile junction so that voltage is generated, which is then measured. Thus, the sensing action takes place in two steps: (1) conversion of radiation to heat and (2) conversion of heat into electrical signal. The radiation signal which is to be detected is usually in the infrared range (wavelength = 10–100 μm) and in the temperature range 100–300 °C. The advantage of this over a conventional bolometer or LED is that dc or optical modes can also be used with them since the only signal transmitted is due to radiation. However, the output signal is also 100–1,000 times lower than that due to bolometer for the same radiation. However, the additional electronic noise, defined by Johnson noise, is also smaller [54]. Secondly, they do not require any external electrical bias, contrary to conventional bolometers. Any small change of this bias due to external reasons could lead to spurious signals in the main signal from the detector [54]. The thermopiles are used not only as radiation detectors, but with minor modifications are used in chemical, flow sensors, etc., where the temperature has to be measured.

The thermopile is made up of polysilicon/aluminium, a few μm in thickness. They are fabricated by using surface and bulk micromachining techniques [55]. One end is coated with a sensitive membrane, while the other end is left uncoated for reference. The temperature difference due to chemical reaction taking place is converted to Seebeck voltage, which is then converted into digital signal by Digital to Analog convertor. There are two common configurations for thermopiles—lateral and vertical. In lateral, the n and p-type legs are joined end to end and the deposition is done on the same substrate. For n- and p-type Si legs, single crystalline Si wafers have been used separately, which were then bonded and connected electrically in series [53]. In the vertical configuration, the n and p-type layers are stacked one above the other.

The basic common parts of a thermopile focussed on detecting radiation are:

- *Absorption of radiation*: The junction is often coated black body for maximum absorption.
- *Isolation from the substrate*: The thermally active junction, i.e. the junction which is exposed to radiation is isolated thermally from the rest of the sensor by an insulating membrane or covering so that the incoming radiation is utilized to the maximum and there are no heat losses due to conduction. This is especially important for materials like silicon whose thermal conductivity is very high and ultimately affects the sensitivity of the sensor. Thermal isolation in bulk solid-backed sensors has been done by [54], in which the thermal isolation of the active junction has been done by depositing a layer of epoxy resin under it. Another method of isolation is by change of design. For example, in micro-machined sensors, suspended membranes have been used, i.e. silicon below the active junction has been etched [56] in order to minimize conduction losses. The additional advantage of this is that with the reduction in mass, the thermal capacitance of the sensor is also reduced, which helps in increasing the response. The length and thickness of the sensor and the heater also needs to be optimized.

Thermopiles deposited by thermal evaporation have also been fabricated [54], in which the substrates like sapphire and anodized aluminium have been used. Bi_2Te_3- and Sb_2Te_3-based thin films have been used as thermopiles by thermal evaporation [57]. In another case [58], n-type Bi_2Te_3 and p-type Sb_2Te_3 thin films (~1.5 μm thick) were used in thermopiles whose dimensions were in millimetres.

The fabrication was done by thermal evaporation using co-evaporation of Bi and Te and Sn and Te, over glass substrates. The heater Ti/Cu/Ti was obtained by sputtering successive layers of 2 μm thick Cu and 0.1 μm thick Ti. This yielded a sensitivity of 2.7 mV/K. In another case, in-plane sensor was fabricated by electrodeposition of thin film thermopiles on glass substrate over which a Ti layer as a seed layer had been deposited initially. Thermopiles of Bi_2Te_3 and Sb_2Te_3 [59], 2–5 μm in thickness, were formed, with 196 pairs in total. Micro-machined sensors are commonly used because of their lower thermal capacity than the bulk thermopiles and have been reviewed excellently in literature [50, 51]. Silicon has been primarily used as a sensor material due to its semiconducting nature which gives higher Seebeck coefficients, than when only metals are used. Thermopiles consisting of n-type and p-type silicon for both the legs have been synthesized [60]. A method to synthesize a large number of thermopiles is to micromachine them and cut by laser cutting.

7.6.3 Chemical Sensor

This sensor works on the principle that heat is evolved or absorbed in chemical reactions, which can be detected. Gas sensors are the chemical sensors that detect the enthalpy change on absorbing/desorbing of a gas on a chemical that is sensitive to it.

They can be detected by changes in the Seebeck coefficient, electrical conductivity, or thermal conductivity.

The detection is usually done by a series of thermocouples (i.e. thermopiles). One end of the thermopile is coated with a catalyst/sensitive chemical that promotes the reaction of the incoming reactants and provides the surface area on which the reaction is supposed to take place. The reactants are made to react on it by bringing through pipes under controlled flow rate and pressure. The other end of the thermopile is kept at a controlled temperature. The heat generated in the reaction creates temperature gradient between the two ends of the thermopile, which gives rise to a voltage due to Seebeck effect. For a thermocouple consisting of 44 thermocouples, voltage of the order of hundreds of μV is obtained [61]. This voltage is measured, amplified, and transduced into the signal, and the enthalpy is calculated by integrating the transient voltage. Alternatively, individual chemicals can also be sensed by coating the thermopiles by enzymes membrane which catalyze the breakdown of the chemical to be sensed, along with the evolution of heat that can be detected. The transport and reaction of chemical inside the membrane for such sensors have been modelled theoretically and verified experimentally [62] for glucose. The chemical can be spray-coated using airbrush and a carrier gas [61].

For testing of volatile organic compounds, they can be synthesized on the spot by mixing saturated vapors of reactant gases sent through computer-controlled mass flow meters onto the coated material [61]. In this method, one requirement is quick measurement (interval ~ms) compared to other sensors in order to account for the fast changes in the enthalpy of the reaction.

Thin film-based chemical sensors based on thermoelectric effect have been used in micro-sized devices like CMOS [63], where both the sensor and the related electronics are integrated on the same chip.

7.6.4 Flow Sensors

These measure the flow rate of the gases or liquids (i.e. fluids) and are based on measurement of change in thermal conductance of the fluid. They are generally used in medical applications like breathing rate of child under normal conditions, breathing rate of adult during exercise, etc. The sensors either are sensitive for only gases or liquids individually or can measure mass flow for both gases and liquids at the same time [64]. Flow of the fluid can either be in one direction (e.g., inhalation and exhalation in spirometric devices, generally used for medical applications [65]), or 2-D, i.e., lamellar (generally used in gas sensors, etc.).

The measurement principle of flow direction is described extensively in [66, 67] and will only be summarized briefly here (Fig. 7.3). In one direction flow, temperature is measured at two opposite ends. For downstream flow, the temperature at one point will be lower than at the other end due to interaction with the fluid and its finite thermal conductivity. The value of temperature at a point downstream is measured

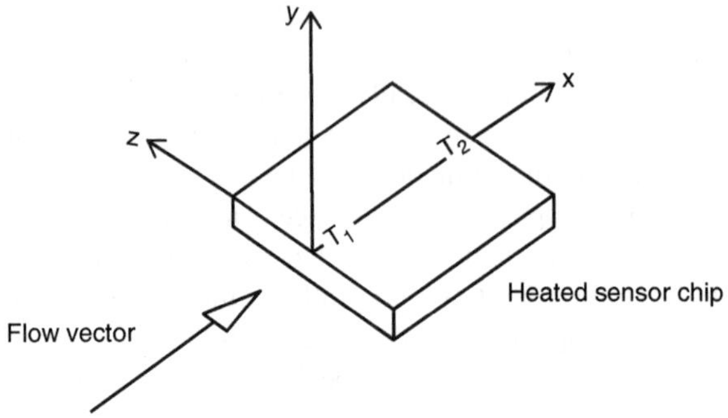

Fig. 7.3 The operating principle of a direction-sensitive thermal flow sensor of a heated chip exposed to a surface flow [66]

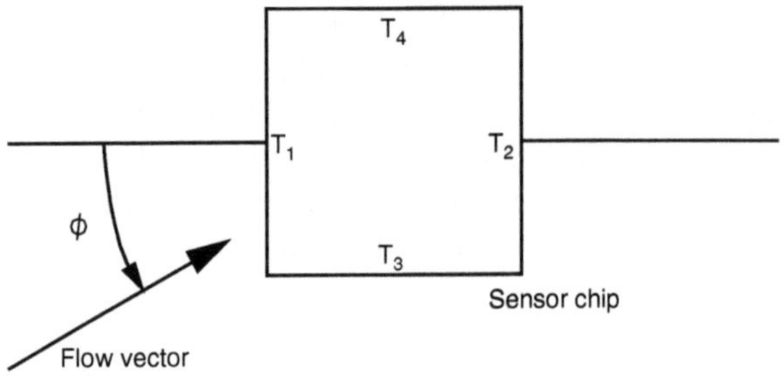

Fig. 7.4 Thermal sensor based on the measurement of flow-induced temperature differences in two-dimensional sensitivity [66]

and whether the temperature gradient is positive or negative gives information about the flow direction.

In 2-D flow (Fig. 7.4), temperature is measured at two other points also and the direction vector is given by [66]

$$\Delta T_{12} = \Delta T_0 \cos \phi, \tag{7.27}$$

$$\Delta T_{34} = \Delta T_0 \sin \phi, \tag{7.28}$$

where $\Delta T_{12} = T_2 – T_1$ and $\Delta T_{34} = T_4 – T_3$ and ϕ is the flow vector.

The temperature measurement is done by thermocouples and the corresponding voltages (V_{12} and V_{34}) are measured.

$$V_{12} = S\Delta TF(U) \cos \phi, \tag{7.29}$$

$$V_{34} = S\Delta TF(U) \sin \phi, \tag{7.30}$$

where S is the Seebeck coefficient and F is a transfer function which includes the material parameters like thermal conductivity, specific heat, viscosity, etc. of the fluid and the sensor. The direction of fluid flow is given by

$$\phi = \tan^{-1}(V_{34}/V_{12}). \tag{7.31}$$

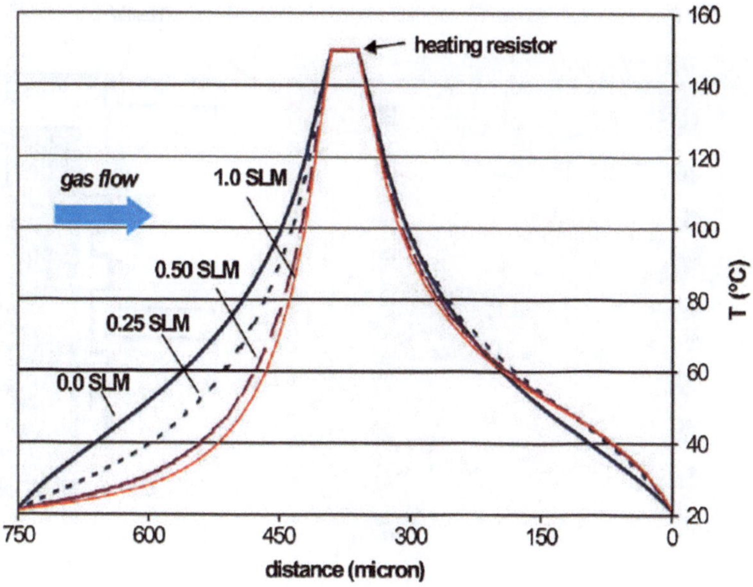

Fig. 7.5 Qualitative FEM results of modification of temperature distribution due to presence of incoming flow [68]

The sensor consists of a heater which is heated by giving electrical power. This heater comes in contact with the fluid and heat exchange takes place between them. This heats up the fluid/or reduces the temperature of the heater. The temperature of the fluid is measured again either by the thermometers placed downstream or the temperature of the heater is measured. The temperature at any point downstream is warmer than upstream (Fig. 7.5) [68].

When there is no air flow, then the distribution of temperature about the heater is symmetric. If there is a fluid flow, then the distribution becomes asymmetric and a temperature gradient is established. The flow velocity of the fluid and the direction of flow are then calculated from the temperature gradient. Here the Seebeck effect is utilized in thermocouples made of thermoelectric thin films as thermometers, and usually a number of thermocouples are used at the same time to get a measurable signal. Final signal is proportional to the square root of fluid's velocity [69]. An advantage of using Seebeck effect in sensing temperature is that there is no electrical voltage offset due to difference in the electronic components. Apart from using Seebeck effect, change in resistance of the element can also be used for detecting change in temperature. Also, by using several resistors together, the flow range can be increased [68]. Thermoelectric flow sensors have also been made based on a wire made of constantan and using gold deposition, which measure the flow based on Peltier effect [70]. An example of a flow sensor is given in Fig. 7.6.

As the heater temperature decreases, two different methods can be used to measure the response: (1) by keeping the heater power constant and measure the

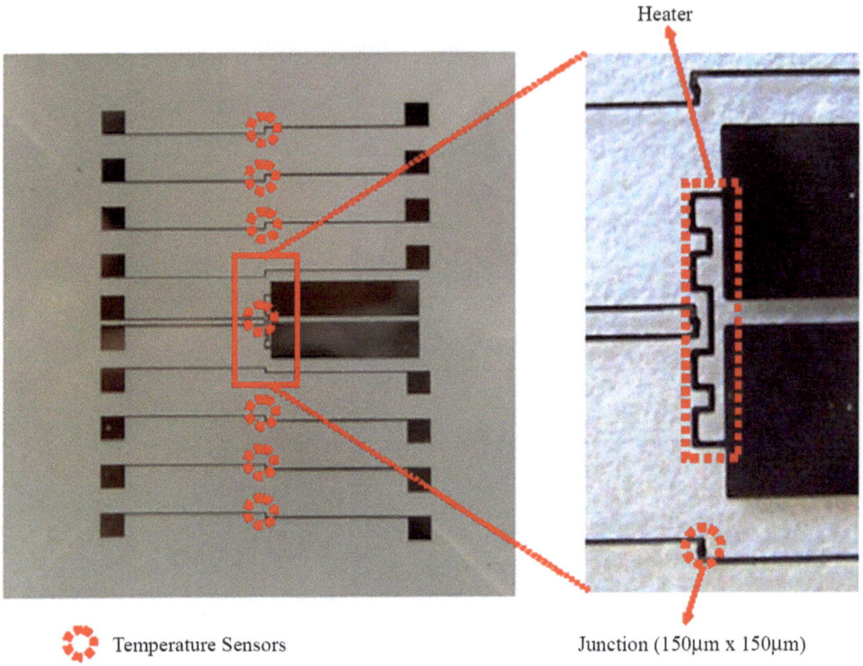

Fig. 7.6 The micro-thermal flow sensor [71]

reduced temperature and (2) by keeping the heater temperature constant by changing the power supplied [68].

In another method of measuring "time of flight" (TOF), the heater is used to send heat pulse into the fluid and the thermocouple detects the time after which it detects that heat pulse that has the higher temperature. The time at which the pulse arrives is measured and is used to find the velocity of the fluid by dividing the distance between the heater and the thermocouple by the time interval. The advantage of this method is that by using both the methods, fluid velocity in a wide range—0.1 to 140 mm/s—can be measured.

The design of flow sensors is dependent to a large extent on the specific requirement of application where it is to be used. However, broadly they follow a common basic design, which is generally modified according to the use, as is discussed in the next paragraph. The sensors essentially have thin film heater and a series of thermocouples. The heater serves also as a thermometer. The thermocouples are placed some distance apart (a few μm) from the heater so that the fluid that is to be sensed is able to exchange heat with the heater. Usually, silicon wafers are used on which the sensor structure is built, but quartz wafers (~200 μm) have also been used [71]. Circular sensors that have four thermopiles arranged symmetrically placed around a heater placed in the center have also been fabricated [72]. Suspended membrane structure has also been attempted [73]. If the direction of the fluid/gas is also to be sensed, the thermocouples are placed symmetrically about the heater [65], so that one set of thermocouples corresponds to upstream, while the other set to downstream. Bi_2Te_3-based sensors have been used to measure flow velocity [74]. Numerical calculations are done, usually by finite element modelling (FEM) method, to simulate the dimensions of sensors, so that when it is placed in the path of gas/liquid flow, it itself does not change the lamellar flow to turbulent flow. Flexible flow sensors have also been developed [75], which use a suspended membrane on which the thermopiles and the heater are deposited; however, the sensors are sensitive to pressure and need to be insulated against the atmosphere.

As thin films, flow sensors have been used in MEMS devices and integrated with IC chips [66], so as to increase their applications and usability. In such devices, the thermocouple is usually made up of silicon (Si) and some metal (e.g. Ni, Au, Al, etc.) due to their compatibility with the rest of the devices and ease of fabrication due to already well-established fabrication procedures. The usual values of Seebeck coefficient from such thermocouples are usually in the range 150–200 μV/K. Here, another advantage is that the value of Seebeck coefficient from the thermocouple can be changed by varying Si doping accordingly (e.g., Si–Al thermopiles give in the range 0.5–1.0 mV/K).

Practically, thin film sensors are used in MEMS device, which utilize the standard procedures of lithography, deposition, patterning, etc. Low pressure chemical vapor deposition (LPCVD) has been used to passivate the sensor layer and deep reactive ion etching (DRIE) has been used to release the membrane made of SiN. The common deposition steps for MEMS-based sensors are:

1. Deposition of membrane materials: A layer of silicon dioxide (thickness of order of a few μm) is grown on Si wafer over which a SiN membrane layer (~150 nm) is deposited on a smaller area (~300 × 600 μm) [65, 68].
2. Deposition of thermopile and heater—Thermopiles and heater are deposited over the SiN membrane. A polysilicon layer (~100 nm) and a 300 nm aluminum layer [65] have been used previously as thermopiles. Platinum has been used as heater [72] as it is found to be stable. In another study [71], chromel and alumel as thermocouples and nichrome heater have been deposited by sputtering over 200 μm quartz substrate. The sensitivity of these thermocouples was measured to be the same as that of bulk K-type thermocouples. The distance between the heater and the thermometer affects the flow rate sensitivity [68].
3. Deposition of protective layer to mechanically insulate fluid/gas and the sensor—Sensing action requires only thermal contact between the sensor and the fluid/gas without the need of any direct contact. Therefore, an integral part of the sensor design is to deposit a protective layer, also called passivating the sensor. Thermocouples can be deposited on membranes or layers of silicon bridge-like structures. Since the entire fabrication procedure and the industry is optimized for silicon, most of the thermocouples are based on silicon. For example, polysilicon and gold deposited over 150 nm thick SiN membrane have been used as thermocouples. The sensor is passivated against the fluid by using polymers, e.g. Si_3N_4 layer by PECVD [65].

4. Removal of the oxide layer below by etching by KOH, so as to have a free standing membrane [65].
5. Enclose in a protective covering—After the fabrication, the sensor is calibrated under well-controlled conditions like controlled flow, temperature, etc., by using mass flow controller before exploiting it in practical applications. The performance depends on the heat transfer between the liquid and the sensing material which depends on the type of flow (lamellar or turbulent) and the velocity of flow. Gas flow induced by adiabatic expansion of gas by bursting a membrane placed at one end of a tube and acoustic waves have also been used to measure response time of the gas [76, 77]. The response time of stagnant gas is higher than that of flowing fluid, e.g. a 62.3 % response time of 0.7 ms, while for flowing gas, it varied from 0.5 to 1.5 ms. The response time has been found to be dependent on the flow velocity also—higher the flow velocity, faster the response [77], and on the distance of the heater from the thermopile. The method of measurement of response time also affects its value.

7.7 Vacuum Sensors

These sensors are used to sense the vacuum level, i.e. the pressure of gases below atmospheric pressure by utilizing the pressure dependence of thermal conductivity of gases. The sensors work on the principle of transport of heat by the gas whose pressure has to be measured [78]. The heater heats up the gas which transports the heat towards a thermopile acting like a heat sink. The heat is transferred by collisions of gas molecules with the heat sink. When the pressure is low and the distance between the gas molecules is greater than distance between two surfaces, the heat is transferred from the gas molecules to the sensor when they hit its surface. The rate at which they hit the surface gives the pressure. If the pressure of the gas changes, i.e. the number of molecules increases or decreases, the amount of heat transported also changes. This means that the temperature of the active end of the thermopile increases, which results in generation of a voltage signal due to Seebeck effect.

These sensors can work only as vacuum sensors and not high pressure as their sensitivity decreases for high pressure. This is because as the number of gas molecules increases, there is more collision between the molecules; the mean free path is reduced and the pressure of the gas is independent of the pressure [78].

Due to the requirement that the gas should transport heat from heater to sink, an empty space between heater and sink is required for the gas to flow. This is fulfilled by designing the sensor either as (1) a cantilever or (2) a membrane (Fig. 7.7).

The sensor basically consists of a heater, an air gap which senses the gas pressure and a heat sink. Micro-machined vacuum sensor on mm-sized IC chips has been modelled and fabricated [79]. A cantilever design was found to be better than a membrane design due to higher output and the possibility of using more number of thermopiles. The gas whose pressure is to be sensed flows below the cantilever.

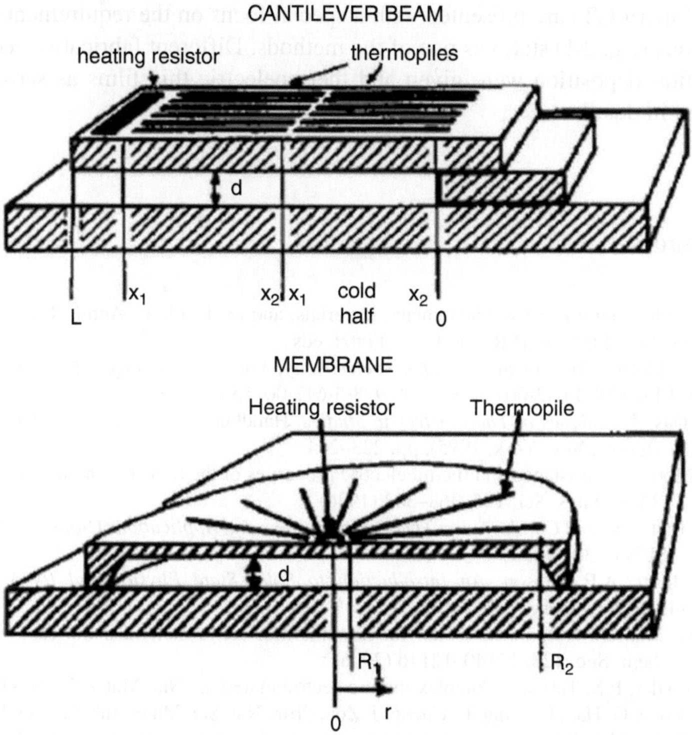

Fig. 7.7 Schematlc diagram of cantilever-beam and membrane vacuum sensor [79]

One end is heated and the heat is carried by both the cantilever and the gas. Since the transfer function (output voltage to input power) of the cantilever is known and does not change, the transfer function of the gas (known as vacuum transfer function) is determined and used to calculate the pressure. This depends upon the height of the cantilever from the base as well as the presence of any supporting bridge under the cantilever. Thinner cantilever beam was found to be much more sensitive than thicker ones and lower heat dissipation. The thermopiles had sensitivity of 22 mV/K with internal resistance of 70 kΩ. The detection of pressure in the range 1 mPa to 1 kPa was done. Vacuum sensors compatible with CMOS technology have also been synthesized by using photolithography and wet etching [80].

7.8 Conclusion

In chapter, thin film materials for thermoelectric applications are described. An introduction to various thermoelectric phenomena is given, followed by a description of transport coefficients—Seebeck coefficient, electrical resistivity, and thermal conductivity. Various strategies to improve the dimensionless thermoelectric figure of merit (zT) are presented with a special focus on the requirement of lower dimensions (e.g. 2-D state) as one of the methods. Different fabrication techniques of thin film deposition were given and thermoelectric thin films as sensors were described in detail.

References

1. T.M. Tritt, Thermoelectric phenomena, materials, and applications. Annu. Rev. Mater. Res. **41**, 433–448 (2011). in D.R. Clarke, P. Fratzl, eds
2. H.J. Goldsmid, *Introduction to Thermoelectricity*, vol. 121 (Springer, New York, 2010). doi:10.1007/978-3-642-00716-3. ISBN 978-3-642-00715-6
3. D.V. Das, *Techniques of Thermoelectric Studies*. Handbook of Advanced Materials Testing (Marcel Dekker, New York, 1995), pp. 223–244
4. N.F. Mott, The resistance and thermoelectric properties of the transition metals. Proc. R. Soc. Lond. A Math. Phys. Sci. **156**, 368–382 (1936)
5. T.M. Tritt, *Thermal Conductivity: Theory, Properties, and Applications* (Springer, New York, 2004). ISBN 0- 306-48327-0
6. R.J. Elliott, A.F. Gibson, *An Introduction to Solid State Physics and Its Applications* (Macmillan, New York, 1974)
7. G.K.H. Madsen, Automated search for new thermoelectric materials: the case of LiZnSb. J. Am. Chem. Soc. **128**, 12140–12146 (2006)
8. G.J. Snyder, E.S. Toberer, Complex thermoelectric materials. Nat. Mater. **7**, 105–114 (2008)
9. Z.-G. Chen, G. Han, L. Yang, L. Cheng, J. Zou, Prog. Nat. Sci. Mater. Int. **22**, 535–549 (2012)
10. L.D. Hicks, M.S. Dresselhaus, Effect of quantum-well structures on the thermoelectric figure of merit. Phys. Rev. B **47**, 12727 (1993)
11. L.D. Hicks, T.C. Harman, M.S. Dresselhaus, Use of quantum- well superlattices to obtain a high figure of merit from nonconventional thermoelectric materials. Appl. Phys. Lett. **63**, 3230–3232 (1993)
12. S. Walia, S. Balendhran, H. Nili, S. Zhuiykov, G. Rosengarten, Q.H. Wang, M. Bhaskaran, S. Sriram, M.S. Strano, K. Kalantar-zadeh, Prog. Mater. Sci. **58**, 1443–1489 (2013)
13. T.C. Harman, P.J. Taylor, M.P. Walsh, B.E. LaForge, Science **297**, 2229 (2002)
14. R. Venkatasubramanian, E. Silvola, T. Colpitts, B. O'Quinn, Nature **413**, 597 (2001)
15. R. Spolenak et al., *Challenges for Thin Film Systems Characterization and Optimization* (Wiley-VCH, Weinheim, 2005)
16. H.O. Pierson, *Handbook of Chemical Vapor Deposition: Principles, Technology and Applications* (Noyes Publication, New Jersey, 1992)
17. J.J. Coleman, Metalorganic chemical vapor deposition for optoelectronic devices, in *Proceedings of the IEEE*, 1997, pp. 1715–1729
18. M.J. Hampden-Smith, T.T. Kodas, Chemical vapor deposition of metals: Part 1. An overview of CVD processes. Chem. Vapor Depos. **1**, 8–23 (1995)
19. A. Giani, A. Boulouz, F. Pascal-Delannoy, A. Foucaran, A. Boyer, B. Aboulfrah, A. Mzerd, Electrical and thermoelectrical properties of Sb_2Te_3 prepared by the metal-organic chemical vapor deposition technique. J. Mater. Sci. Lett. **18**, 541–543 (1999)
20. H. You, S.-H. Bae, J. Kim, J.-S. Kim, C. Park, Deposition of nanocrystalline Bi_2Te_3 films using a modified MOCVD system. J. Electron. Mater. **40**, 635–640 (2011)
21. A. Boulouz, S. Chakraborty, A. Giani, F.P. Delannoy, A. Boyer, J. Schumann, Transport properties of V–VI semiconducting thermoelectric BiSbTe alloy thin films and their application to micromodule Peltier devices. J. Appl. Phys. **89**, 5009–5021 (2001)

22. S. Do Kwon, B.-K. Ju, S.-J. Yoon, J.-S. Kim, Fabrication of bismuth telluride-based alloy thin film thermoelectric devices grown by metal organic chemical vapor deposition. J. Electron. Mater. **38**, 920–924 (2009)
23. R. Venkatasubramanian, T. Colpitts, E. Watko, M. Lamvik, N. El-Masry, MOCVD of Bi_2Te_3, Sb_2Te_3 and their superlattice structures for thin-film thermoelectric applications. J. Cryst. Growth **170**, 817–821 (1997)
24. I. Chowdhury, R. Prasher, K. Lofgreen, G. Chrysler, S. Narasimhan, R. Mahajan, D. Koester, R. Alley, R. Venkatasubramanian, On-chip cooling by superlattice-based thin-film thermoelectrics. Nat. Nanotechnol. **4**, 235–238 (2009)
25. I. Gurrappa, L. Binder, Electrodeposition of nanostructured coatings and their characterization—a review. Sci. Tech. Adv. Mater. **9**, 043001 (2008)
26. A. Mallik, B.C. Ray, Evolution of principle and practice of electrodeposited thin film: a review on effect of temperature and sonication. Int. J. Electrochem. **2011**, 16 (2011)
27. F. Xiao, C. Hangarter, B. Yoo, Y. Rheem, K.-H. Lee, N.V. Myung, Recent progress in electrodeposition of thermoelectric thin films and nanostructures. Electrochim. Acta **53**, 8103–8117 (2008)
28. F.-H. Li, Q.-H. Huang, W. Wang, Investigations on the electrodeposition behaviors of $Bi_{0.5}Sb_{1.5}Te_3$ thin film from nitric acid baths. Electrochim. Acta **54**, 3745–3752 (2009)
29. D. Tang, J. Yang, F. Liu, Y. Lai, M. Jia, J. Li, Y. Liu, One-step electrodeposition and annealing of $CuSbSe_2$ thin films. Electrochem. Solid State Lett. **15**, D11–D13 (2012)
30. W. Kern, *Thin Film Processes II* (Academic, San Diego, 1991)
31. H. van Kranenburg, C. Lodder, Tailoring growth and local composition by oblique-incidence deposition: a review and new experimental data. Mater. Sci. Eng. R Rep. **11**, 295–354 (1994)
32. K. Reichelt, X. Jiang, The preparation of thin films by physical vapour deposition methods. Thin Solid Films **191**, 91–126 (1990)
33. V.D. Das, R.C. Mallik, Study of scattering of charge carriers in thin films of $(Bi_{0.25}Sb_{0.75})_2Te_3$ alloy with 2 % excess Te. Mater. Res. Bull. **37**, 1961–1971 (2002)
34. V.D. Das, R.C. Mallik, Determination of the nature of principal scattering mechanism in well-annealed vacuum deposited thin films of the ternary thermoelectric material $Bi_2(Te_{0.8}Se_{0.2})_3$. Thin Solid Films **424**, 75–78 (2003)
35. R.C. Mallik, V.D. Das, Size and temperature-dependent thermoelectric and electrical properties of $Bi_{88}Sb_{12}$ alloy thin films. Vacuum **77**, 275–285 (2005)
36. M. Takashiri, T. Shirakawa, K. Miyazaki, H. Tsukamoto, Fabrication and characterization of bismuth–telluride-based alloy thin film thermoelectric generators by flash evaporation method. Sens. Actuat. A Phys. **138**, 329–334 (2007)
37. D. Zhao, M. Zuo, H. Geng, Enhanced thermoelectric performance of Ga-added $Bi_{0.5}Sb_{1.5}Te_3$ films by flash evaporation. Intermetallics **31**, 321–324 (2012)
38. X.K. Duan, Y.Z. Jiang, Microstructure and thermoelectric properties of Sn-doped $Bi_2Te_{2.7}Se_{0.3}$ thin films deposited by flash evaporation method. Thin Solid Films **519**, 3007–3010 (2011)
39. R. Suryanarayanan, The co-evaporation technique—a potential tool for basic and applied research. Thin Solid Films **50**, 349–355 (1978)
40. H. Zou, D.M. Rowe, S.G.K. Williams, Peltier effect in a co-evaporated Sb_2Te_3(P)-Bi_2Te_3(N) thin film thermocouple. Thin Solid Films **408**, 270–274 (2002)
41. R.A. Baragiola, Sputtering: survey of observations and derived principles. Phil. Trans. R. Soc. Lond. A **362**, 29–53 (2003)
42. P. Sigmund, Theory of sputtering. I. Sputtering yield of amorphous and polycrystalline targets. Phys. Rev. **184**, 383–416 (1969)
43. P.J. Kelly, R.D. Arnell, Magnetron sputtering: a review of recent developments and applications. Vacuum **56**, 159–172 (2000)
44. Y. Furuta, K. Kato, T. Miyawaki, H. Asano, T. Takeuchi, Fe_2VAl-based thermoelectric thin films prepared by a sputtering technique. J. Electron. Mater. **43**, 2157–2164 (2014)

45. H. Anno, K. Matsubara, Y. Notohara, T. Sakakibara, K. Kishimoto, T. Koyanagi, Thermoelectric Properties of rf-Sputtered $CoSb_3$ Films, in *Thermoelectrics, Fifteenth International Conference on, 1996*, 1996, pp. 435–439
46. K. Kishimoto, M. Tsukamoto, T. Koyanagi, Temperature dependence of the Seebeck coefficient and the potential barrier scattering of n-type PbTe films prepared on heated glass substrates by rf sputtering. J. Appl. Phys. **92**, 5331–5339 (2002)
47. Y.H. Shing, Y. Chang, A. Mirshafii, L. Hayashi, S.S. Roberts, J.Y. Josefowicz, N. Tran, Sputtered Bi_2Te_3 and PbTe thin films. J. Vac. Sci. Technol. A **1**, 503–506 (1983)
48. H. Le-Quoc, A. Lacoste, E.K. Hlil, A. Bès, T.T. Vinh, D. Fruchart, N. Skryabina, Thin films of thermoelectric compound Mg_2Sn deposited by co-sputtering assisted by multi-dipolar microwave plasma. J. Alloys Compd. **509**, 9906–9911 (2011)
49. S.V. Herwaarden, Physical principles of thermal sensors. Sens. Mater. **8**, 373–387 (1996)
50. U. Dillner, E. Kessler, H.-G. Meyer, Figures of merit of thermoelectric and bolometric thermal radiation sensors. J. Sens. Sens. Syst. **2**, 85–94 (2013)
51. A. Graf, M. Arndt, M. Sauer, G. Gerlach, Review of micromachined thermopiles for infrared detection. Meas. Sci. Technol. **18**, R59–R75 (2007)
52. J. Fraden, *Handbook of Modern Sensors: Physics, Designs and Applications*, 4th edn. (Springer, New York, 2010). ISBN 978-1-4419-6465-6
53. S.C. Allison, R.L. Smith, D.W. Howard, C. Gonzalez, S.D. Collins, Sens. Actuat. A **104**, 32–39 (2003)
54. R.W. Astheimer, S. Weiner, Solid-backed evaporated thermopile radiation detectors. Appl. Opt. **3**(4), 493–500 (1964)
55. T. Huesgen, P. Woias, N. Kockmann, Design and fabrication of MEMS thermoelectric generators with high temperature efficiency. Sens. Actuat. A **145–146**(2008), 423–429 (2008)
56. M. Boutchich, K. Ziouche, M.A.-H. Yala, P. Godts, D. Leclercq, Package-free infrared micro sensor using polysilicon thermopile. Sens. Actuat. A **121**, 52–58 (2005)
57. U. Birkholz, R. Fettig, J. Rosenzweig, Fast semiconductor thermoelectric devices. Sensors Actuators **12**, 179–184 (1987). doi:10.1016/0250-6874(87)85015-1
58. J.-H. Kim, J.-Y. Choi, J.-M. Bae, M.-Y. Kim, T.-S. Oh, Thermoelectric characteristics of n-type Bi_2Te_3 and p-type Sb_2Te_3, thin films prepared by Co-evaporation and annealing for thermopile sensor applications. Mater. Trans. **54**, 618–625 (2013)
59. M.-Y. Kim, T.-S. Oh, Thermoelectric characteristics of the thermopile sensors with variations of the width and the thickness of the electrodeposited bismuth-telluride and antimony-telluride thin films. Mater. Trans. **51**, 1909–1913 (2010)
60. H. Muro, T. Mitamura, S. Kiyota, Sen. Mater. **18**, 433–444 (2006)
61. A. Koll, A. Schaufelbuhl, N. Schneeberger, U. Munch, O. Brand, H. Baltes, C. Menolfi, Q. Huang, Micromachined CMOS calorimetric chemical sensor with on-chip low noise amplifier, in *Micro Electro Mechanical Systems*, 1999. MEMS '99. *Twelfth IEEE International Conference on IEEE, Orlando, FL, USA*, 1999, pp. 547–551
62. M.J. Muehlbauer, J.E.J. Guilbeau, B.C. Towe, Model for a thermoelectric enzyme glucose sensor. Ann. Chem. **61**, 77–83 (1989)
63. A. Hierlemann, D. Lange, C. Hagleitner, N. Kerness, A. Koll, O. Brand, H. Baltes, Sens. Actuat. B **70**, 2–11 (2000)
64. N.T. Nguyen, R. Kiehnscherf, Low cost silicon sensors for mass flow measurement of liquids and gases. Sens. Actuat. A **49**, 17–20 (1995)
65. F. Hedrich, K. Kliche, M. Storz, S. Billat, M. Ashauer, R. Zengerle, Thermal flow sensors for MEMS spirometric devices. Sens. Actuat. A **162**, 373–378 (2010)
66. B.W.V. Oudheusden, J.H. Huijsing, Direction-sensitive thermal IC flow sensors. Flow Meas. Instrum. **1**, 157–164 (1990)
67. B.W.V. Odheusden, J.H. Huijsing, Integrated silicon flow direction sensor. Sens. Actuat. **16**, 109–119 (1989)
68. N. Sabaté, J. Santander, L. Fonseca, I. Gràcia, C. Cané, Multi-range silicon micromachined flow sensor. Sens. Actuat. A **110**, 282–288 (2004)

69. M. Elwenspoek, Thermal flow micro sensors, in *Semiconductor Conference, 1999. CAS '99 Proceedings. International*, v2, Oct 1999, pp. 423–435. doi:10.1109/SMICND.1999.810580
70. H. Stachowiaka, S. Lassue, A. Dubernard, E. Gaviot, A thermoelectric sensor for fluid flow measurement. principles, calibration and solution for self temperature compensation. Flow Meas. Instrum. **9**, 135–141 (1998)
71. T.H. Kim, S.J. Kim, Development of a micro-thermal flow sensor with thin-film thermocouples. J. Micromech. Microeng. **16**, 2502–2508 (2006)
72. S. Kim, T. Nam, S. Park, Measurement of flow direction and velocity using a micromachined flow sensor. Sens. Actuat. A **114**, 312–318 (2004)
73. B.W.V. Oudheusden, A.W.V. Herwaarden, High-sensivity 2-D flow sensor with an etched thermal isolation structure. Sens. Actuat. A **A21–A23**, 425–430 (1990)
74. M.A. Khalfiou, A. Michez, A. Giani, A. Boyer, A. Foucara, Anemometer based on Seebeck effect. Sens. Actuat. A **107**, 36–41 (2003)
75. R. Buchner, K. Froehner, C. Sosna, W. Benecke, W. Lang, Toward flexible thermoelectric flow sensors: a new technological approach. J. Microelectromech. Syst. **17**(5), 1114–1119 (2008). doi:10.1109/JMEMS.2008.926143
76. C. Sosna, T. Walter, W. Lang, M. Ashauer, Response time of thermal flow sensors. Procedia Eng. **5**, 524–527 (2010)
77. C. Sosna, T. Walter, W. Lang, Response time of thermal flow sensors with air as fluid. Sens. Actuat. A **172**, 15–20 (2011)
78. S.V. Herwaarden, Sensor applications of thermoelectric thin films, in *16th International Conference on Thermoelectrics IEEE*, 1997, pp. 47–55
79. A. Herwaarden, P.M. Sarro, H.C. Meijer, Integrated vacuum sensor. Sens. Actuat. **8**, 187–196 (1985)
80. O. Paul, H. Baltes, Novel fully CMOS-compatible vacuum sensor. Sens. Actuat. A **46–47**, 143–146 (1995)

Chapter 8
Electroluminescent Thin Film Phosphors

Samuel Paul David and Romain Gaume

Abstract In the later part of the twentieth century, classic cathode ray tubes (CRTs) are replaced by flat panel displays (FPDs) with a thinner, smaller size and with a significant reduction in power consumption. Such displays are currently used in several applications including mobile phones, laptops, and televisions. Among other FPDs, electroluminescent displays perform better than its counterparts such as liquid crystal, field emission, and plasma displays. Electroluminescent (EL) display is a common term for devices such as light-emitting diodes (LEDs), powder phosphor devices, thin film, and thick dielectric electroluminescent devices. Thin film EL devices (TFEL) show advantages such as quick response, high resolution, wide viewing angles and wide operating temperatures and so on. TFEL displays are all solid state and based on electroluminescence in which luminescent material gives light under high electric field.

Phosphors act as luminescent materials both in powder as well as in thin film forms in these devices. ZnS doped with Mn is one of the earliest and still in use electroluminescent phosphor. Monochrome display was developed using ZnS:Mn sandwiched between single or double multi-stack dielectric layers. An efficient yellow emission was generated when a high voltage was applied between front indium tin oxide (ITO) and rear aluminium (Al) electrodes. Multiple color displays were then achieved by combination of different rare earth-doped ZnS layers emitting different colors. Similar displays were developed more efficiently with other materials. Mostly oxides, sulphides, and fluorides are highly preferred host for such phosphors because of wider bandgap that allows the transmission of visible light. Oxides have been found to have more stability compared to sulphides which suffer degradation under high electron bombardment due to sulphur depletion.

TFEL phosphors have been grown by physical vapour deposition techniques such as pulsed laser deposition, sputtering, and chemical reaction techniques such as spray pyrolysis, chemical vapour deposition, sol–gel process followed by spraying, dipping, or spinning. Over the years, extensive research has been carried out to develop multi-color TFEL devices using combination of red, green, blue emitting phosphors. Optimum growth techniques, dopant concentration, choice of

S.P. David (✉) • R. Gaume
College of Optics and Photonics (CREOL), University of Central Florida,
Orlando, FL 32816, USA
e-mail: samuelpaul83@gmail.com

S. Babu Krishna Moorthy (ed.), *Thin Film Structures in Energy Applications*,
DOI 10.1007/978-3-319-14774-1_8

rare earths, and hosts have been investigated and led to the production of efficient, high luminance display devices. This chapter introduces basics of phosphors, minimum requirements to be an efficient phosphor, different fabrication techniques, and a review of different color electroluminescent phosphors for display devices.

8.1 Introduction

Solid state white lighting such as LEDs requires UV LED chips covered by a combination of red, green, and blue downconverting phosphors [1]. A field emission display (FED) comprises of cathode field emitter that emits electrons which are then accelerated towards the phosphor screen that generates light [2]. Thin film electroluminescent displays use double insulating structure in which a thin phosphor layer is sandwiched between them with a metal film (Al film) as top electrode and transparent conductive film as back electrode. Under irradiation of electrons, these thin film phosphors generate electroluminescence and hence the light. Phosphors are luminescent materials and are being used for several decades in various applications such as fluorescent tubes, X-ray detectors, televisions, cathode ray tubes, and so on. Inorganic phosphors are generally insulating materials such as ZnS or YAG doped with small concentration of rare earth ions (Eu^{3+}, Tb^{3+}, Ce^{3+}, and so on) or transition metal ions (Mn^{2+}, Cr^{3+}, and so on) that generate light on excitation either with photons or electrons. These activator ions are in general directly excited by photons or electrons and are responsible for the luminescence [3]. If the absorption cross-section of activators is not strong enough, excitation is carried out by an indirect energy transfer from an additional ion which acts as a sensitizer for the activators. A common example of sensitized phosphors is the most popular calcium halophosphate ($Ca_5(PO_4)_3(Cl,F):Sb^{3+}$, Mn^{2+}) white phosphor in which Sb^{3+} acts as sensitizer by absorbing Hg emission at 254 nm and then transferring its blue emission to Mn^{2+} ion. Mn^{2+} gives yellow emission by absorbing blue light generating white light.

The phosphors are based on two prominent materials:

1. Lanthanides or lanthanoids (Ln^{3+})—elements of atomic number from 57 (Lanthanum) to 71 (Lutetium). They are also called as rare earth metal ions.
2. Transition metal ions—elements in the d-block of the periodic table

Transitions happen differently in these two activator ions. In lanthanoids, optical transitions take place purely between 4f energy levels which are well shielded by outer electrons. This leads to weak interactions by the crystal field of the host lattice, whereas transition metals strongly experience the host lattice field. Transition metals involve incomplete d-shell for the optical transitions and it is affected by crystal field. Transitions between 4f levels are forbidden in free 4f ions, which makes 4f–4f transitions substantially long-lived in the range of milliseconds, whereas 5d–4f transitions are parity allowed and are much faster (10^{-5} s). Different dopant ions generate different emission wavelengths corresponding to their energy level transitions.

Desirable properties of host materials are larger band gap, crystallinity, and a lower refractive index. Wider band gap avoids reabsorption of the emitted light and also reduces slow electrons generated at low electric fields when these phosphors are electrically excited. These slow electrons degrade the efficiency of EL devices. A much wider band gap can also reduce the generation of electron–hole pairs and therefore materials having bandgap to fit the activator band gap is desirable. Higher crystallinity limits grain boundaries and defects and hence the scattering.

Inorganic phosphors are prepared in several forms including polycrystalline powders, nanocrystals, nano/microcrystalline powders, and as thin films. Size-dependent luminescent properties in nanocrystalline phosphors make nanophosphors attractive for optical, biological, optical applications. The quantum confinement effect on such materials gives rise to novel optoelectronic properties. Phosphors in powder form have found several applications in fluorescent lamps, LEDs, and so on. In field emission displays (FEDs), ZnS doped with Ag and Cu were used as blue and green phosphors. Unfortunately, lifetime of these phosphors are shorter and the performance goes down due to degradation by electron bombardment due to the formation of nonluminescent layer such as ZnO or $ZnSO_4$, which leads to the degradation of brightness in such materials [4]. Such environmental degradation can be avoided by thin film phosphors. Table 8.1 lists few examples of inorganic phosphors for different applications.

Thin film phosphors are materials with thicknesses that range from few nanometers to several micrometres. These thin film phosphors are found to be more attractive alternatives in display technologies which require high resolution and long-term operation. These luminescent thin films also show better thermal stability, insensitivity to surface contamination, reduced outgassing rate, and better adhesion to the solid surface than that of powders. Under electron beam bombardment, powders were found to degas CO, CO_2, H_2 ions, whereas thin films outgassing rates are significantly smaller to that of the powders. Due to these

Table 8.1 List of phosphors and their emission colors for respective devices

Phosphor	Color of Emission	Applications
ZnS:Cu, Al	Green	Television screens
ZnS:Ag, Cl	Blue	Display tubes
(Zn,Cd)S:Ag	Yellow	Display tubes
Zn_2SiO_4:Mn	Yellowish-green	Display tubes
MgF_2:Mn	Orange	Radar screens
YAG:Tb	Yellow–green	Projection tubes
SiO_2:PbS	Red	Television screens
Y_2SiO_5:Ce	Blue	Field emission displays
$SrAl_2O_4$:Eu^{2+}, Dy^{3+}	Green	Long afterglow

Table 8.2 Comparison of thin films over powders as phosphors

Property	Thin film phosphor	Phosphor powder
Efficiency	Poor	Excellent
Resolution	<1 μm	5–10 μm
Screen contrast	Excellent	Good
Lifetime	Good	Poor
Mechanical stability	Excellent	Good
Thermal stability	Excellent	Good

advantages, thin film phosphors are replacing conventional powder-based display screens. Several luminescent thin films have been investigated over the years for flat panel displays, field emission displays, backlights for display panels, and so on. Electroluminescent thin films gained much attraction because they consume low voltage for operation and help to produce low-cost flat-panel display. Laser active ion doped thin films act as diode pumped waveguide lasers with lower lasing threshold and with high efficiency. Codoped thin films such as Yb/Ce:YAG have been used for quantum cutting applications to enhance the Si solar cell efficiency. Thin films can be prepared in powder form and then deposited either electrophoretically, or it can be directly deposited as thin film from their gaseous precursors such as in MOCVD. In general, thin film-coating techniques are divided into two: (1) physical vapour deposition and (2) chemical vapour deposition. Comparison of powder and thin film phosphors has been tabulated in Table 8.2 [5].

8.2 Characteristics of Phosphors

Activators in luminescent thin film phosphors should have a large absorption cross-section for the excitation mechanisms, either by high energy photons or high energy electrons. Ionic radius and valence state of the activator ion should match with that of an ion in the host lattice for better solubility and stability of the system. For example, in Ce:YAG, Ce^{3+} having an ionic radius of 1.14 Å substitutes Y^{3+} in YAG of ionic radius 1.02 Å. Apart from these initial requirements, phosphors should have other important properties such as high luminous efficacy, quantum efficiency, color rendering index, and color chromaticity coordinates.

8.2.1 Luminous Efficacy

Light sources are mainly assessed by its efficiency to convert radiant flux to luminous flux known as luminous efficacy of radiation (LER) and efficiency of conversion from electrical power to optical power known as luminous efficacy of the source (LES). Former one measures the optical efficiency of the light emitted by a source and is given by the Eq. (8.1) [6]

$$\mathrm{LER} = 683\ \frac{\mathrm{lm}}{\mathrm{W}} \frac{\int_{\lambda} P_{\mathrm{white}}(\lambda) V(\lambda) \mathrm{d}\lambda}{\int_{\lambda} P_{\mathrm{white}}(\lambda) \mathrm{d}\lambda} \tag{8.1}$$

where 683 lm/W is a normalization factor, $P_{\mathrm{white}}(\lambda)$ is the spectral power density of the white light source, and $V(\lambda)$ is the eye sensitivity function with 555 nm being the most sensitive wavelength to the human eye. LES is the ratio of luminous flux to input electrical power. For instance GaInNwhite LEDs has LES of 249 lm/W at 20 mA which is two- to three times higher than standard fluorescent sources and an order of magnitude better than incandescent sources. Luminous efficacy of phosphors is size-dependent and an optimum phosphor size gives a better performance. Ce:YAG phosphors in InGaN LED chips with mean phosphor size of 15 μm show about 3.2 % higher lumen output than that for the phosphors having size of 6–8 μm with 20 μm being an optimum phosphor size [7].

8.2.2 *Quantum Efficiency*

Apart from luminescence efficiency, the luminescent devices such as LEDs should have a good external quantum efficiency which is the product of internal quantum efficiency and light extraction efficiency.

$$\eta_{\mathrm{EQE}} = \eta_{\mathrm{int}} \times \eta_{\mathrm{Lext}} \tag{8.2}$$

η_{int} is the ratio of number of photons emitted from the active region per second to the number of exciting electrons/photons per second. For an LED, η_{int} is P_{int}/I where P_{int} is the number of photons emitted for the given number of electrons (current) per second. η_{Lext} is the extraction efficiency given by the ratio of number of photons that make it out of the material to the number of incident photons ($\eta_{\mathrm{Lext}} = P/P_{\mathrm{int}}$). Phosphor external quantum efficiency is given by

$$\eta_{\mathrm{Phosphor\ EQE}} = \eta_{\mathrm{absorption\ QE}} \times \eta_{\mathrm{int}} \tag{8.3}$$

External quantum yield (η_{ext}) of a thin film electro luminescent emitters is defined as the ratio of the number of photons emitted by the device to the number of charge carriers that passed through the phosphor layer during luminescence. Assuming the device's surface is monochromatic and direction-dependent, η_{ext} is given by [8]

$$\eta_{ext} = K_0\eta_{int} = K_0 A\frac{L_e T}{Q_p} = K_0 A\frac{\int_0^T L(t)\mathrm{d}t}{\int_0^T I_P(t)\mathrm{d}t} \quad (8.4)$$

where K_0 is the coefficient of emission extraction from the device, η_{int} is the internal quantum yield, $A = (\pi S_e q)/(h\nu f_\lambda)$ where S_e is the emitting surface area; q is the charge of an electron, $h\nu$ is the photon energy, f_λ is the spectral luminous efficiency of the emission, L_e is the electroluminescence brightness averaged over the exciting voltage period T given by

$$L_e = \frac{1}{T}\int_0^T L(t)\mathrm{d}t \quad (8.5)$$

and Q_p is the charge passed through the device for the period T

8.2.3 Color Rendering Index

Luminous efficacy and color rendering index are the most widely used metrics to evaluate the light sources. Color rendering index (CRI) is a unitless measurement of how an artificial light source represents to a perfect illuminant, for example, daylight. International Commission on Illumination (CIE) defines CRI as follows: "Color rendering: Effect of an illuminant on the color appearance of objects by conscious or subconscious comparison with their color appearance under a reference illuminant"; "CRI: under certain conditions, when an object is illuminated by the light source and illuminated by the reference lighting body (i.e. the standard light source D^{65}), the measurement of the feeling match level is the color-rendering index of the light source". It is determined by the following equation

$$R_i = 100 - 4.6\Delta E_i$$

It is sometimes called "average color-rendering index" which is calculated by

$$R_a = 100 - 4.6\Delta\overline{E}_a \quad (8.6)$$

$$\Delta\overline{E}_a = \frac{1}{8}\sum_{i=1}^{8}\Delta E_i$$

In simple words, CRI is defined as a quantitative measure of the ability of a light source to reproduce the colors of various objects faithfully in comparison with an ideal or natural light source. Reference source can either be a blackbody below 5,000 K or a CIE daylight source above 5,000 K and will have a CRI typically of 100. High quality lamps for interior lighting show R_a values greater than 80, whereas for visual inspection purposes, $R_a > 95$ is desirable. CRI values of white LEDs vary depending on the type of RGB LED combinations employed. When 640 nm red LED was replaced by 615 nm one, CRI value improved from 23 to 70 [9]. So a high CRI is desirable for an ideal thin film phosphors for desirable display applications.

8.2.4 Degradation Characteristics

Phosphors experience degradation due to electron-stimulated surface chemical reaction which leads to a loss in intensity over time due to electron beam exposure. Degradation of sulphide phosphors have been extensively studied using different models [10, 11]. Formation of dead layer by chemical reaction inhibits the emission intensity. Holloway proposed a mathematical model to explain electron-stimulated surface chemical reaction (ESSCR) that leads to degradation of phosphors [10].

Other than the mentioned requirements of luminescent thin films, an increased phosphor life-times, brightness saturation, and excitation density are all other required parameters for phosphors in different applications.

8.3 Preparation of Thin Film Phosphors

Phosphors are prepared in both powder forms and thin films. Thin film phosphors find attractive than powders because of some unique properties, especially for field emission display (FED) environment. Thin films reduce light scattering with good thermal contact between the screen and the faceplate. In addition, luminescent thin films show lower outgassing and high uniform properties across the covered area with high resolution and contrast with less material. These films also have few disadvantages such as low light output and low efficiency when irradiated with electron beam [12]. Luminescent thin films are fabricated by several thin film growth techniques such as pulsed laser deposition, sputter deposition, electron beam deposition, metal-organic chemical vapour deposition, spray pyrolysis, sol–gel technology, and so on. We will discuss each technique separately.

8.3.1 *Pulsed Laser Deposition Technique*

Pulsed laser deposition (PLD) is a well-known fast and effective technique to grow thin films of complex-oxide materials with a sharpness that was earlier achieved only with molecular-beam epitaxy methods. With physical principle of PLD being very complex, this technique uses short and intensive pulsed laser to transfer the material from a solid source to a substrate by evaporation without losing its stoichiometry. When high power laser pulses are focused onto a target inside a vacuum chamber, evaporation of the target material forms a plasma plume. As the plume expands and flows from the target material, the particles in the plume condense on a heated substrate and forms a thin film. After PLD was successfully used for the deposition of titanate thin films for the first time in 1969 [13], several other complex oxides, cuprates, nitrides, metallic multilayers have been developed including materials that cannot be otherwise obtained by an equilibrium route. High temperature superconductor ($YBa_2Cu_3O_7$) thin film was grown by Dijkkamp and Venkatesan in 1987 with a superior quality than films grown with other techniques.

8.3.1.1 Description of PLD Technique

PLD process can be divided into four stages:

1. Rapid removal of material from the surface, i.e., formation of plasma.
2. Propagation of plume—travel of atoms, ions, electrons from target to the substrate
3. Deposition of the ablation material on the substrate
4. Nucleation and growth of the film on the substrate surface.

Photons are converted into electronic excitations and then into thermal, chemical, and mechanical energy, resulting in the evaporation, ablation, and formation of the plasma. The penetration depth is dependent on the laser wavelength and on refractive index of the target at the incident wavelength. Since this depth decides that most of the energy is absorbed in a very shallow layer near the target surface, determination of penetration depth (mostly in the region of 10 nm) is critical. Excimer laser (KrF at 248 nm) or frequency quadrupled Nd:YAG (266 nm) are the most commonly used lasers for PLD. Wavelength below 200 nm is not generally used due to the possibility of absorption of molecular oxygen (Schumann–Runge bands). Pulse duration of the order of tens of ns is mostly used. Long pulse durations help to achieve thermalization and create very dense plasma with a plume of all species with an identical angular distribution. This increase in laser heating of the plasma develops high quality thin films. Laser fluence in the order of 1–3 J/cm^{2}is another critical parameter to exceed a certain threshold to form a plume. Temperature of the generated plasma plume is around 10,000 K.

The plume contains a variety of particles such as neutral atoms, ions, and electrons which travel at different velocities from the target's surface towards the

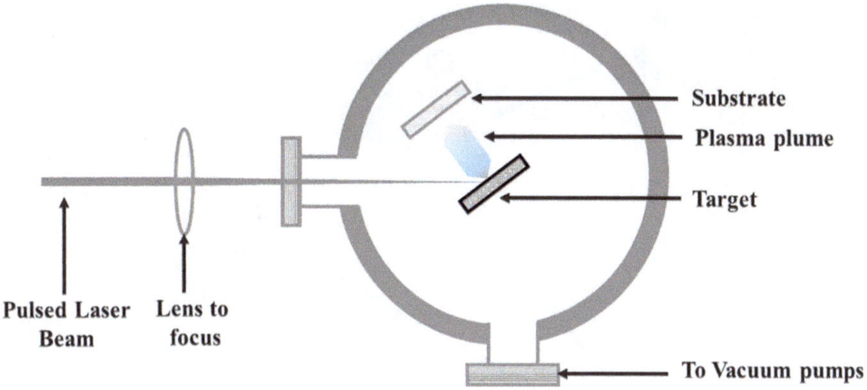

Fig. 8.1 Schematic diagram of PLD setup

substrate. Considering the thermal energy to be high enough to ionize all the atoms in the plume, all the ions reach the substrate with equal lateral and forward velocities. Inside the PLD chamber, under vacuum plume is very narrow and directed without any scattering with the background gasses. Under these conditions of thermalized plume, the optimal growth rate is around 1 Å per pulse. Plume can be seen inside the chamber due to fluorescence and recombination process occurring in the plasma. Stoichiometric removal of the material from the target does not always lead to stoichiometric deposition in the substrate since different elements in the plume get deposited in a different rate. Some compounds having volatile elements need to be compensated by an excess target material to supply the volatile species [14].

Well-known blue-emitting SrS:Cu thin films have been grown by PLD technique using KrFexcimer laser. SrS:Cu solid target is irradiated with by 248 nm, ns pulses at an angle of 22.5° in H_2S/Ar mixture atmosphere. Substrate such as Al_2O_3/ITO/glass was rotated and maintained at 500 °C [15]. A schematic diagram of PLD setup is shown in Fig. 8.1.

8.3.2 Metal Organic Chemical Vapour Deposition (MOCVD)

Chemical vapour deposition uses chemical reaction of precursors in vapour phase to form thin films on a substrate. Semiconductor industry has extensively used MOCVD technique for the deposition of high pure epitaxial thin films for applications such as laser- and light-emitting diodes (LEDs). MOCVD uses metal-organic precursors that make chemical reaction both in the gas phase and on the substrate. MOCVD process is extremely complex and can be divided into several steps:

1. Evaporation and transport of precursors
2. Pyrolysis of precursors followed by deposition of the material
3. Removal of residues from the reactor after the reaction

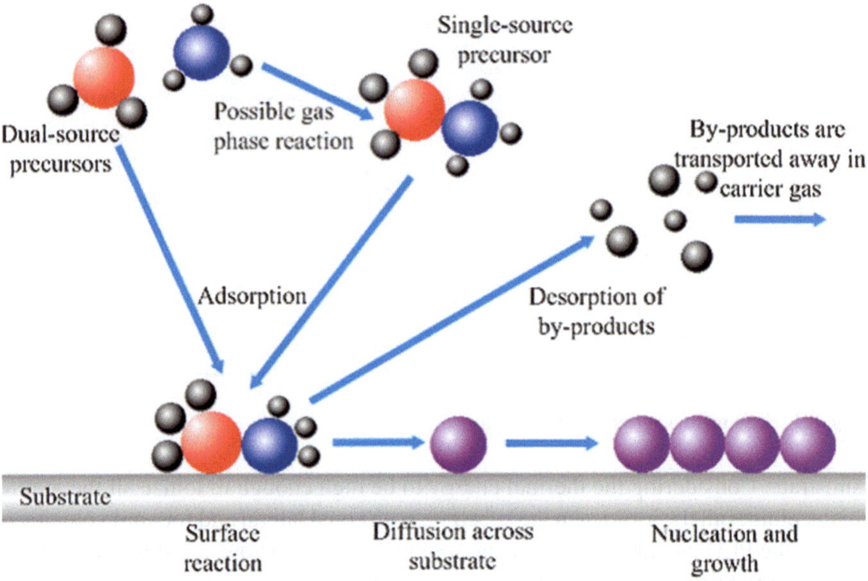

Fig. 8.2 General processes in MOCVD technique

Possible processes that can take place in an MOCVD chamber are represented in Fig. 8.2. MOCVD uses metal organics as source materials with a general growth temperature of 550–700 °C and low pressure for the deposition process. In 1971, first MOCVD ZnS phosphor layers were prepared by Manasevit using dimethyl zinc (DMZ) with hydrogen sulphide, H_2S [16]. Deposition of ZnS layers from metal-organic sources can be carried out in two different ways: the additive method which is usually called as MOCVD and the decomposition method such as spray pyrolysis. MOCVD uses widely used DMZ metal organic and H_2S as the precursors and forms ZnS as per the following chemical reaction

$$(CH_3)_2Zn_{(g)} + H_2S \rightarrow ZnS_{(s)} + 2CH_{4(g)}$$

Step-by-step growth of ZnS by MOCVD is given as follows

1. DMZ is introduced in the vapour phase over the substrate and forms Zn bonds on the substrate surface.
2. DMZ turned off and unreacted excess DMZ is removed from the reactor
3. H_2S is then introduced and reacts with DMZ and then purged.

Gases like $(C_2H_5)_2Zn$, Cs_2, $(CH_3)_2S$ have also been used as precursors for ZnS to have better stability over the reaction. Advantages of this technique are its low deposition temperature, control over stoichiometry and dopant, and its ability to deposit large area. Doping Mn in ZnS was done by thermal diffusion using thin Mn layer as a diffusion source or by in situ doping using tricarbonyl methylocyclopentadienyl manganese. Modified techniques of MOCVD have also been invented to deposit continuously rather than a large number of repeated cycles. In addition to that, mercaptan such as *t*-butyl mercaptan has been preferred as a precursor for H_2S because of its less toxicity [17].

8.3.3 Sputter Deposition

Electroluminescent phosphors are primarily developed by sputter deposition. Sputtering is a plasma process that uses ionized gas atoms which are accelerated in an electric field and hit a target. Material from the target is removed as a gaseous phase and then condenses on a substrate. When a strong negative electric field is applied to the substrate, all the positive ions are attracted and settled over the substrate. First sputtering process was done by Grove in 1852 when he found the formation of oxide layer on a silver plate when a voltage was applied by a battery in a gas discharge tube [18]. The technique has been well developed over the years and is being used for a variety of materials for different applications.

The ion that is accelerated by the electric field carries a momentum. As these accelerated ions hit the target, this momentum is transferred and causes a series of collisions between the atoms of the target leading to the removal of surface atoms. This ejection process is known as sputtering. The efficiency of momentum transfer is quantified by sputtering yield which is given by the number of atoms ejected from the surface per incident ion (*S* atoms/ion).

$$S \propto (E_{\text{ion}} - E_{\text{thres}}) \tag{8.7}$$

where E_{ion} is the incident ion energy, E_{thres} is the threshold energy.

Threshold energy is related to the mass number of the incident ion and the target and is given by

$$E_{\text{thres}} = 8\,U_{\text{s}}(M_1/M_2)^{2/5} \tag{8.8}$$

where M_1 and M_2 are the mass number of the ion and the target, respectively, U_{s} is the surface potential barrier. The maximum energy transferred between the two colliding particles is also related to their masses and the angle of collision, θ, and is given by

$$\frac{E_2}{E_1} = \frac{4M_1M_2}{(M_1 + M_2)}\cos^2\theta \tag{8.9}$$

where E_1 is the energy of the incident ion and E_2 is the energy that are transferred to remove the surface atoms which can also be called as surface binding energy (E_{b}). Using these energy equations, sputtering yield is given by

$$S = \frac{3\alpha}{4\pi_2} \frac{4M_1M_2}{(M_1 + M_2)^2} \frac{E_1}{E_b} \tag{8.10}$$

where α is the efficiency with which the momentum is transferred between the incident ions and the surface atoms.

There are various sputtering techniques available such as DC, RF, magnetron, and reactive sputtering based on the sources that are used to accelerate ions. DC sputtering is widely used because of its high sputtering rate with improved film properties. Material to be deposited is first prepared as a pellet and then sputtered as a thin film. For instance, Eu:YVO_4 is first formed by stoichiometric amounts of Y_2O_3, NH_4VO_3, and Eu_2O_3. Powders are cold pressed and fired at 1,000 °C for 2–3 h to form the compound and is followed by sintering at 1,200 °C for 48 h. Sputtering takes place in an argon/oxygen atmosphere at a background pressure. Sputtering techniques have also been efficiently used to deposit multi-component oxide thin films.

8.3.4 Spray Pyrolysis

Compared to physical vapour deposition techniques such as PLD and sputter deposition, spray pyrolysis is a chemical deposition technique using solution precursors. Reduced cost and relaxation on usage of precursors make this technique more advantageous and was the predecessor of CVD techniques. This process also can be done in an open atmosphere and the films fabricated from this technique exhibit more durability at a lower cost than other deposition techniques such as chemical vapour deposition that requires vacuum. Other advantages include relatively uniform and high quality coatings and also for substrates with multi-layer and complex geometries. Spray pyrolysis deposition process involves spraying metal salt solution onto a heated substrate and is schematically given in Fig. 8.3 [19]. Sequence of deposition process includes three steps.

1. Atomization of the precursor solution
2. Transport of the aerosol droplet
3. Decomposition of the precursor and thin film growth

The first and an important step in the spray pyrolysis process is atomization which produces droplets from a spray solution. Different types of atomizers such as air blast, ultrasonic, and electrostatic atomizers have been employed based on applications, liquid properties, and operating conditions. Different atomizers yield different droplet size, atomization rate, and the initial velocity of the droplets. Compared to ultrasonic atomizers, electrostatic and air blast atomizers are being used extensively because of ease of production. Controlling the pressure in air blast atomizers helps achieve desirable droplet size, whereas precursor solution is atomized using an electric field in an electrostatic atomizers. The solution is exposed to

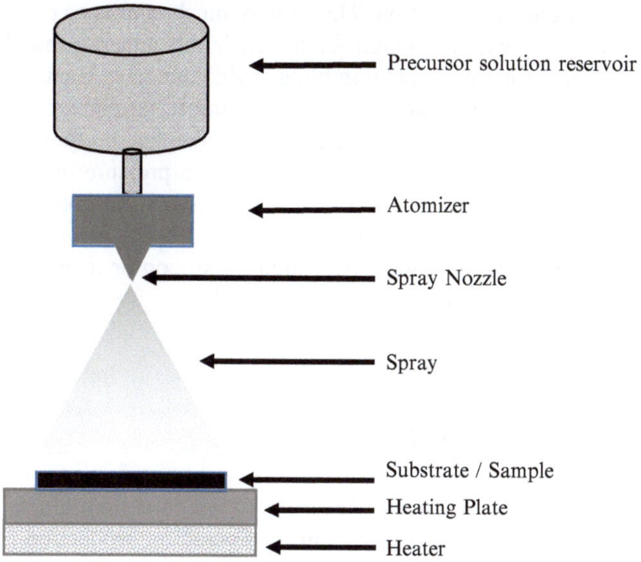

Fig. 8.3 Schematic representation of a spray pyrolysis process

an intense electric field between the charged atomizer and grounded work piece in electrostatic atomizers. Various spraying modes such as cone-jet and multi-jet modes are employed based on electric field strength, fluid flow rate, and fluid properties.

Aerosol droplets produced in atomizers need to be transported without evaporation and any formation of powder or salt particles. In general, solvent evaporates before the droplet reaches the substrate and reduces the droplet size. This causes concentration gradient within the droplet and leads to formation of precipitate on the surface when the surface concentration is above the solubility limit. If solvent evaporation is fast enough, thin films obtained from such precipitates will have poor quality due to formation of hollow particles. Large droplet sizes at low concentration also cause hollow particles. Smaller droplet size at a higher concentration of droplet number ends up in uniform concentration distribution within the droplet and leads to uniform particles on the substrate [20]. Carrier gas flow rate and nozzle-to-substrate distance play a significant role in the efficiency of deposition in this spray pyrolysis process.

Decomposition of the precursor on the substrate is the final step in this thin film deposition process. By varying the processing parameters and with the control of droplet size, residual solvent can be evaporated well ahead and precipitate can be formed well before the droplet reaches the substrate. When the precipitate approaches the immediate vicinity of the substrate maintained at high temperature, it is converted into a vapour state. Lowest temperature causes the droplets to splash onto the substrate, whereas the highest temperature leads to the complete evaporation of the precursor before reaching the substrate. At an optimum temperature, solid state precipitate vaporizes without decomposition. The vapour diffuses and forms solid particles on the substrate after chemical reaction. This is very much similar to chemical vapour deposition (CVD) and results in a high quality and highly adherent thin films.

Well-known electroluminescent thin films of ZnS:Mn have been prepared using ultrasonic spray pyrolysis technique. Precursor solution was prepared by dissolving 0.1 M zinc chloride, 0.1 M thiourea, and 0.1 M manganese chloride in distilled water. Purified air was used as the carrier gas with a pressure of 10^5 Pa and the substrate was maintained at 350–400 °C [21]. Similarly Eu-doped alumina thin films were prepared using Al and Eu acetilacetonates as precursors diluted in dimetylformamide. Thin films were deposited on silicon or quartz substrate at a deposition temperature of 500–600 °C.

8.3.5 Sol–Gel Technology

Sol–gel process is gaining much attraction to prepare thin films, fibers, and nanopowders because of its low processing temperature, production of high pure materials, and ease in compositional and microstructural control. A sol is a colloidal suspension of very fine particles in a liquid. A gel is formed when a sol containing colloidal particles from inorganic or organic precursors is converted into highly viscous mass and extends throughout the solution. This consists of network of polymer chains formed by hydrolysis and condensation reaction of metal-organic compounds. Metal alkoxide with general formula $M(OR)_x$ is the most commonly used precursor where R is an alkyl group C_nH_{2n+1}. By addition of water and with appropriate alcohol, hydrolysis occurs by removal of a proton to form a hydroxyl (MOH) ligand by aqueous reaction. Metal alkoxides are characterized to undergo hydrolysis easily in which alkoxide groups (OR) are replaced by hydroxyl groups (OH). An example of one such hydrolysis reaction is given by

$$M(OR)_x + H_2O \rightarrow M(OR)_{x-1}OH + ROH$$

Hydrolysis of such metal alkoxides depends on the characteristics of the metal and that of the alkyl group.

Condensation follows the hydrolysis reaction to form the macromolecular oxide network. Hydroxy metal alkoxide forms polymer chains of M–O–M bonds with alcohol (ROH) or water as a by-product. Such condensation reactions can be given as follows

$$M(OH)(OR)_{x-1} + M(OR)_x \rightarrow (OR)_{x-1}M - O - M(OR)_{x-1} + HOR$$
$$2M(OH)(OR)_{x-1} \rightarrow (OR)_{x-1}M - O - M(OR)_{x-1} + H_2O$$

The end product formed after the condensation reaction is polymerized together by cross-linking and interlinking polymer chains. Eventual growth of these polymer chains by the condensation of polymers or particles leads to the formation of high viscous gel with a continuous solid network and shear modulus. The gel undergoes aging process during which its properties such as viscosity and elastic modulus change with time due to continued chemical reactions even after condensation.

Conventional process of powder synthesis from sol–gel technology involves drying and sintering of the gel following the aging process. But for thin film fabrication for electroluminescent displays, the fluid sol or solution before the polymerization process is suitable. Sol–gel thin film processes are generally carried out by dip coating and spin coating from the solution. Compared to sputtering, MOCVD, these techniques are cheaper with an ability to control the microstructure of deposited film. In dip coating, the substrate to be coated is immersed into the sol and withdrawn vertically at a constant speed. After the drainage and evaporation of the remaining solvent, thin solid polymer film is deposited with a thickness which is dependent on several factors such as substrate pulling speed, concentration and viscosity of the solution, temperature of the substrate and solution, and so on. Spin coating is another thin film deposition technique that involves four stages: deposition, spin-up, spin-off, and evaporation. In this case, the solution on the spinning substrate forms a thin film by balancing the centrifugal and inward viscous force. Thickness of the deposited thin film depends on the time between spin-up and spin-of, evaporation, liquid viscosity, angular velocity, and so on. Other coating methods such as electrophoresis and thermophoresis also have been employed for forming thin films from solution.

Well-known electroluminescent material such as ZnS:Mn has been developed as thin films by sol–gel thin films for displays. Flowchart shows the step by step preparation of ZnS thin films in Fig. 8.4.

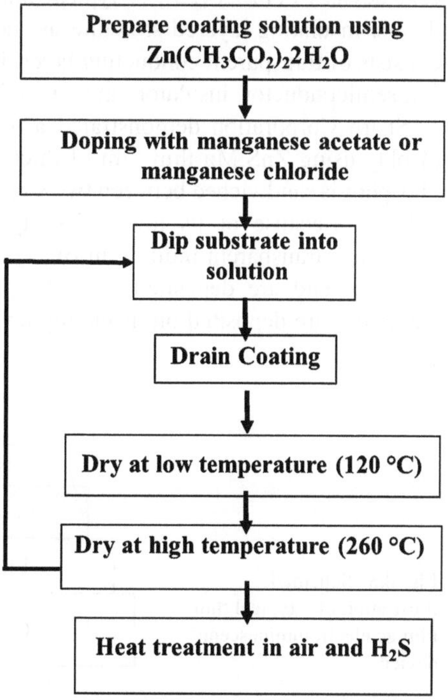

Fig. 8.4 Flow chart of fabrication of ZnS thin films sol–gel and dip coating processes

Zinc acetate ($Zn(CH_3CO_2)_2{\cdot}2H_2O$) is dissolved in anhydrous ethanol or methanol to form a solution of ZnO precursor. Manganese acetate ($Mn(CH_3CO_2)_2{\cdot}4H_2O$) or manganese chloride ($MnCl_2$) is added to the solution for doping. Thin film is deposited from this solution by dipping the substrate and withdrawing slowly. For thicker films, multiple coatings are needed. Deposited substrate is heat-treated in air. Any unreacted ZnO films are annealed in a mixture of H_2S, H_2, and N_2 above 500 °C to achieve the complete conversion to ZnS.

8.4 Electroluminescent Phosphors

Flat panel display (FPD) finds more attractive over-conventional cathode ray tubes because of its thickness which is a small fraction of its viewing area. Several major techniques such as gas-plasma display, liquid crystal display, electroluminescent display, flat CRT display, and electrochromic display have been used for the development of FPD technologies. Among these techniques, Electroluminescence is completely a solid-state technique and it does not require any gas or backlite. Destriau demonstrated electroluminescence for the first time in 1936 when he applied a high field to a ZnS powder dispersed in an insulating liquid [22]. Electroluminescent thin films gained much interest over powders after the development of thin film technology. In 1960, Vlasenko and Popkov discovered the practical form of electroluminescent displays, which are well-known as thin film electroluminescent devices (TFEL) [23]. A typical thin film ac electroluminescent device is a double insulating layered structure as shown in Fig. 8.5. This structure generally consists of transparent conducting layer, insulator, and phosphor, which in general is a semiconductor, insulator, and a metal electrode [24].

Sharp Corporation demonstrated a monochrome thin film electroluminescent display using ZnS:Mn thin film of thickness 0.5–1 μm. This polycrystalline ZnS phosphor is sandwiched between two single or stacked dielectric layers of thickness 0.2 μm. Because of its low resistivity, high transparency, and better thermal resistance, transparent indium tin-oxide column electrodes are chosen as the front electrode and are deposited onto the glass substrate, whereas aluminium row electrodes are deposited on insulating layer and act as rear electrode. When an a.c

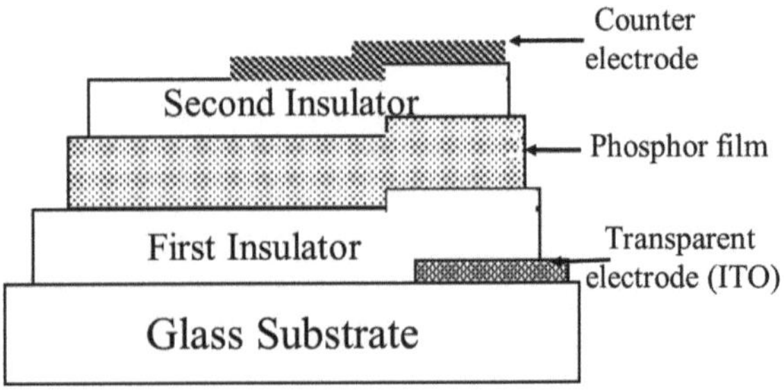

Fig. 8.5 Schematic illustration of a typical thin film ac electroluminescent device

voltage of 200 V (~1.5 MV/ cm) is applied between these electrodes, electrons tunnel into the conduction band of the phosphor from the electronic states and ionize the crystal lattice and excite the luminescent centers, Mn^{2+}. These excited Mn^{2+} ions produce an efficient yellow emission due to intra d-shell $^4T_1(G)$–$^6A_1(S)$ transitions. Since high voltages are used, thin film electroluminescent devices experience instability due to degradation or possible dielectric breakdown. The essential requirements of these TFEL phosphors are that (1) the luminescence must not be quenched by the high electric field (greater than or equal to 10^8 V/m), (2) a current of hot electrons must run through the phosphor film, (3) the phosphor must not be degraded by the high electric field. Multicolor TFEL displays became more attractive after the first multicolor EL display demonstration by William Barrow in 1984 using ZnS:Mn and ZnS:Tb phosphor layers [25]. In the early stages of multicolor TFEL displays, oxides were not primarily considered as TFEL phosphor because of its difficulty to achieve hot electron density to obtain sufficient excitation. Certain binary and ternary sulphides with dopants were promising for multicolor TFEL displays because of light emission from internal transitions. In general, multicolor TFEL displays can be either made using a patterned phosphor film structure of different colors (mostly red, blue, and green) or by using a single phosphor with a wide emission spectrum.

As discussed earlier in this chapter, a phosphor consists of a host material doped with a light emission center which is an activator. A host material should have a band gap large enough to emit visible light without absorption such as large band gap semiconductors (>2.5 eV) and insulators. An additional requirement for these phosphors to be electroluminescent host materials, they also should provide a medium for efficient transfer of high energy (>2 eV) electrons. Sulfides such as ZnS, CaS, SrS were found to satisfy these requirements. When doped with Cu, these materials perform very poorly because of unstable light-emitting centers in high fields because of shallow donor and acceptor levels in these host materials. Most classical stable dopants were found to be Mn, Tb, Eu, Ce, and Pr with respective yellow, green, red, blue, and white colors [26]. Table 8.3 shows some of the important electroluminescent phosphors being used in several display devices. We will discuss in detail some of the primary color phosphors needed for thin film electroluminescent display applications.

Table 8.3 Important sulphide and oxide EL phosphors having primary emission colors

Phosphor	Color	Efficiency, lm/W
Ga_2O_3:Eu	Red	0.5–1
$BaAl_2S_4$:Eu	Blue	0.5–1.5
SrS:Ce	Blue–green	0.5–1.5
ZnS:Tb	Green	0.5–2
$SrGa_2S_4$:Eu	Green	1–2
$ZnSiO_4$:Mn	Green	0.5–2
$ZnGa_2O_4$:Mn	Green	1–2
ZnS:Mn	Yellow	3–10
$Y_xGa_yO_3$:Mn	Yellow	10

8.4.1 Red Emitting Electroluminescent Phosphor

ZnS:Mn yellow phosphor with a red filter was used as the first red emitting EL phosphor with a luminous efficacy of 1 lm/W. Broad emission of ZnS:Mn from the green to red part of the spectrum helps achieve this red emission, but complicates fabrication process with a loss of luminance efficiency [27]. CaS:Eu became an efficient red EL emitter without any color filter and thin films of Eu:CaS were prepared by electron beam evaporation technique under sulphur atmosphere to compensate sulphur deficiency. Such thin films suffer oxygen contamination with low brightness that requires H_2S addition for better color purity [28]. Also it lacks in luminance and faster response upon the application of electric field. Such low response was reduced by addition of Sr to get rid of shallow traps due to Eu-clusters [29]. These $Ca_{1-x}Sr_xS$:Eu still suffers with low response at a higher above threshold voltage. Therefore, traditional sulphide phosphors such as ZnS, CaS, and SrS have been found to be inefficient candidates as red emitting EL phosphors. Gallium oxide (Ga_2O_3) gained much attraction as a phosphor host material after it has been demonstrated as emitting layers when doped with Mn, Cr, and Eu for TFEL devices because of its high luminance and stable operation [30]. Europium has several advantages such as high quantum efficiency and high-speed radiation luminescence such as organic materials. Ga_2O_3:Eu has been found to be a bright red-emitting electroluminescent phosphors with an efficiency of 0.38 lm/W [31] and hence found to be an attractive EL phosphor for multicolor TFEL devices. β-Ga_2O_3 is the most stable phase among five crystal structures of gallium oxide. Wide band gap of this material (~4.8 eV) with high thermal and chemical stability makes this material a good host of rare earth (RE) ions with high luminescence efficiency. Because of smaller ionic radius of Eu compared to other RE, it is compatible with Ga_2O_3. Interestingly, this phosphor works even in an amorphous state and proves that crystallization is not essential for all electroluminescent phosphors [32]. Thin films of Eu:Ga_2O_3 have been fabricated by several different techniques, including spray pyrolysis, atomic layer deposition, electron beam evaporation, sol gel process, and pulsed laser deposition. Wellenius et al. demonstrated a low voltage high brightness Eu:Ga_2O_3 TFEL devices fabricated by pulsed laser deposition technique [33]. The device consists of 250 nm thick Eu:Ga_2O_3 phosphor layer followed by a 100 nm thick indium gallium zinc oxide top contact and indium tin oxide (ITO)/ATO nano aluminate on Corning 7,059 glass plates as transparent bottom conducting layer (Fig. 8.6b). A maximum luminance of 220 cd/m^2 was achieved when excited by a 1 kHz square wave with rms voltage of 100 V. Threshold voltage for the emission is low at approximately 60 V compared to many other TFEL devices, which require a threshold voltage of 100–250 V. Figure 8.6 shows the schematic of the device and EL spectrum illustrating the characteristic red emission of Eu^{3+} ions at 611 nm due to $^5D_0 \rightarrow {}^7F_2$ transition.

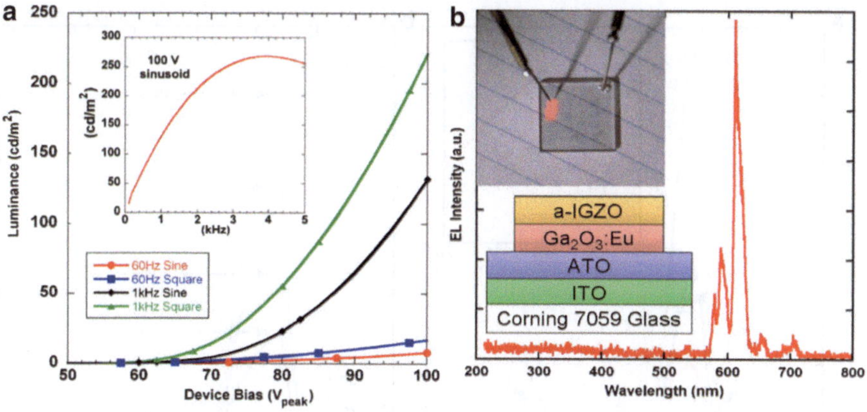

Fig. 8.6 (**a**) Light vs voltage curves for a device under varying excitation waveforms. The *inset* shows the frequency dependence of light emission for a 100 V sinusoidal waveform. (**b**) Electroluminescence spectrum illustrating the characteristic red emission of Eu^{3+} ions. A device schematic and a picture of a working device are the inset. Reprinted with permission from [33], Lee et al., *ACS Appl. Mater. Interfaces,* 6 (2008). © 2008, American Institute of Physics

Monoclinic β-Ga_2O_3 with a space group of C2/m has tetrahedral and octahedral sites for Ga^{3+} which are occupied by Eu^{3+} when doped. These two different sites produce two similar but distinct emission lines as can be seen in figure. Emission and efficiency are dependent on the frequency and shape of the wave. Maximum luminance of 221 cd/m^2 was achieved at 100 V for a 1 kHz square wave, whereas for sinusoidal excitation, luminance was 131 cd/m^2. Threshold for this emission is near 60 V for 1 kHz excitation, whereas a threshold of 80 V is needed for 60 Hz excitations (Fig. 8.6a). Maximum emission intensity was reached for 100 V excitation for a 3.9 kHz sinusoidal wave. Because the emission intensity is dependent on concentration of dopant ions, finding an optimal concentration with high brightness and efficiency without concentration quenching is important for better performance of such devices. Optimum concentration of Eu content in Ga_2O_3 was found to be 5–10 mol% after which nonradiative quenching mechanisms dominate that lead to drop in emission intensity [34]. As the concentration is increased to 20 mol%, average ion–ion distance decreases that leads to non-radiative energy transfer from Eu^{3+} ions either to nearby ions or to quenching centers. Difference in emission intensity of 2.4 and 20 mol% in continuous wave photoluminescence (CWPL) and time-integrated photoluminescence (TIPL) is due to change in decay constants as concentration is increased (Fig. 8.7).

In TIPL, 20 mol% sample is quenched because of drop in slow decay constant ($\tau_s < 90$ μs) compared to 2.4 mol% sample ($\tau \simeq 1\,\text{ms}$), whereas in CWPL, quenching is compensated by photo-excited carrier distribution during excitation.

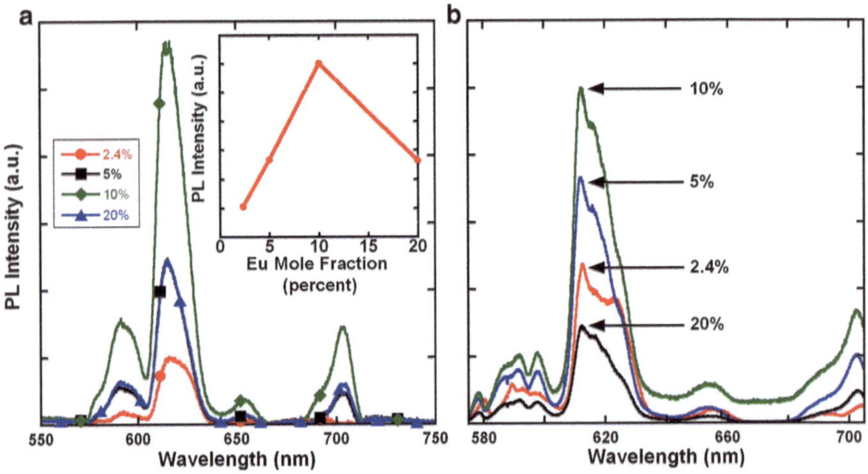

Fig. 8.7 CWPL (**a**) and TIPL (**b**) spectra for $(Eu_xGa_{1-x})_2O_3$ with Eu content between 2.4 and 20 mol%. The peak intensity of the $^5D_0 \rightarrow {}^7F_2$ transition is used to compare the relative intensity of the various films (*inset*). Reprinted with permission from [34], Wellenius et al., *J. Appl. Phys.* **107** (2010). © 2010, American Institute of Physics

8.4.2 Blue Emitting Electroluminescent Phosphor

A need for an efficient blue phosphor was challenging for many years and delayed the development of full color thin film EL display devices. Earlier, full color TFEL devices used color filters by filtering the yellow emission of ZnS:Mn and blue–green emission of SrS:Ce. Use of filters reduces the luminescence efficiency and luminance of the device. For instance, ZnS:Mn/SrS:Ce device showed a luminance of 97 and 220 cd/m^2 for filtered red and green colors, respectively, whereas for filtered blue, value was only 11 cd/m^2 under the same operation conditions of 244 V excitation voltage at 60 Hz frequency [35]. Among the rare earths, Eu^{2+}, Ce^{3+}, Cu^+, and Ag^+ are few ions that are mostly used for the generation of blue luminescence on such phosphors. Compared to rare earths involving well-shielded 4f electrons, Eu^{2+} involves $5d^14f^6$ excited state which is not well-shielded and experiences significant crystal field due to the interaction of 5d electrons. So, the emission of Eu^{2+} is host lattice-dependent and its wavelength varies from near-UV to red due to $4f^65d^1$–$4f^7$ transitions. Similar case with cerium as Ce^{3+} ions produces broad emission around bluish green region. Among the hosts, II–VI sulphides such as CaS, BaS, and SrS have been investigated for blue thin film phosphors. Sulfides have advantages over oxides because they provide a sufficient electron acceleration below the breakdown field and are found to be attractive for TFEL devices. Among them, SrS:Ce was the first blue phosphor and was studied intensively because of its relatively good efficiency of 0.8–1.6 lm/W. MOCVD grown thin films of SrS:Ce showed better performance with a luminescence brightness of 51 cd/m^2, as measured at 40 V above the threshold at a frequency of 60 Hz [36]. But as pointed out earlier, SrS:Ce suffered from stability, chromaticity, luminescence efficiency, and charge mismatch. Fabricating thin films of SrS:Ce with good crystallinity is complex. Compensating the charge by incorporating single or multiple impurity ions and requirement of a filter to isolate blue color raised practical difficulties for it to be considered for commercial display devices. Cu-doped SrS was found to be better than SrS:Ce with 40 % of the emission from wavelengths below 460 nm compared to only 5 % of SrS:Ce. Enhancement of blue emission was also observed by codoping with monovalent cation ion such as Ag^+ which also provides a significant blue shift from 480 to 430 nm [37]. In addition to binary sulphides, ternary sulphides involving II–III–VI group elements in the periodic table were investigated for blue phosphors. Combination of II group (Mg, Ca, Sr, and Ba), III-group (Al, Ga, In), and VI-group S4 hosts was found to be attractive materials for EL devices. Strontium thiogallates ($SrGa_2S_4$) doped with rare earth prepared by sputtering technique had good cathodoluminescence property and a potential candidate for EL devices [38].

These ternary sulphides when doped with Ce^{3+} or Eu^{2+} show a broad emission compared to its binary counterparts. Cerium-doped $CaGa_2S_4$ and $SrGa_2S_4$ generate blue emission but not good enough for full color displays [39]. In addition, Eu^{2+} in these ternary sulphides emit blue wavelength, whereas binary sulphides such as MgS, CaS, SrS, BaS when doped with Eu^{2+} have peak emission wavelength at 591, 651, 616, and 572 nm, respectively [40, 41]. Table 8.4 shows emission wavelength of Eu^{2+} in selected ternary sulphides [42].

As can be seen from the list, only $BaAl_2S_4$ emits blue wavelength at 467 nm due to dipole-allowed transition from the $4f^65d^1$ excited state to $4f^7$ ground state of Eu^{2+} ions and this makes it the most promising blue-emitting thin film electroluminescent material.

Even though powders of $BaAl_2S_4$:Eu with an emission wavelength of 475 nm were reported for the first time by Donohue and Hanlon [43], it was until 1999 that

Table 8.4 List of ternary sulphides and their emission wavelengths

Host	Emission wavelength (nm)
$BaGa_2S_4$	495
$MgGa_2S_4$	543
$CaGa_2S_4$	562
$SrGa_2S_4$	538
$EuGa_2S_4$	546
$BaIn_2S_4$	663
$SrIn_2S_4$	614
$CaIn_2S_4$	731
$BaAl_2S_4$	467
$CaAl_2S_4$	516
$MgAl_2S_4$	499
$SrAl_2S_4$	495
$EuAl_2S_4$	508

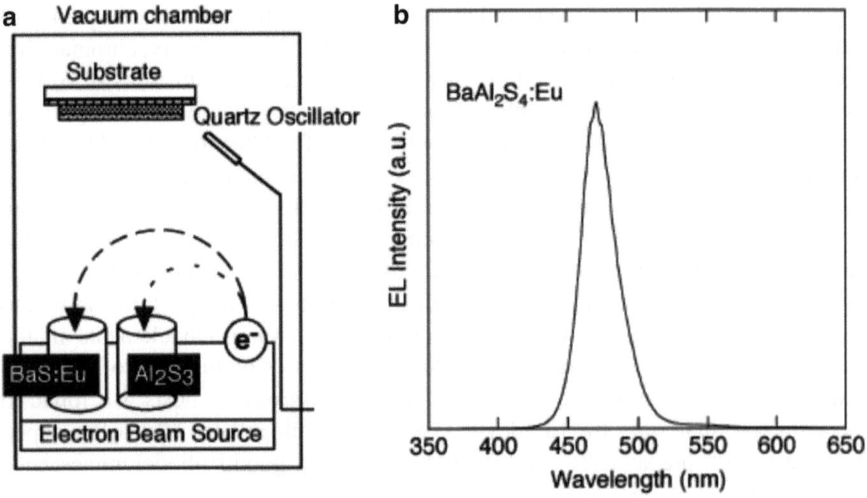

Fig. 8.8 (**a**) Two targets pulse-electron beam evaporation technique. (**b**) Electroluminescent spectrum from $BaAl_2S_4$:Eu phosphor. Reprinted with permission from [44], Miura et al., *Jpn. J. Appl. Phys*. **38** (1999). © 1999, The Japan Society of Applied Physics

$BaAl_2S_4$:Eu were found to be a good EL phosphor in the form of thin films by Miura et al. [44]. The first thin films of $BaAl_2S_4$:Eu were fabricated by dual source pulsed electron-beam evaporation using two targets (Fig. 8.8a).

$BaAl_2S_4$ belongs to a cubic phase with a bandgap of 4.10 eV. They showed a luminance of 65 cd/m^2 at 80 V above the threshold at a frequency of 50 Hz. Color coordinates satisfy the requirements of blue color television displays and allowed the fabrication of full-color EL devices. A full-color 34 in. flat-panel display was developed using color-by-blue scheme by company iFire [45]. This display demonstrated a brightness of 1,000 cd/m^2 using high luminescence $BaAl_2S_4$:Eu blue phosphor in combination with red and green color phosphors. By varying the molar ratio of BaS to Al_2S^3, emission of Eu^{2+} can be varied from 467 to 540 nm in this barium thioaluminate thin films. This is because of the strong crystal field interaction on 5d–4f transition of Eu^{2+}[41]. Because of close ionic radii match, Eu^{2+} ions ($R_{Eu}{}^{2+} = 117$ pm) are likely to replace the barium sites ($R_{Ba}{}^{2+} = 135$ pm). Addition of cations was found to enhance the luminescence efficiency of several phosphors. Mg^{2+} as codopants seems to enhance the PL intensity by 2.5 times in $BaAl_2S_4$ compared to single-doped $BaAl_2S_4$:Eu under the excitation of 395 nm. Comparable ionic radii ($R_{Mg}{}^{2+} = 35$ pm) and charge neutrality allows solid solution possible in this case. As mentioned earlier, Eu^{2+} experiences crystal field strength and an addition of Mg^{2+} increases the effect. Increasing the concentration of Mg^{2+} shifts the emission wavelength from 474 nm to 476 and 477 nm when the concentration is increased to 6 and 15 mol%, respectively. The maximum intensity was achieved at 6 mol% after which there is a drop in intensity because of concentration quenching effect [46] (Fig. 8.9).

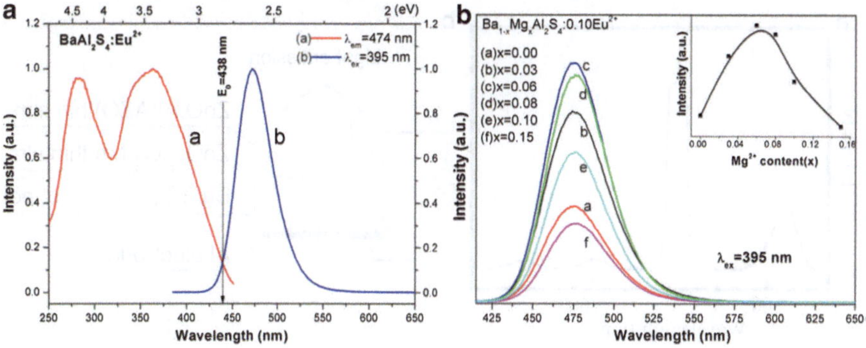

Fig. 8.9 (**a**) Excitation (λ_{em}—474 nm) and emission (λ_{ex}—395 nm) spectra of $BaAl_2S_4$:0.10Eu^{2+}. (**b**) Emission spectra of $Ba_{1-x}Mg_xAl_2S_4$:0.10Eu^{2+} with varying Mg^{2+} concentrations. *Inset* shows the influence of Mg^{2+} content on the emission intensity of the $Ba_{1-x}Mg_xAl_2S_4$:0.10Eu^{2+} phosphors (λ_{ex}—395 nm). Reprinted with permission from [46], Yu et al., *Ceram. Int.* **39** (2013). © 2013, Elsevier

8.4.3 Green Emitting Electroluminescent Phosphor

Since the outermost 3d orbitals of Mn^{2+} are not shielded, host lattice plays a major role on the optical properties of Mn. So manganese atom can emit red, green, or yellow light depending on the host lattice. Energy levels of this transition metal are highly affected by crystal-field effect that results in broad emission range. Green emission was obtained earlier with the well-established ZnS:Mn phosphor by filtering out the green component from its broad yellow–green luminescence. Among the unfiltered green EL phosphors, ZnS:Tb and ZnS:TbOF have been very attractive and found to produce green emission with a luminous efficiency of more than 1 lm/W and a luminance of 100 cd/m^2. It has a strong emission at 545 nm due to $^5D_4$–$^7F_5$ electron transition of Tb^{3+} ion as shown in Fig. 8.10a [47]. But these sulphide-based phosphors experienced stability problems and also a limited color purity because of its multiple emission peaks. Zn-based oxide phosphors have been found to be a good green emitting EL phosphor because of its higher stability over sulphides under electron irradiation. Zinc gallate ($ZnGa_2O_4$) having a band gap of 4.4 eV and with a normal spinel structure was found to be a promising oxide phosphor both as a thin film and powder for EL devices [49]. Typical device structure using zinc gallate is shown in Fig. 8.10b.

Zinc ions with +2 charge occupy tetrahedrally coordinated sites, whereas Ga ions with +3 charge are octahedrally coordinated. When doped with Manganese, Mn^{2+} substitutes Zn^{2+} sites and gives an intense green emission at 505 nm due to $^4T_1$–$^6A_1$ transitions of Mn^{2+} ions. Thin films of Mn-doped zinc gallate have been prepared by different techniques including pulsed laser and RF sputtering deposition techniques from stoichiometric pellets of ZnO and Ga_2O_3. In addition to such Zn-based gallates, stoichiometric (100 mol% activator ions) europium thiogallate ($EuGa_2S_4$) thin films were grown by molecular beam epitaxy technique.

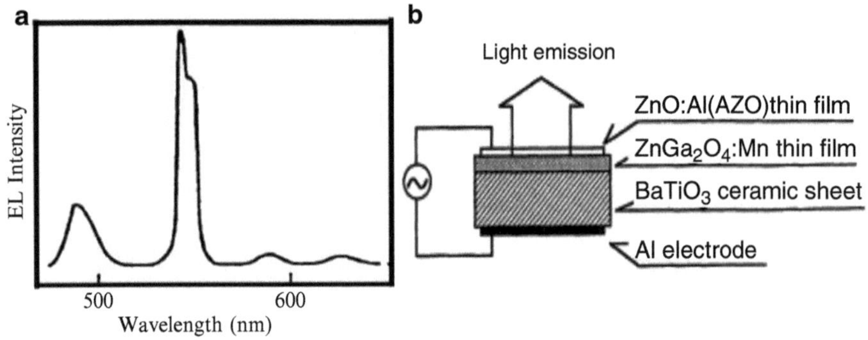

Fig. 8.10 (**a**) EL emission spectrum of ZnS:Tb, F. Reprinted with permission from [47], Ono, *Ann. Rev. Mater. Sci.* **27** (1997). © 1997, Annual Reviews. (**b**) TFEL device structure. Reprinted with permission from [48], Minami et al., *Jpn. J. Appl. Phys.* **34** (1995). © 1995, Japan Society of Applied Physics

This orthorhombic phosphor thin film has weak interactions between Eu^{2+} ions and makes stoichiometric doping possible with only a small concentration quenching. Pure green luminescence due to intra-atomic (5d–4f) transition of Eu^{2+} ion is observed at 536 nm [50].

Zinc aluminate ($ZnAl_2O_4$) with a band gap of 3.9 eV is another attractive spinel-structured EL phosphor because of its transparent and electroconductive properties. Green emitting manganese-doped $ZnAl_2O_4$ thin film phosphors have been prepared by spray pyrolysis technique with good quality. An aqueous solution of nitrate salts of Zn and Al was sprayed as aerosol using ultrasonic nebulizer on a heated substrate. Consecutive pyrolysis of aerosol forms thin films on the substrates such as quartz or aluminosilicate ceramic plates maintained at a higher temperature. An intense green emission with an efficiency of 4.5 lm/W and a luminance of 540 cd/m^2 was achieved at an excitation voltage of 4 kV and current density of 38 μA/cm^2. Among the oxides for green emitting phosphors, Zn_2SiO_4 doped with Mn has also been widely investigated and produced bright green electroluminescence with an efficiency of 0.78 lm/W. For better crystallization, zinc silicate requires annealing process at a higher temperature which in turn requires high temperature-resistant substrates such as $BaTiO_3$, which in turn requires a high driving voltage. To avoid higher processing temperature, Si is substituted by Ge to form Zn_2GeO_4 with a smaller band gap and lower processing temperature. Mn-doped Zn_2GeO_4 thin films have been prepared by sputtering techniques using stoichiometric targets of ZnO, GeO_2, and MnO. Deposited thin films were further annealed at around 600–650 °C. Germanate thin films showed an intense green electroluminescence at 540 nm due to transition of Mn^{2+} ions sitting on tetrahedrally coordinated Zn sites. An efficiency of 0.33 lm/W and a luminance of 105 cd/m^2 have been observed at an optimum concentration of 2 % Mn^{2+} ions at 60 Hz. These values are comparable to standard green TFEL phosphors such as ZnS:Tb and ZnS:TbOF.

8.5 Conclusions

Conversion of electrical energy into luminous energy is used in electroluminescent displays which gain much interest recent days. The integral part of thin film electroluminescent devices is thin film phosphor that emits light under high electric field. It consists of a host material with activator ions that act as luminescent centers. II–VI sulphides such as ZnS and SrS were well studied in the early stages of electroluminescence and are still being used in display devices. Later, oxides also gained equal interest because of its better thermal stability and relatively ease of fabrication. Full color thin film EL displays require efficient phosphors for primary colors such as red, blue, and green. Various materials for these primary colors have been discussed in this chapter. Recent investigations on several oxides yield new materials for red and green colors, whereas sulphides dominate for efficient blue color thin film phosphor. In addition to host materials, dopant ions, their concentration and introduction of codopants play a major role in achieving the desirable emission wavelengths necessary for different applications. This makes the researchers continue investigating new host matrices as well as activator ion combinations for more efficient, reliable, and stable thin film electroluminescent displays.

References

1. N.C. George, K.A. Denault, R. Seshadri, Phosphors for solid-state white lighting. Ann. Rev. Mater. Res. **43**, 481–501 (2013)
2. M. Leskela, Rare earths in electroluminescent and field emission display phosphors. J. Alloys Compd. **275–277**, 702–708 (1998)
3. T. Justel, H. Nikol, C. Ronda, New developments in the field of luminescent materials for lighting and displays. Angew. Chem. Int. Ed. **37**, 3084–3103 (1998)
4. A. Lakshmanan, *Luminescence and Display Phosphors: Phenomena and Applications* (Nova Science, New York, 2008)
5. S.L. Jones, D. Kumar, K.-G. Cho, R. Singh, P.H. Holloway, Pulsed laser deposition of Y_2O_3: Eu thin film phosphors. Displays **19**(4), 151–167 (1999)
6. R. Mirhosseini, M.F. Schubert, S. Chhajed, J. Cho, J.K. Kim, E. Fred Schuber, Improved color rendering and luminous efficacy in phosphor-converted white light-emitting diodes by use of dual-blue emitting active regions. Opt. Express **17**(13), 10806–108013 (2009)
7. N.T. Tran, J.P. You, F.G. Shi, Effect of phosphor particle size on luminous efficacy of phosphor-converted white LED. J. Lightwave Tech. **27**(22), 5145 (2009)
8. N.T. Gurin, A.V. Shlyapin, O.Y. Sabitov, Quantum yield and luminous efficacy of thin-film ZnS electroluminescent emitters. Tech. Phys. **48**(4), 479–491 (2003). Translated from ZhurnalTekhnichesko Ï Fiziki **73**(4), 100–112, 2003
9. N. Narendran, N. Maliyagoda, L. Deng, R. Pysar, Characterizing LEDs for general illumination applications: mixed-color and phosphor-based white sources. SPIE Proc. **4445** (2001). 137–147
10. P.H. Holloway, T.A. Trottier, J. Sebastian, S. Jones, X.-M. Zhang, J.-S. Bang, B. Abrams, W.J. Thomes, T.-J. Kim, Degradation of field emission display phosphors. J. Appl. Phys. **88**(1), 483–488 (2000)

11. H.C. Swart, J.S. Sebastian, T.A. Trottier, S.L. Jones, P.H. Holloway, Degradation of zinc sulfide phosphors under electron bombardment. J. Vac. Sci. Technol. A **14**(3), 1697–1703 (1996)
12. K.T. Hillie, H.C. Swart, Electron beam induced degradation of a pulsed laser deposited ZnS: Cu, Au, Al thin film on a Si (1 0 0) substrate. Appl. Surf. Sci. **183**(3–4), 304–310 (2001)
13. H. Schwarz, H.A. Tourtellotte, Vacuum deposition by high-energy laser with emphasis on barium titanate films. J. Vac. Sci. Technol. **6**, 373 (1969)
14. H.M. Christen, G. Eres, Recent advances in pulsed-laser deposition of complex oxides. J. Phys. Condens. Matter **20**(26), 264005 (2008)
15. J.Y. Choe, S.M. Blomquist, D.C. Morton, Characteristics of SrS:Cu thin-film electroluminescent device fabricated by pulsed-laser deposition. Appl. Phys. Lett. **80**(22), 4124–4126 (2002)
16. H.M. Manasevit, W.I. Simpson, The use of metal-organic in the preparation of semiconductor materials. J. Electrochem. Soc. **118**(4), 645 (1971)
17. R. Hiskes, RP. Helbing, SA. DiCarolis, Growth of Electroluminescent Phosphors by MOCVD, Patent Number 5942284, 1999
18. W.R. Grove, On the electro-chemical polarity of gases. Phil. Trans. R. Soc. **142**, 87–101 (1852)
19. L. Filipovic, S. Selberherr, G. Mutinati, E. Brunet, S. Steinhauer, A. Kock, J. Teva, J. Kraft, J. Siegert, F. Schrank, Modeling spray pyrolysis deposition, in *Proceedings of the world congress on engineering*, vol. II (WCE, London, 2013), ISBN: 978-988-1952-8-2
20. D. Perednis, L.J. Gauckler, Thin film deposition using spray pyrolysis. J. Electroceram. **14**, 103–111 (2005)
21. F. Ozutok, K. Erturk, V. Bilgin, Growth, electrical and optical study of ZnS:Mn thin films. Proceedings of the international congress on advances in applied physics and materials science. Acta Phys. Polonica A **121**(1), 221–223 (2012)
22. G. Destriau, Recherchessur les scintillations des sulfures de zinc aux rayons. J. Chim. Phys. **33**, 587–625 (1936)
23. N.A. Vlasenko, Y.U. Popkov, A study of electroluminescence of a sublimated ZnS-Mn phosphor. Opt. Spectrosk. **8**, 81 (1960)
24. H. Ohnishi, Electroluminescent display materials. Ann. Rev. Mater. Sci. **19**, 83–101 (1989)
25. A.N. Krasnov, Electroluminescent displays: history and lessons learned. Displays **24**, 73–79 (2003)
26. C.N. King, Electroluminescent displays. J. Vac. Sci. Technol. A **14**, 1729 (1996)
27. R. T. Tuenge, J. Kane, Bright red EL using a thin-film filter, *Digest of 1991 SID International Symposium*, 279–281 (1991)
28. D. Poelman, R. Vercaemst, R.L. Van Meirhaeghe, W.H. Laflkre, F. Cardon, Influence of the growth conditions on the properties of CaS:Eu electroluminescent thin films. J. Lumin. **75**, 175–181 (1997)
29. J.E. Van Haecke, P.F. Smet, D. Poelman, The influence of source powder composition on the electroluminescence of $Ca_{1-x}Sr_xS$:Eu thin films. Spectrochim. Acta B Atom. Spectr. **59**(10–11), 1759–1764 (2004)
30. T. Miyata, T. Nakatani, T. Minami, Gallium oxide as host material for multicolor emitting phosphors. J. Lumin. **87–89**, 1183–1185 (2000)
31. A.H. Kitai, Oxide phosphor and dielectric thin films for electroluminescent devices. Thin Solid Films **445**, 367–376 (2003)
32. A.H. Kitai, *Luminescent Materials and Applications* (Wiley, Chichester, 2008)
33. P. Wellenius, A. Suresh, J.F. Muth, Bright, low voltage europium doped gallium oxide thin film electroluminescent devices. Appl. Phys. Lett. **92**, 021111 (2008)
34. P. Wellenius, E.R. Smith, S.M. LeBoeuf, H.O. Everitt, J.F. Muth, Optimal composition of europium gallium oxide thin films for device applications. J. Appl. Phys. **107**, 103111–103111-5 (2010)
35. W.-M. Li, M. Ritala, M. Leskela, R. Lappalainen, J. Jokinen, E. Soininen, B. Hüttl, E. Nykänen, L. Niinisto, Elemental characterization of electroluminescent SrS:Ce thin films. J. Appl. Phys. **84**(2), 1029–1035 (1998)

36. K.W. Barth, J.E. Lau, G.G. Peterson, D. Endisch, A.E. Kaloyeros, R.T. Tuenge, C.N. King, Metallorganic chemical vapor deposition of SrS:Ce for thin film electroluminescent device applications. J. Electrochem. Soc. **147**(6), 2174–2180 (2000)
37. S.-S. Sun, A new blue emitting TFEL phosphor: SrS:Cu. Displays **19**, 145–149 (1999)
38. P. Benalloul, C. Bat-thou, J. Benoit, L. Eichenauer, A. Zeinerta, HA-Ill & ternary compounds: new host matrices for full color thin film electroluminescence displays. Appl. Phys. Lett. **63**, 1954 (1993)
39. Y. Inoue, I. Tanaka, K. Tanaka, Y. Izumi, S. Okamoto, M. Kawanishi, D. Barada, N. Miura, H. Matsumoto, R. Nakano, Atomic composition and structural properties of blue emitting $BaAl_2S_4$:Eu electroluminescent thin films. Jpn. J. Appl. Phys. **40**, 2451–2455 (2001)
40. S. Shionoya, W.M. Yen, *Phosphor Handbook* (CRC, Boca Raton, 1999)
41. H. Kasano, K. Megumi, H. Yamamoto, Cathodoluminescence of Ca1 − xMgxS: A (A = Eu or Ce). J. Electrochem. Soc. **131**(8), 1953–1960 (1984)
42. Philippe SMET, Study of $BaAl_2S_4$:Eu and SrS:Cu,Ag as blue emitting materials for thin film electroluminescence, Ph.D. Dissertation, Universiteit Gent, 2005
43. P.C. Donohue, J.E. Hanlon, The synthesis and photoluminescence of MnM:n(S, Se). J. Electrochem. Soc. **121**, 137 (1974)
44. N. Miura, M. Kawanishi, H. Matsumoto, R. Nakano, High-luminance blue-emitting $BaAl_2S_4$: Eu thin-film electroluminescent devices. Jpn. J. Appl. Phys. **38**, L1291–L1292 (1999)
45. X. Wu, A. Nakua, D. Cheong, Color-by-blue: a novel method to achieve full-color inorganic EL displays. J. Soc. Inf. Disp. **12**, 281–286 (2004)
46. R. Yu, H.M. Noh, B.K. Moon, B.C. Choi, J.H. Jeong, K. Jang, H.S. Lee, S.S. Yi, Enhanced photoluminescence of $BaAl_2S_4:Eu^{2+}$ phosphor by Mg^{2+} ions doping. Ceram. Int. **39**(8), 9709–9713 (2013)
47. Y.A. Ono, Materials for full-color electroluminescent displays. Ann. Rev. Mater. Sci. **27**, 283–303 (1997)
48. T. Minami, T. Maeno, Y. Kuroi, S. Takata, High-luminance green-emitting thin-film electroluminescent devices using $ZnGa_2O_4$:Mn phosphor. Jpn. J. Appl. Phys. **34**, L684–L687 (1995)
49. G. Anoop, K. Mini Krishna, M.K. Jayaraj, Photoluminescent studies in co-sputtered $ZnGa_2O_4$: Mn phosphor thin films, in *Proceedings of ASID*, New Delhi, 2006, pp. 280–282
50. K. Tanaka, S. Okamoto, Green electroluminescence of $EuGa_2S_4$ thin films. Appl. Phys. Lett. **83**, 647 (2003)

Chapter 9
Thin Films for Energy-Efficient Mechanical Tools

Hisayuki Suematsu and Tsuneo Suzuki

Abstract Chromium oxynitride-based thin films have been developed for application as hard coatings on cutting tools. In this work, the hardening mechanisms of crystalline solids were examined and a variety of thin films were fabricated based on these mechanisms. Methods of controlling the compositions of such films as well as the resulting microstructures and properties are discussed.

9.1 Introduction: Demands from Machining Industry

Machining industries have been developed with the improvement of properties of cutting tool such as life and hardness. Results of cost analysis in machining are shown in Fig. 9.1 [1]. Seventy-five percentage of the costs depends on the time of the machining so that the cost is suppressed with the increase of the cutting speed. To reduce the heat in the high speed cuttings, cutting tools with high toughness and hardness is required. However, it was highly difficult because they are contradictory properties [2].

First attempt was to use bulk materials with higher hardness. High speed tool steels and WC-based materials were developed. Then, surface treatments and thin film coatings were applied on the tools. Nitrization and oxidation by annealing the tools in NH_3 gas and water vapor, respectively, were the early methods to form nitrides and oxides on the surface [2]. For obtaining higher hardness, carbide and nitride coatings were developed using chemically vapor deposition and other coating methods. With addition of third elements, many thin films including Ti–Al–N were developed [3].

Different from the cation additions, anion additions were very rare except for Ti–C–N. Furthermore, it was widely believed that the increase of the oxygen content in the nitride thin films caused hardness decrease because of the formation of soft oxide grains.

H. Suematsu • T. Suzuki (✉)
Extreme Energy-Density Research Institute,
Nagaoka University of Technology, Nagaoka, Japan
e-mail: suzuki@vos.nagaokaut.ac.ip

S. Babu Krishna Moorthy (ed.), *Thin Film Structures in Energy Applications*,
DOI 10.1007/978-3-319-14774-1_9

Fig. 9.1 Cost analysis in machining

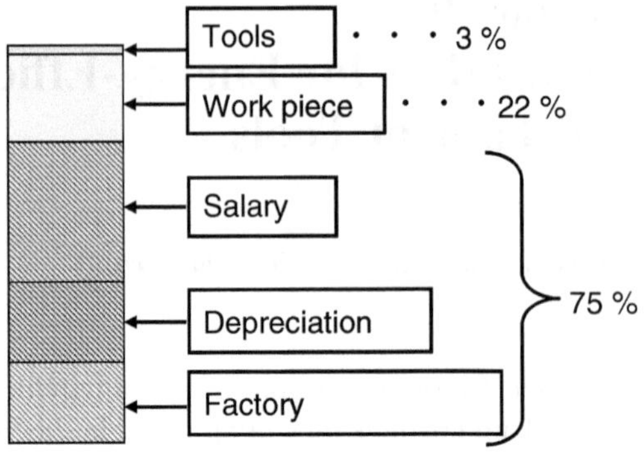

Utilizing the concept of nanocomposite [4], some attempts on preparations of thin films with nanocomposite structure [5] including Cr–Si–N [6, 7] were reported. However, in some reports, the microstructural observation was very limited so that it was difficult to know whether the microstructure contributed the hardness or not.

Among them, CrN has been known as a material which has slightly lower hardness and higher oxidation resistance [8]. With these properties, CrN coatings have been applied on drills and end mills for cutting Cu-based alloys [9].

Similar to TiN-based hard coatings, various CrN-based thin films were prepared to improve the hardness and other properties. Most of them are cation addition or substitution [8]. In order to improve the hardness of CrN-based coatings, other novel methods or combination of the conventional and novel methods have been required. In the next section, hardening mechanisms are described.

9.2 Hardening of Solid

Hardness of metals has been widely studied. In Knoop indentation anisotropy, the experimental data was qualitatively explained by assuming tensile stress parallel to the surface of the indentation to move dislocations in the primary slip system at the testing temperature [10, 11]. Thus, drag of the dislocation motion is a key factor to determine the hardness. Even though the brittle oxides and nitrides, an isostatic stress component induced by the indentation prevents the crack propagation and a high shear stress component from the tensile stress beneath the indentation drives the dislocations. Although it was difficult to introduce dislocations in nanosized grains, this empirical rule is still believed to be applicable to many materials.

The drag of dislocations known as Peierels force is shown below [12, 13],

$$\tau_{\mathrm{p}} = \frac{2G}{1-\nu}\exp\left[-\frac{2\pi}{(1-\nu)}\frac{d}{b}\right], \tag{9.1}$$

where, τ_{p} is the shear stress, G is the bulk modulus, ν is the Poison's ratio, d is the interplanar distance of the slip plane, and b is Burger's vector.

In order to increase the hardness of solids, the following methods have been known:

1. Increasing the bulk modulus by changing the chemical bonds.
2. Increasing dislocation pinning force by introducing inclusions in grains.
3. Preventing stress concentration from the neighboring pilled-up dislocations by decreasing the grain size described in the following Hall–Petch relation [14, 15],

$$\sigma_y = \sigma_0 + \frac{k}{\sqrt{d}}, \tag{9.2}$$

   where σ_y is the yield stress with dislocation pile up, σ_0 is the yield stress without dislocation pile up, k is a constant, d is the grain size.
4. Decreasing the external stress.

We have carried out the above (1–3) to tailoring the hardness. Details are described in the following sections.

9.3 Anion Substitution of Transition Metal Nitrides to Change the Chemical Bonds

In order to replace partial N sites by O, a pulsed laser deposition (PLD) method was utilized to deposit as Cr(N,O) thin films. Plates of Cr or Cr_2N were rotated in a vacuum chamber and used as targets. By changing the base pressure to control O_2 partial pressure and introducing NH_3 gas at various pressures in the chamber, thin films were deposited on Si(100) single crystal substrates. Details of the deposition are shown elsewhere [16, 17]. A typical example of Rutherford backscattering spectroscopy (RBS) data is shown in Fig. 9.2 [16]. By fitting simulated results to the experimental data, the composition of the thin film was determined. X-ray diffraction (XRD) patterns for the thin films are shown in Fig. 9.3 [18]. For thin

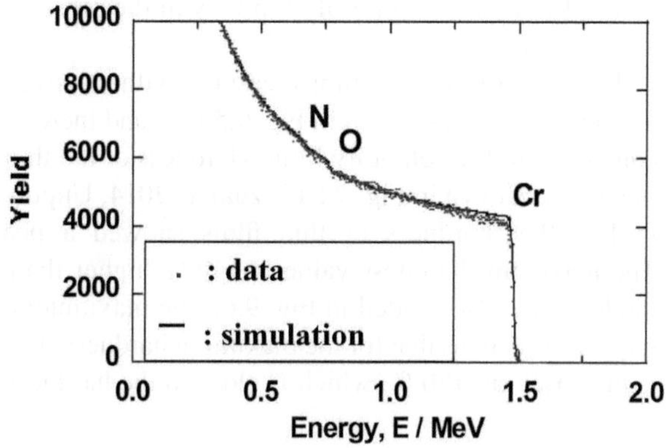

Fig. 9.2 RBS experimental and simulated data for a Cr(N,O) thin film

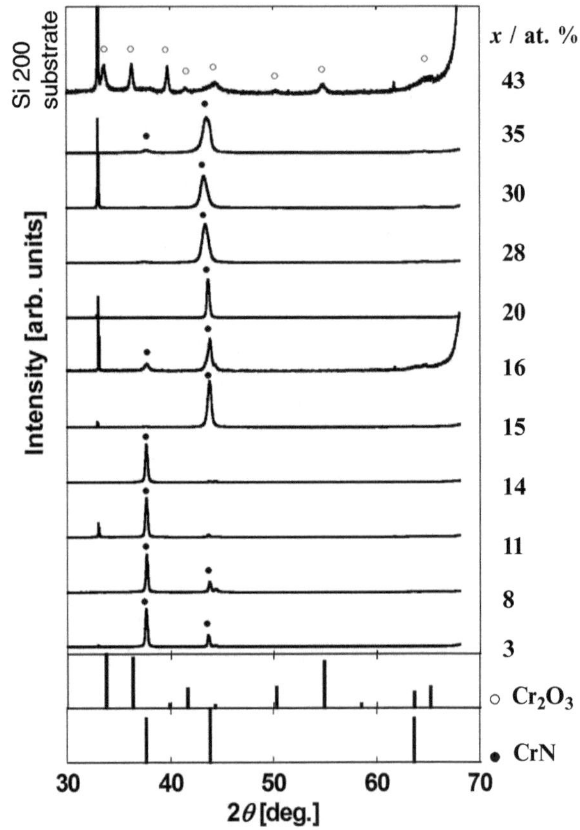

Fig. 9.3 XRD patterns for Cr(N,O) thin films

films with the oxygen content (x) = 3–35 at.%, all peaks correspond to those for CrN structure. This indicated that the thin films were of an NaCl type structure. Lattice constants of the thin films are plotted in Fig. 9.4 (Suzuki T, 2014, Unpublished work). From these results, Cr(N,O) thin films with various anion compositions were successfully deposited. At $x = 43$ at.%, peaks for Cr_2O_3 were seen. The solubility limit of O in CrN in the present experimental conditions is in $x = 35$–43 at.%.

Hardness of the thin films measured with a conventional hardness tester using a Vickers indenter is shown in Fig. 9.5 [19] and increases with increasing the oxygen content up to the solubility limit. Hardness of the thin films measured with a nano-indenter is shown in Fig. 9.6 (Suzuki T, 2014, Unpublished work). Similar to those in Fig. 9.5, hardness of thin films showed a peak near the solubility limit. The maximum hardness value is 140 % higher than that with $x = 8$ at.%. Elastic modulus are also plotted in Fig. 9.6. The maximum elastic modulus is at the same oxygen content as that for the maximum hardness. Furthermore, the elastic modulus is increased by 150 %, which is close to the hardness increase.

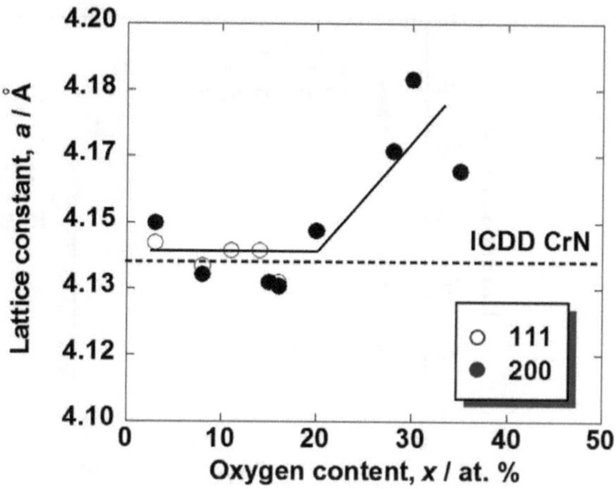

Fig. 9.4 The lattice constants of Cr(N,O) thin films

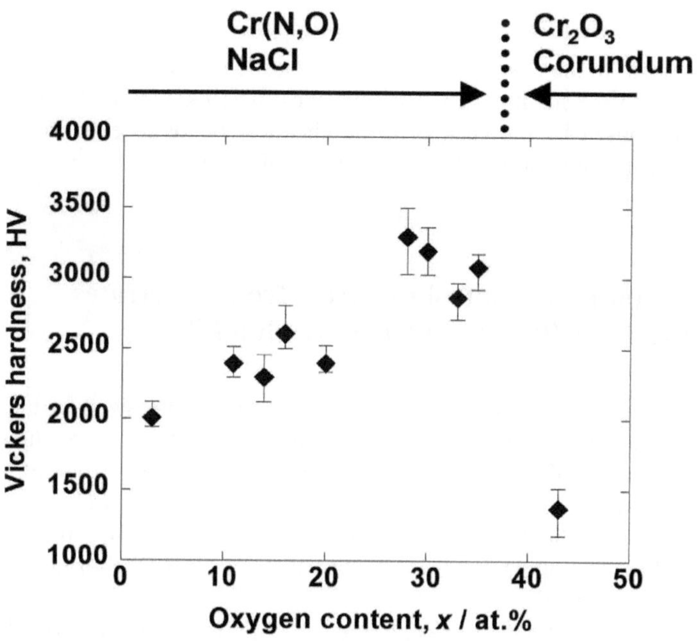

Fig. 9.5 Hardness in Cr(N,O) thin films

For these thin films, grain size measurements by transmission electron microscopy (TEM) were carried out and no significant changes were detected [20]. Furthermore, residual stress of the thin films was estimated by the bending curves and was below 1 GPa.

From XRD results (Fig. 9.3), Cr(N,O) with an NaCl type structure is the only phase in the thin films with $x \leqq 35$ at.%. From Eq. (9.1), the drag of dislocation must

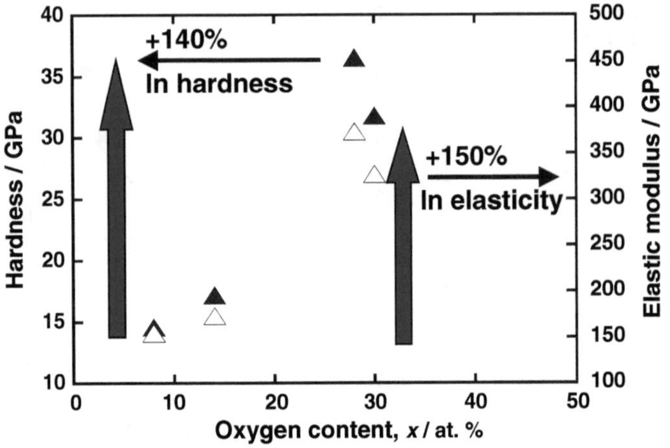

Fig. 9.6 Hardness and elastic modulus in Cr(N,O) thin films measured with a nanoindentation

be proportional to the elastic modulus. From elastic modulus change being almost proportional to the hardness change (Fig. 9.6), it is likely that the hardening of Cr(N,O) is caused by the change in the chemical bonds to enhance the elastic modulus change and not by other microstructural factors.

9.4 Composition Control by Stress from Substrates in Epitaxially Grown Cr(N,O) Thin Films

In the last section, novel compounds of Cr(N,O) were synthesized as thin films by PLD. The anion compositions were controlled by the partial gas pressures. On the other hand, compositions of thin films can be affected by stress from the substrate, which can reach as high as a few GPa. In particular, by epitaxial growth, lattice mismatch between the thin film and the substrate is a measure to estimate and tune the stress. For controlling the anion composition of Cr(N,O) thin films, epitaxial growth was attempted.

Thin films of Cr(N,O) were deposited by PLD on MgO(100), $MgAl_2O_4$(100) single crystal substrates. The misfit strain between CrN to the substrates are 1.7 and −2.4 %, respectively. Si(100) single crystals were also used for the substrates for comparison. The MgO, $MgAl_2O_4$, and Si (100) single crystal substrates were place in a deposition chamber at the same time. After the deposition chamber was evacuated to 1.0×10^{-6} Pa, oxygen gas was introduced at 2.0×10^{-5} Pa. Then, a N radical source was turned on at 1.5×10^{-2} Pa. These three thin films were prepared in a single experiment to keep the environment and plasma conditions identical. Details of experiments are shown elsewhere [21].

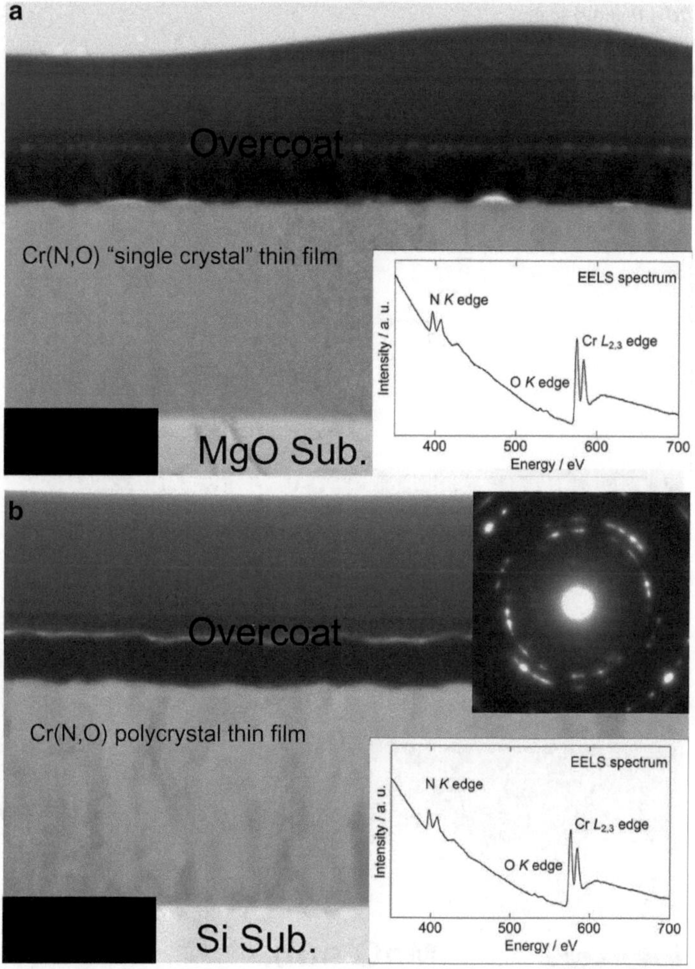

Fig. 9.7 Bright field images, SAD patterns, and EELS data for Cr(N,O) thin films on (**a**) MgO and (**b**) Si (100) single crystal substrates

A bright field cross-sectional TEM micrograph for the Cr(N,O) thin film deposited on the MgO (100) substrate is shown in Fig. 9.7a [21]. There are no grain boundaries. A selected area electron diffraction (SAD) pattern is also shown in Fig. 9.7a. The crystal in the thin film has a clear orientation to the substrate. An electron energy loss spectroscopy (EELS) data shown in Fig. 9.7a indicates that the thin film consists of N, O, and Cr. TEM, SAD, and EELS data for a thin film deposited at the same conditions on the Si (100) single crystal substrate are shown in Fig. 9.7b for comparison. This is clearly polycrystalline Cr(N,O) thin film.

An XRD pattern for the deposited Cr(N,O) thin film on the MgO (100) substrate is shown in Fig. 9.8 [21]. In the $\theta - 2\theta$ scan, a weak peak at Cr(N,O) 200 is visible. In the ϕ scan results using a 111 peak shows a clear fourfold symmetry. A high

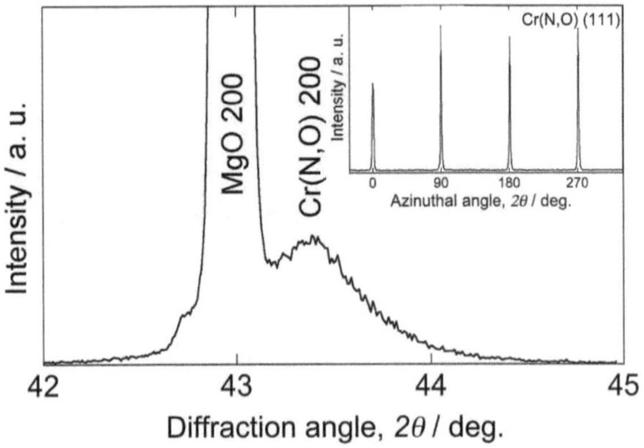

Fig. 9.8 $2\theta - \theta$ and ϕ scan XRD patterns for Cr(N,O) thin film deposited on a MgO (100) single crystal substrate

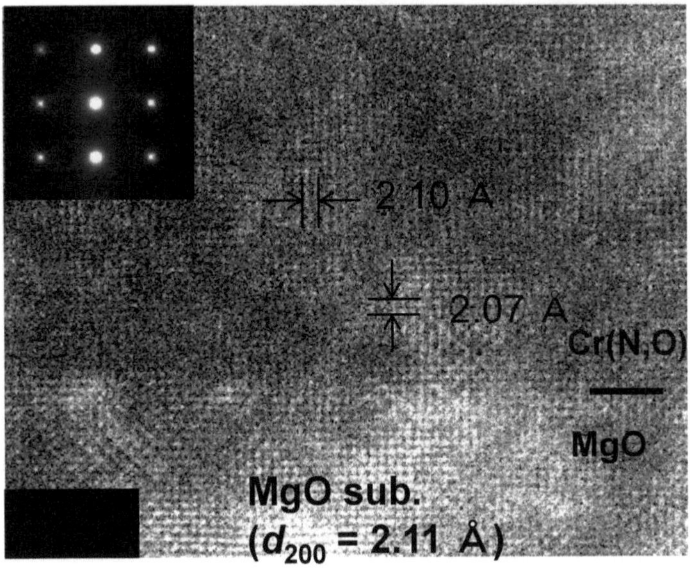

Fig. 9.9 HRTEM image for Cr(N,O) thin film deposited on a MgO (100) single crystal substrate

resolution transmission electron microscopy (HRTEM) image for the thin film is shown in Fig. 9.9 (Suzuki T, 2014, Unpublished work). From these results, it was concluded that epitaxially grown Cr(N,O) thin film was successfully obtained by using a MgO substrate. Similar observations were carried out for a thin film deposited on $MgAl_2O_4$ (100) single crystal substrate and it was found that a epitaxially grown Cr(N,O) was also deposited on this substrate.

Since the O–K edge and the Cr–L edge are very close in Fig. 9.7, only anion composition was estimated from EELS data. By also using RBS data, composition of the Cr(N,O) thin films deposited was determined. Compositions of the thin films

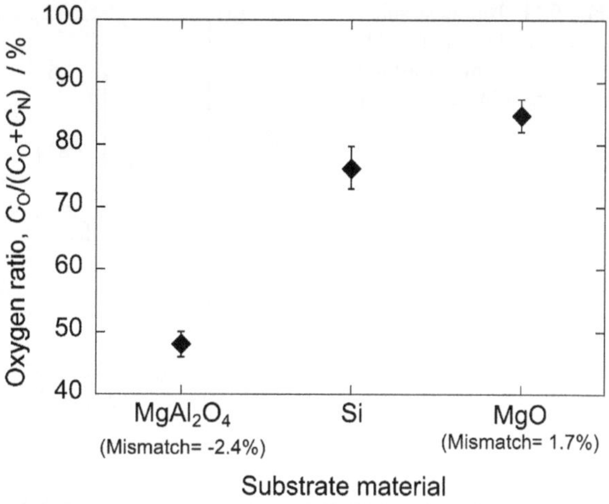

Fig. 9.10 Effect of lattice mismatch on the oxygen content in epitaxially grown Cr (N,O) thin films

deposited on $MgAl_2O_4$, Si, and MgO (100) single crystal substrates are shown in Fig. 9.10 (Suzuki T, 2014, Unpublished work). By changing the lattice mismatch, composition of the thin films was controlled.

Free energy change (ΔG) induced by the stress is estimated. Assuming the elastic modulus of CrN being 328 GPa, stresses caused by the mismatch of 1.7 and −2.4 % are 5.5 and −7.9 GPa, respectively.

$$\Delta G = \int_{p_0}^{p} V \mathrm{d}p = V(p - p_0), \tag{9.3}$$

where p_0 and p are initial and final pressures, V is the volume. Thus, epitaxially grown Cr(N,O) thin films on MgO and $MgAl_2O_4$ (100) single crystal substrates have 60.8 and −86.6 kJ/mol than that for stress free CrN, respectively. Although free energy change between CrN and Cr(N,O) is not known, these high free energy change is thought to alter the composition. This is the reason why composition of Cr(N,O) is affected by the substrate material.

9.5 Hardening in (Cr, Al) (N,O) Thin Films by Introducing Inclusions

Thin films of (Cr1-*x*Al*x*)(N,O) were deposited by PLD using rotating Cr2N/AlN targets with various surface area ratios. Details of the deposition conditions are shown elsewhere [22]. Hardness measured by a nanoindentation tester is shown in Fig. 9.11. The maximum hardness was obtained at $x = 0.35$. Elastic modulus estimated by the same method is also shown in Fig. 9.11 [22]. Comparing to

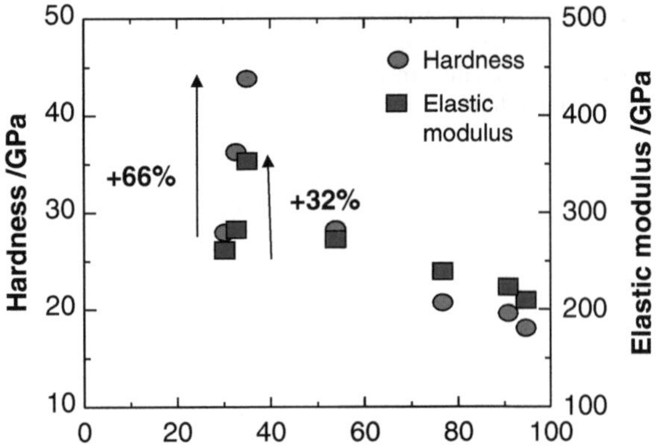

Fig. 9.11 Hardness and elastic modulus in (Cr,Al)(N,O) thin films measured with a nanoindentaion tester

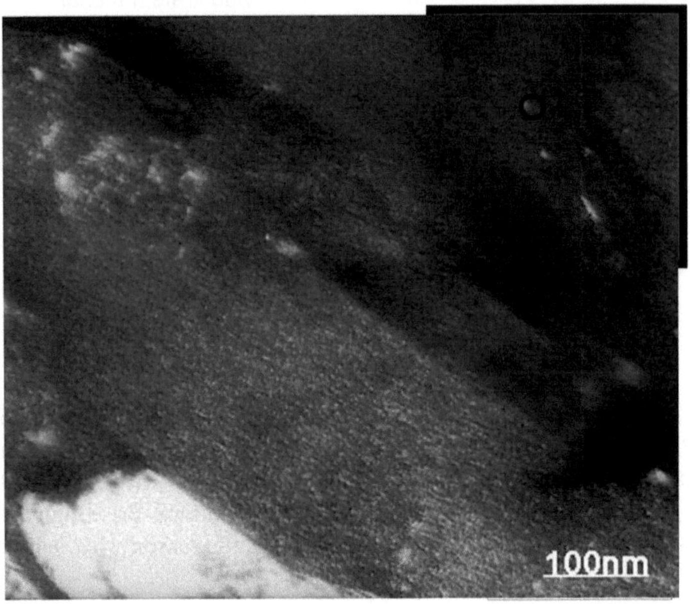

Fig. 9.12 Bright field image and SAD pattern for a (Cr,Al)(N,O) thin film with $x = 0.35$

these values in the thin film with $x = 0.30$, the hardness enhances by 66 %, while the elastic modulus increases only by 32 %. A dark field TEM image is shown in Figs. 9.12 and 9.13. In the grain, fine white spots with sizes of approximately 3 nm are seen. After stress estimations and grain size measurements, there are no significant difference among the thin films. Thus, it was concluded that these white spots are inclusions to pin down dislocations to increase the hardness.

9.6 Hardening by Cr–Si–N–O Thin Films by Controlling the Grain Size

A compositionally gradient thin film of Cr–Si–N–O was deposited in a single experiment on a Si single crystal substrate by magnetron sputtering using a Si–Cr target. Then, the thin film was cut into samples. Details of the compositionally gradient thin film were shown elsewhere [23]. Cation composition variation of the thin film determined by RBS is plotted in Fig. 9.13 [24]. Patterns of XRD for the samples are shown in Fig. 9.14 [23]. All peaks except for the substrate correspond to those for a NaCl type structure, which indicates crystalline phase is only of (Cr,Si)(N,O). HRTEM images of thin films with silicon content (C_{Si}) = 3 and 5 at. % are shown in Fig. 9.15 [23]. In the thin film with C_{Si} = 3 at.%, only crystalline grains are seen, while grain boundary layer without lattice fringes is visible in the thin film with C_{Si} = 5 at.%. It is likely that, above C_{Si} = 3 5 at.%, amorphous grain boundary layer was formed. The grain size distribution of the samples is shown in Fig. 9.16 [23]. With the increase in C_{Si}, the grain size is decreased. Compositional analysis of grains and the grain boundary layers are shown in Fig. 9.17 (Suzuki T, 2014, Unpublished work). The grains consist mainly of Cr, Si, and N, while the

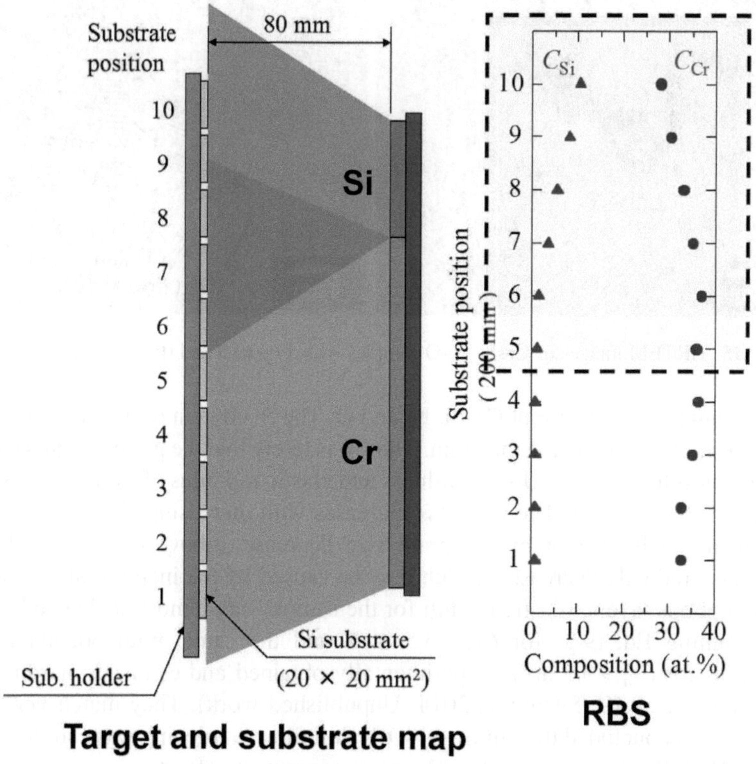

Fig. 9.13 Target and substrate configuration (*left*) and composition (*right*) of Cr–Si–N–O samples

Fig. 9.14 XRD patterns for Cr–Si–N–O samples

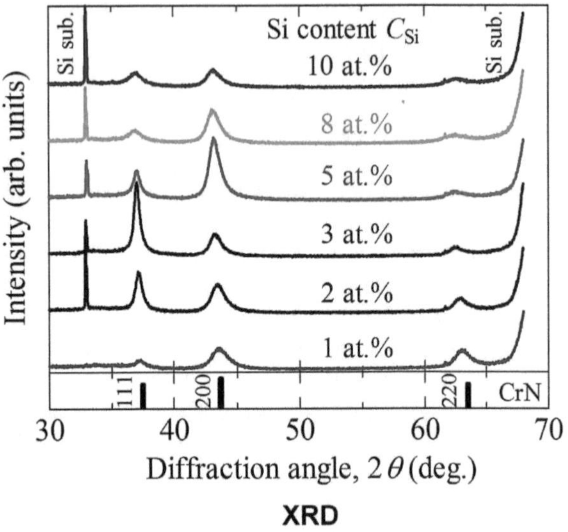

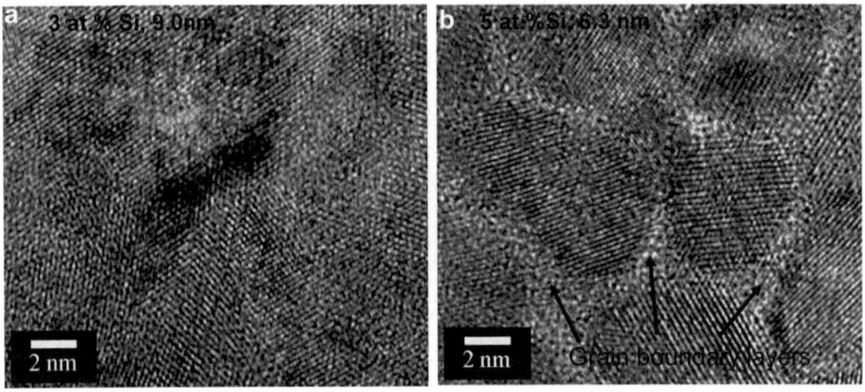

Fig. 9.15 HRTEM images of Cr–Si–N–O samples with x = (**a**) 3 and (**b**) 5 at.% [23]

grain boundary layers are of Cr, Si, N, and O. The Si content of the grain boundary layers is higher than that in the grain. Thus, it is likely that the grain boundary layers are of amorphous Cr–Si–O–N. Hardness and elastic modulus of samples are shown in Figs. 9.18 and 9.19. The hardness increases with increasing C_{Si} up to 3.5 at.%, which means hardening by the grain size decrease. Above $C_{Si} = 3.5$ at.%, the hardness gradually decreases, which may be caused by the increase of the volume of amorphous layers. Elastic moduli for the samples are almost unchanged.

Assuming Eq. (9.2) for $C_{Si} \leqq 3.5$ at.% and the same grain boundary layer thickness for $C_{Si} > 3.5$ at.%, experimentally obtained and estimated hardness are shown in Fig. 9.19 (Suzuki T, 2014, Unpublished work). They match very well. Thus, it was concluded that, in Cr–Si–O-N thin films, hardening was caused mainly by the Hall–Petch relation and not by a nanocomposite effect.

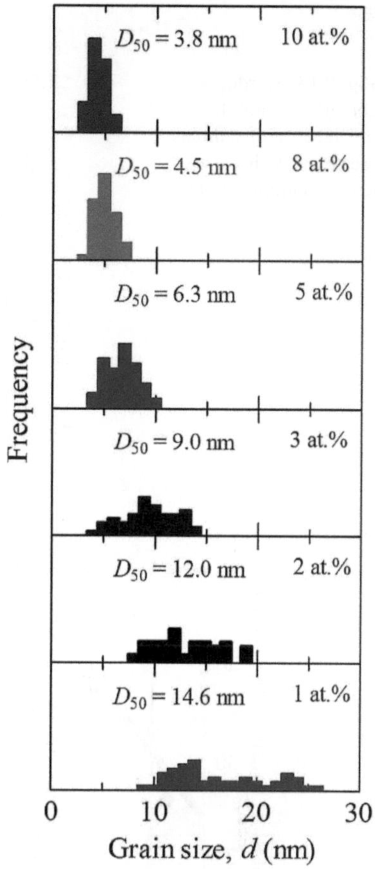

Fig. 9.16 Grain size distributions in Cr–Si–N–O samples

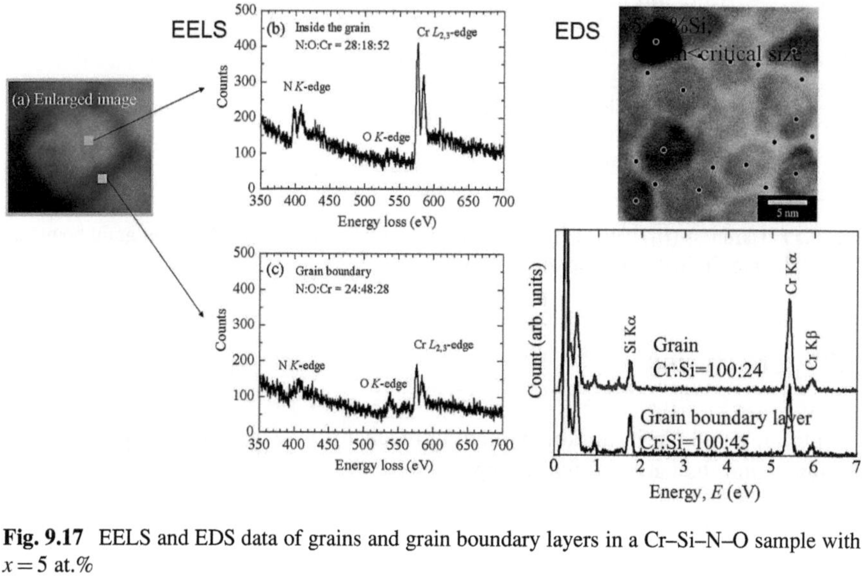

Fig. 9.17 EELS and EDS data of grains and grain boundary layers in a Cr–Si–N–O sample with $x = 5$ at.%

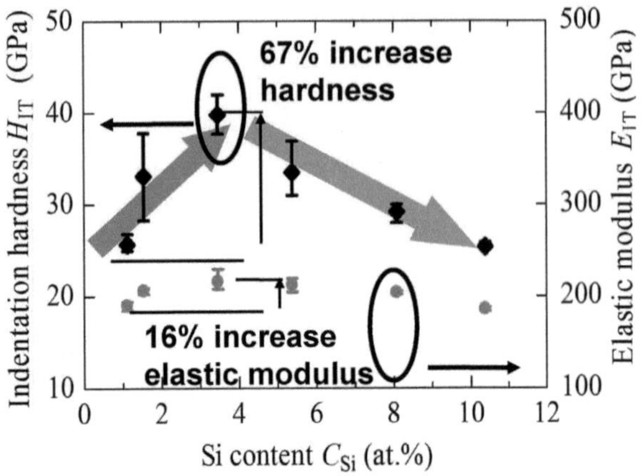

Fig. 9.18 Hardness and elastic modulus in Cr–Si–N–O samples measured with a nanoindentation tester

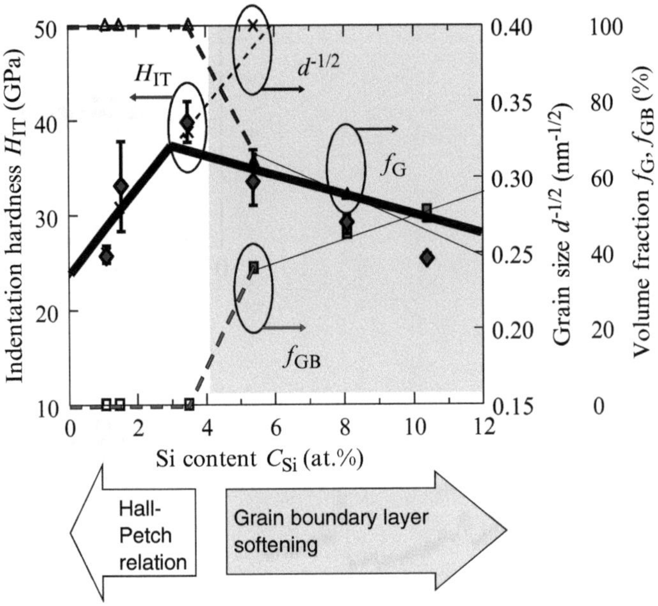

Fig. 9.19 Hardness (H_{IT}), grain size (d), and volume fraction of grains (f_G) and grain boundary layers (f_{GB}) in Cr–Si–N–O samples

9.7 Conclusion

By utilizing and clarifying the hardening mechanisms, CrN-based thin films were designed and the hardness was tailored.

References

1. N. Komiya, Kikaitokougu **46**, 35 (2002)
2. K. Karino, *New Technologies in Machining* (Kogyo Chosakai Publishing, Tokyo, 1992)
3. O. Knotek, T. Leyendecker, J. Solid State Chem. **70**, 318 (1987)
4. K. Niihara, J. Ceram. Soc. Jpn. **99**, 974 (1991)
5. S. Veprek, S. Reiprich, Thin Solid Films **268**, 64 (1995)
6. H.Y. Lee, W.S. Jung, J.G. Han, S.M. Seo, J.H. Kim, Y.H. Bae, Surf. Coat. Technol. **200**, 1026 (2005)
7. J. Musil, Surf. Coat. Technol. **125**, 322 (2000)
8. J. Vetter, E. Lugscherider, S.S. Gurreiro, Surf. Coat. Technol. **98**, 1233 (1998)
9. V.K. Sarin, Comprehensive Hard Materials, Volume 3 (Elsevier, Amsterdam 2014)
10. C.A. Brookes, J.B. O'Neill, B.A. Redfern, Proc. R. Soc. Lond. A **322**, 73 (1971)
11. F.W. Daniels, C.G. Dunn, Trans. Am. Soc. Metal. **41**, 419 (1949)
12. F.R.N. Nabarro, Proc. Roy. Soc. **59**, 256 (1947)
13. R.E. Peierls, Proc. Phys. Soc. **52**, 34 (1940)
14. E.O. Hall, Proc. Phys. Soc. Lond. **64**, 747 (1951)
15. N.J. Petch, J. Iron Steel Inst. Lond. **173**, 25 (1953)
16. T. Suzuki, J. Inoue, H. Saito, M. Hirai, H. Suematsu, W. Jiang, K. Yatsui, Thin Solid Films **515**, 2161 (2006)
17. T. Suzuki, H. Saito, M. Hirai, H. Suematsu, W. Jiang, K. Yatsui, Thin Solid Films **407**, 118 (2002)
18. T. Suzuki, J. Inoue, H. Asami, T. Ibi, T. Nakayama, H. Suematsu, K. Niihara, Key Eng. Mater. **373–374**, 122 (2008)
19. J. Inoue, H. Saito, M. Hirai, T. Suzuki, H. Suematsu, W. Jiang, K. Yatsui, Trans. Mater. Res. Soc. Jpn. **28**, 421 (2003)
20. J. Inoue et al., J. Jpn. Soc. Powder Powder Metal. **54**, 194 (2007)
21. K. Suzuki, T. Endo, A. Sato, T. Suzuki, T. Nakayama, H. Suematsu, K. Niihara, Mater. Trans. **54**, 1957 (2013)
22. M. Hirai, T. Suzuki, H. Suematsu, W. Jiang, K. Yatsui, J. Vac. Sci. Technol. A **21**, 947 (2003)
23. J. Shirahata, T. Suzuki, T. Nakayama, H. Suematsu, K. Niihara, J. Jpn. Inst. Metal. **75**, 97 (2011)
24. J. Shirahata, T. Ohori, H. Asami, T. Suzuki, T. Nakayama, H. Suematsu, Y. Nakajima, K. Niihara, Thin Solid Films **519**, 3497 (2011)

Index

S. Babu Krishna Moorthy (ed.), *Thin Film Structures in Energy Applications*,
DOI 10.1007/978-3-319-14774-1

MIX
Papier aus verantwortungsvollen Quellen
Paper from responsible sources
FSC® C105338

If you have any concerns about our products,
you can contact us on
ProductSafety@springernature.com

In case Publisher is established outside the EU,
the EU authorized representative is:
Springer Nature Customer Service Center GmbH
Europaplatz 3, 69115 Heidelberg, Germany

Printed by Libri Plureos GmbH
in Hamburg, Germany